GB8 $14.95
D0483959

A time capsule's view of Earth history

Adapted from a drawing in *Geologic Time*, 1976, a publication of the United States Geological Survey.

COSMOS, EARTH, AND MAN

COSMOS, EARTH, AND MAN

A SHORT HISTORY OF THE UNIVERSE

Preston Cloud

New Haven and London
Yale University Press
1978

*Published with assistance from the
Kingsley Trust Association Publication Fund
established by the Scroll and Key Society of Yale College.*

*Copyright © 1978 by Yale University.
All rights reserved. This book may not be
reproduced, in whole or in part, in any form
(beyond that copying permitted by Sections 107
and 108 of the U.S. Copyright Law and except by
reviewers for the public press), without written
permission from the publishers.*

*Designed by Sally Harris
and set in VIP Janson type.
Printed in the United States of America by
The Vail-Ballou Press, Binghamton, N.Y.*

*Published in Great Britain, Europe, Africa, and
Asia (except Japan) by Yale University Press,
Ltd., London. Distributed in Australia and
New Zealand by Book & Film Services, Artarmon,
N.S.W., Australia; and in Japan by Harper & Row,
Publishers, Tokyo Office.*

Library of Congress Cataloging in Publication Data

Cloud, Preston Ercelle, 1912-
Cosmos, Earth, and man.

Includes index.
1. Cosmology. 2. Earth sciences. 3. Life
sciences. I. Title.
QB981.C66 523.1 78-2666
ISBN 0-300-02146-1

For Karen, Lisa, and Kevin

CONTENTS

PART FOUR: MAN

ILLUSTRATIONS

TABLES

PREFACE

This book is a distillation, in simple terms, of some aspects of what I think I have learned from a lifetime of research, reading, and reflection about our planet, its history and cosmic connections, the development of life on it, and its capability to sustain our descendants. It is meant for the educated layman—the kind of person who reads a good weekly news magazine, who might subscribe to a scholarly journal in some field, and who likes to ponder an issue. It might also be used as a textbook in courses that aim to present a broad introductory view of material covered.

It was written as a labor of love by one who has always had trouble confining his interests or finding good reason why he should. It was only natural for such a person to become a geologist. In this profession I was enabled, without fear of excommunication, successively to study fossils of many kinds and ages, recent as well as ancient coral reefs, ancient sediments on land and the depositional environments of their modern counterparts at sea, and the evolution of the primitive earth in all its aspects. Such investigations have taken me to many parts of the nation and the world and provided opportunities to further long-standing interests in the broader generalities of history, geography, biology, chemistry, and astronomy.

Thus it seemed that it might be fun, and perhaps appeal to others, if I could get it all together as a sequence of interconnected vignettes focusing on my favorite subject, the earth. But I am easily diverted and the gestation period has been long. At first I thought to write for my young children, wanting to tell them about the wonderful, complex, and sometimes intransigent world they were about to inherit. While I was mulling this over, however, and trying to write the first chapters, that generation grew up, inspiring the present book.

What I try to do here is to relate the story of the co-development of Earth's air, water, crustal rocks, climate, and life in cosmic perspective as a vantage point from which to view the present state and future prospects of mankind. Earth has come to be what it is as the result of a sequence of historical interactions stretching over billions of years. What it is likely to become in the future is affected to an extraordinary degree by the actions of its now dominant organism—man (of course including woman). Man, however, is himself a product of planetary evolution, and it behooves him to understand his habitat as a complexity of evolving and interacting systems that have changed in the past and will change in the future.

Following a brief introductory chapter, the bulk of the book deals with the nature and evolution of matter; cosmic evolution; the reckoning of planetary and cosmic time; the history of air, water, and the solid earth; and the development of life on Earth to the present. The last three chapters view Earth's present state and future prospects, with special reference to the nature of and constraints on its ever more industrialized and urbanized society. I would hope that most readers might cover all of the chapters in sequence, as they are intended to hang together as a consistent story. Because some, however, are necessarily more sophisticated scientifically than others, I have tried to write each in such a way, and with sufficient overlap of basic concepts and data, that continuity would not be unduly impaired by skipping over parts that may seem difficult or uninteresting to some readers.

I like to think of the story I have tried to tell as a kind of historical and, to some extent, predictive novel about the earth, its ancestry, and its geologically near-term future. The central characters are elements, stars, planets, and life. On planet Earth a local population of elements takes the form of air, water, the great rocky sphere beneath, and a variety of living forms, culminating in man himself.

The processes that shaped the events related are of great generality and apply equally to past, present, and future. With understanding of these balancing processes, and of the historical significance, the uniqueness, and the vulnerability of life on our planet comes the recognition that man is a product of Earth's historical development. He has shaped and been shaped by the larger ecosystem of which he is a part and his actions have foreseeable consequences for that ecosystem and the future generations that will inherit it. When man fully grasps those truths, he must see compassion for other humans, and for nonhuman forms of life, as consistent with his own best interests and those of his descendants. And when he acts thoughtfully on such convictions there will be better management of the planet and its precious natural resources, fewer descendants, and a freer and more ample world for them to live in.

I have not hesitated to interrupt this account to tell some relevant bit of scientific history, introduce alternative explanations, or take notice of controversy. Much of what we know or think we know has not yet been satisfactorily explained. The creative friction generated by competition among hypotheses is as much a part of science as observations, measurements, and experiments. At the same time I do not shrink from expressing personal views and preferences.

As to procedural matters, I try to tell the story in simple language but without going to great lengths to avoid those occasional technical terms for which there are no good, brief, alternate expressions. Those whose meanings are not clear from context or to be found in an ordinary dictionary are explained as they appear. The metric system of measurements is used throughout the book except in figures 46 and 47, where currently conventional United States commercial notation seemed more appropriate. Conversion tables are given on the back endpapers. The book being primarily oriented to an American audience, I use the term billion to refer to units of 10^9 or 1,000,000,000. In selecting references for further reading I considered accessibility, comprehensibility, reliability, and currency. Compromise was necessary and often agonizing.

It must also be acknowledged that I have at places described unconscious biological and inanimate processes or events as if they were conscious, directed acts. Although such expressions are taboo in proper scientific writing, they serve to dramatize and to avoid roundabout constructions. I beg the reader, therefore, to understand that when I say, for instance, that some unknown ancestral microbe invented photosynthesis, I don't really mean to suggest that it applied for a patent.

The production of this book in its present form owes much to the help and inspiration of others. First among these is my wife, Janice, who served as lay critic and typist through two complete revisions of the original 1973 draft. For review of the whole first draft I am also grateful to James Gilluly, Winifred W. Gregory, and A. O. Woodford; and for individual chapters to S. W. Awramik, John Crowell, W. W. Murdoch, E. C. Olson, Stanton Peale, George Tilton, and Robert Warner. David Crouch prepared the line drawings from my rough sketches and revisions of published copy. And Catherine Skinner directed me to the Yale Press, where I was fortunate enough to have Edward Tripp and Maureen Bushkovitch for editors. Finally, I wish to thank those anonymous critics of the several publishers who rejected an earlier draft of this book for their often cutting but nevertheless helpful comments. Should any of them come upon the present version I trust they will find that all that blood was not let in vain.

Mammoth Lakes, California Preston Cloud
June 1977

1

TO MAKE A BEGINNING

The planet called Earth is a very unusual cosmic body. Its sun, the billions of other stars in our galaxy, the myriads of other galaxies in the visible universe, interstellar and intergalactic space, and even interplanetary space and the planets beyond the asteroid belt consist mainly of hydrogen and helium. Earth, however, like the other inner or "terrestrial" planets of our solar system, is anomalous for its relative scarcity of these light elements and its relative abundance of heavier ones.

Earth has also been regarded as unique in other respects that are not borne out by scientific inquiry. It was long supposed, from the observed motions of the visible heavenly bodies, and quite reasonably too from a commonsense point of view, that Earth stands at the center of the universe, while everything else revolves around it. To earlier generations also Earth seemed awesome and illimitable in its capabilities. Although we have known differently for more than 400 years, the term *infinite* is still commonly but incorrectly used, even by highly intelligent people, to refer to the life-supporting capabilities of our planet.

Now that the entire earth has been photographed from afar in a single field of view from a variety of spaceships, it is hard to defend such a view. At last our planet can be seen by all as it really is: a relatively tiny habitable sphere, suspended in the black emptiness of the heavens, hurtling at 30 kilometers a second in captive and delicately balanced orbit around the life-giving but potentially lethal sun.

Planet Earth is of surpassing interest to us because it is the only object in the solar system, indeed in the universe, that is *known* to be capable of supporting life. More to the point for us, it also happens to be the only planet in our solar system on which we or our descendants can sustain life under natural conditions. And, with the important exception of its daily quota of energy from the sun, it is very nearly a closed system. Except in science fiction, no great quantity of matter can be added to or taken away from it in its present state or in any foreseeable future state. Only a thin atmospheric blanket shields us from the lethal radiation beyond, and from equally lethal extremes of heat and cold. Only a little film of water moderates our climates and provides the solvent in which our life stuff is suspended. Earth's interior is forever sealed to entry beyond a few kilometers depth, whether we start our penetration on land or at sea.

It is worth our while, therefore, to try to understand our unequivocally finite planet, its limitied and fragile life-support system, and the necessity of treating it with care.

These pages reach toward such an understanding. They aim to reduce the mystery, so that Earth's limitations may be grasped and dealt with. Our home planet is seen in cosmic perspective as a unique and evolving configuration of matter on which has arisen a sometimes noble and creative, sometimes foolish and destructive creature who is affecting that evolution in unprecedented ways. If we can appreciate our antecedents, it should help us to understand how things got to be as they are and how we might best proceed to affect the future favorably.

Many of us, to be sure, are very much concerned about our world these days. We can see many places where improvement would be desirable. But we are not sure what improvements are possible or how best to go after them—least of all how to motivate others to go with us. We talk vaguely about coming into balance with nature, as if natural balance were some permanent ideal which has only to be restored and maintained. We seek specific individual actions thought to be good. Yet the effects we would like to correct are often themselves the cumulative consequences of earlier actions, also seen as good when taken, but taken without thinking hard enough about possible unintended feedbacks. It behooves us, therefore, to keep ever in mind as we lay our plans for the future John Muir's passionate insistence that everything on Earth is connected with everything else, and with the rest of the universe. We are concerned, however remotely, with the entire complex system.

Nature is indivisible, constantly changing, historical. Man lives in a man-altered environment. It can be no other way. We are part of the environment; we interact with it even when at complete rest, indeed even after death. Such interactions are what ecology is all about. A space occupied by living things displays a multitude of interactions not only between the different kinds of organisms that live within it but also between them and its inanimate parts. Every part of our global ecosystem, animate and inanimate, interacts in some sense with every other part. All animals eat and are eaten, live, give birth, and die. Every change, anywhere, leads to a sequence of other changes that ends only when a new balance is attained. Whatever that new balance may be, it is always temporary.

It is one of the fundamental rules of nature that any disturbance of a balanced state, be that state within a single organism or the global ecosystem, leads to reactions, sometimes catastrophic, that tend eventually to restore balance—a fever, for instance, or a hurricane. And all changes are irreversible in the sense that nothing can ever be restored exactly as it was before disturbance. The wheat fields of Kansas can never be returned to the same original mix of native grasses. The three-toed horse of yore will never roam the plains again. A river, once diverted, will never return to exactly its previous channel.

Nature evolves. The world was different yesterday and it will be more different still tomorrow. It will change as a result of what we do, even if we seem to do nothing at all. But we are not helpless pawns of fate. We can foresee, however dimly, the consequences of our actions. Although all our yesterdays may have lighted fools the way to dusty death, it is not required that all generations take such an inglorious route to the grave. It is harder to behave foolishly when one reflects on the consequences of one's actions.

In thinking about the present and the future, there is no better foundation than a historical one—the longer the better. It is germane to an assessment of our problems that Earth evolved from stardust, that it was not always capable of supporting life, that before life as we know it could have arisen there had to be air and water, and that air and water did not always exist. Furthermore, there could have been no significant amount of free oxygen in the air when life arose. For not only does oxygen prevent the formation of or burn up the basic organic molecules from which living things can be made, it is also lethal to all forms of life in the absence of advanced systems of enzymes that harness it to advanced life functions (see chapters 9-10). The primitive atmosphere would have seemed to us to be hopelessly polluted. Indeed, before even the lowest animal could appear and survive, there had to be photosynthesizing plants to produce plenty of free oxygen and a revolution in cellular chemistry to facilitate its controlled burning in oxidative metabolism. Before man could arise, there had to be a long preceding course of animal evolution. And the level of man's ecological dominance has increased in proportion to his unparalleled success in controlling ever larger amounts of Earth's total nonbiological energy supply.

Many books tell a part of the story. In this book I have tried to weave the main parts together briefly in order to show how our world got to be the way it is and what might be done to improve its potentiality for gracious living by all mankind. In trying to understand our world, it is important to stick as closely as possible to ideas that can be tested, and to look for the simplest explanations that are consistent with all observed facts. As these are the methods of science, it is important to say a few words at this point about the nature of scientific inquiry.

It has been said that the scientist is like a dog, sniffing in aimless ecstasy at a thousand trees and hedges. That simile applies more appropriately to human curiosity in general, without which, of course, no science would exist. The analogy with science would be better were the ecstatic canine on the trail of some particular scent, especially were he to discover, in the process, an even more delectable scent—perhaps one that he was not even looking for.

Our world, indeed, is so intricate that if we were just to describe it in detail the way it is, it would not make much sense. We would have forgotten where we started before we got to the end. Besides, no one could remember and recite such a welter of uncoordinated detail.

Although facts are essential to science, a mere recital or catalog of facts is not science any more than a pile of bricks is a cathedral. Science begins where connections are made between facts in such a way that larger generalizations emerge.

Stated more "scientifically," the goal of science is to discover what order there may be in nature; to find the connections between seemingly unconnected processes and events; to seek simple statements about the complex; and to devise tests by means of which the validity of such statements may be checked. Aldous Huxley finds the distinction between the man of science and the man of letters in the former's focus on the generality of experience and the latter's focus on its uniqueness. And I would add that it takes a healthy balance of both to make a whole culture or even a whole person.

Science, however, achieves its highest expression in statements that are precise, simple, and very comprehensive. Albert Einstein put it that: "Science seeks to reduce the connections discovered to the smallest possible number of independent elements." The most extreme reduction of connections is seen, for example, in Isaac Newton's law of universal gravitation and Einstein's own formulation of the relations between matter and energy. Newton saw that the gravitational attraction between any two bodies in the universe must vary as the product of their masses, divided by the square of the distance between them. Einstein's famous equation, $E=mc^2$, simply says that matter and energy are interchangeable, and that, when matter is converted to energy, the energy generated is equal to the mass of the matter converted times the square of the speed of light (indicated by *c*, for celerity)—but what a sweeping simplicity!

Such insights never arise from the simple collection and listing of observations and measurements alone, although data are necessary before they can arise. Moreover, the insights that emerge at any level of scientific development are strongly conditioned by those that have gone before. Einstein's intellectual contributions were as dependent on Newton's laws of motion and the laws of thermodynamics as the invention of the cart was on that of the wheel. Nevertheless, there is a strong intuitive and artistic quality to creative scientific thought. Ideas often spring seemingly full-blown to the prepared mind, sometimes even during sleep, or when the mind is not in any way consciously engaged.

Science progresses, because scientists make up explanations, *hypotheses*, that fit or seem to fit existing observations. The author of a hypothesis is often himself not fully aware of how it came to mind. However, and this is where science differs from other systems of explanation, in order to be considered *scientific* hypotheses must be testable. That is to say, they must include verifiable consequences that if not found to be the case would require their rejection or modification. To proceed differently in any instance is not science, no matter how brilliant the performer or how great his scientific contributions in other respects. Consistency between interpretations and available data and testability of hypotheses proposed are as essential to fruitful science as intellect itself. The miraculous is excluded by definition. The goal being to explain natural phenomena by naturalistic processes, the introduction of supernatural postulates amounts to an abandonment of science.

When data are few, few limits are placed on the kinds of hypotheses that may be proposed. As information increases, the permissible range of interpretations usually decreases. The credibility of a hypothesis increases as it survives opportunities to be discredited—as Einstein's induction survived the practical tests of the atomic and hydrogen bombs. A hypothesis or a group of hypotheses may also be called a *model* (or in some sense a paradigm, but not quite in the senses of T. S. Kuhn). When a hypothesis or a model has survived many opportunities for disproof, it may be elevated to the status of a *theory* (although hypotheses and models are often, and improperly, called theories before they are adequately tested). If, in due time, a relation is found both to obtain in all of many observed circumstances and to have a high degree of generality, it may then be called a *law*. Or, if it is highly general but not invariable, the term *rule* may be used. It is well to remember, however, that the laws and rules of science are written by men as an effort to express their perception of the laws and rules of nature.

Man's understanding of nature often turns out to be less complete than he thought. Models and paradigms may be upset by new data or new perceptions of old data. Even though we may speak of proving this or that, science progresses more by disproof, or by survival of opportunity for disproof, than by proof. Ideas may progress from hypotheses to theories to laws, usually with modification, as they continue to withstand tests. Most do not survive. Scientists must expect to be wholly wrong part of the time and partly wrong much of the time. Intellectual discipline is essential. Should it flag, it will be revived or replaced by peer pressure. The work of science is never finished.

Such rigorous rules of thought are not, of course, unique to science any more than they are invariably followed to the letter by scientists, but they express the ideal. The same methods of thought can be and often are applied with profit to other aspects of life—including the area of scholarly theology (research on the Dead Sea Scrolls and extracanonical scriptures, for instance).

Our reach toward the ultimate is no less sublime in concept because it is constrained by facts. Poetry, theology, and science occasionally do blend, as they do in the felicitous words of William Cowper, poet and hymnist, written two centuries ago:

> Some say that in the origin of things,
> When all creation started into birth,
> The infant elements receiv'd a law
> From which they swerve not since . . . That under force
> Of that controlling ordinance they move,
> And need not his immediate hand who first
> Prescrib'd their course, to regulate it now.
>
> *The Task*, Book 6

In the chapters that follow we will explore the many ramifications of that controlling ordinance, beginning with the infant elements themselves.

For Further Reading

Beck, W. S. 1961. *Modern science and the nature of life*. Anchor Books. 334 pp.

Conant, J. B. 1967. Scientific principles and moral conduct. *American Scientist* 55. 311-28.

Platt, John R. 1966. *The step to man*. John Wiley and Sons, Inc. 216 pp.
Popper, Karl R. 1959. *The logic of scientific discovery* (translation from the original German edition of 1934). Hutchinson of London. 480 pp.
Weisskopf, V. F. 1977. The frontiers and limits of science. *American Scientist* 65: 405–11.

I

COSMOS

2

ON THE NATURE OF MATTER

Why bother with the nature of matter when our purpose is to discern the history of our planet and contemplate its future? What relevance can things so esoteric and remote possibly have for dreaming, scheming man?

The nature of matter is deeply germane because we and our planet are made of matter. If it were different we would be different. We and our planet are products of its evolution. Except for some rare productions of the atomic age, Earth and everything in it, the seas of water and the air around it, all consist of combinations of the ninety elements that occur naturally in our part of the universe. Elements and their compounds are made of atoms. Atoms consist of elementary particles. Their structure determines how Earth is put together and what we are.

Like life, and the planets, the elements did not always exist. They were cooked in stars under circumstances that do not occur naturally on Earth. The heaviest elements are produced only in dying stars, ones that have imploded as novae or supernovae, flinging a shock wave of new elements into space for reincarnation as successor stars and planets. Stars have died that we might live. It might seem enough that they have done so, that they guide the wayfarer, stir our emotions, and pique our curiosity. But there is much more.

Stargazers and other thinking men and women have long been fascinated with the minute and the gigantic. It is a part of being human to seek relevance in the seemingly irrelevant. It is the key to practical progress that the search is often successful. The most successful searchers are often those who operate out of simple curiosity and the joy of problem solving. It may be important to their success that they frequently do not care what the solution is, so long as they find one. Relevance in the practical sense is likely to be an accident of history. To overemphasize it is to distort the intellectual process.

Understanding the elements and the compounds made from them is the key to the time-blurred mystery of how Earth originated in circumstances precisely suited to the appearance and sustenance of life as we know it, from the first living thing until now. For Earth, third planet outward in our solar system from its central star, the sun, is of just the right size and composition, at just the right distance from a central star of just the right properties, to be one of those unusual (but probably not rare) planets in the seemingly illimitable universe capable of supporting the evolutionary development whose acme man fancies himself to be. And the story of how scientists think that came about, although it may be told in the unemotional words of science, has an enchanting quality to match the most gripping flights of literary fancy.

Elements, molecules, and other arrangements of atoms themselves have a history, even a kind of life-style. They grow and decay, depending on their surroundings, in a manner reminiscent of living things, although with highly variable time scales. Particular kinds of compounds are found in the particular physical and chemical environments suitable to their formation. We can no more comprehend the balance of nature without understanding something about such matters than Shakespeare could have produced *Hamlet* without mastering the English language.

The Structure of Matter

The elements of all pure substances are arranged in a characteristic or systematically varying order, the nature of which determines the ways in which they combine with other substances to make more complex matter, including the cells, tissues, and organs of living things. With rare exceptions, elements exist only as components of molecular or other compounds, which may consist either of two or more atoms of the same element, like molecules of oxygen and hydrogen, or of combinations of atoms, like simple carbon dioxide or the giant molecules of the various proteins that sustain us and regulate our life processes. The idea that atoms are the basic components of matter arose long before the Christian Era with the musings of Democritus and Lucretius. It remained in very crude form, however, until around the turn of the eighteenth century, when an English schoolteacher named John Dalton, a Frenchman, Joseph Gay-Lussac, and an Italian, Amedeo Avogadro, introduced major refinements. The realization that atoms are not in fact the ultimate stuff of the universe, however, had to await the serendipitous discovery of radioactivity by the Frenchman Henri Becquerel in 1896 and the subsequent invention of instruments and methods that have allowed science to look inside the atom.

It is now known, from the work of particle physicists in this century, that the supposedly indivisible atom actually consists of around a hundred different *elementary particles*, of which the main ones are protons, neutrons, and electrons. The entering wedge to the partitioning of the atom was driven in 1919 by the brilliant New Zealander Ernest Rutherford, later professor of physics at Cambridge University. The foundation for Rutherford's researches had been laid by Becquerel's discovery of radioactivity and the elegant subsequent investigations of Marie and Pierre Curie.

The early alchemists, although unable to make gold from lead, had not been entirely mistaken about the transmutability of the elements. Although most elements are immutable under conditions that prevail on Earth, a few, like uranium and its daughter product radium, spontaneously change to different elements as a result of their emission of radiation. Becquerel observed that this *radioactivity* could darken shielded photographic film and produce other unusual effects. The Curies also found that, whatever the state of combination of the radioactive elements or the physical conditions surrounding them, they continued to radiate spontaneously at seemingly unvarying rates.

Rutherford discovered that when radiation emitted was passed through a strong magnetic or electric field it split into three different kinds. Some rays were deflected as if they were the paths of particles carrying a positive charge, others were deflected as if negatively charged, and still others passed through the field undeflected. Rutherford called them alpha, beta, and gamma particles or rays, respectively.

The nature of this radiation became the subject of intensive study. The positively charged particles were eventually found to be the charged nuclei of the inert gas helium, without its two neutralizing, negatively charged electrons. You cannot see these particles in a literal sense, even with the most powerful microscope, but you can confirm their existence by examining, with an ordinary magnifying glass, the luminous dial of a watch in a dark room. Like Fourth-of-July sparklers, the beautiful glow consists of innumerable tiny flashes, each of which represents a single helium nucleus, emitted with great energy by the material of the watch dial and impinging on a substance that momentarily glows on contact with it.

The ray paths that are followed by the positively and negatively charged particles emitted by radioactive substances can be observed in an experimental device called a cloud chamber. Tiny water droplets condense on other charged particles produced by their energized passage through the moist atmosphere of the cloud chamber. The track so defined shines under oblique lighting against a dark background. The prominent tracks made by the positively charged alpha particles are for the most part not deflected by collisions with gaseous atoms in the cloud chamber nor even when they penetrate sheets of metal foil. Rutherford interpreted this to mean that, outside their relatively heavy but extremely tiny central cores or nuclei, atoms were mostly open space. This space is occupied only by orbiting, sparsely scattered, negatively charged particles of far smaller mass, called *electrons*, which are present in exactly the number required to neutralize the positive charge in the central nucleus. The alpha particles simply pass through the mainly unoccupied space surrounding atomic nuclei the way Magellan sailed across the Pacific without sighting most of its tiny oceanic islands.

The first elementary particle to be identified was the *proton*, the most distinctive component of the nucleus, having a positive charge. In 1919, on examining a photograph of the tracks made by alpha particles (helium nuclei) zooming through nitrogen gas in a cloud chamber, Rutherford noticed a curious thing. From the end of one conspicuous, broad track that stopped abruptly where its positively charged alpha particle plowed into a nitrogen nucleus, a dimmer track headed off in an oblique direction. Characteristics of the two tracks enabled Rutherford to calculate that the dim, oblique one was made by a new particle of about one-fourth the weight of the helium nucleus. It had a positive charge, equal and opposite to that of the electron. It was called a *proton*. Although blasted out of a nitrogen nucleus, its weight and charge were identical to those of the hydrogen nucleus, which, as we now know, consists of a single proton. Later work by Rutherford, involving the passage of alpha particles through different substances, showed that protons could be jarred loose from many kinds of atoms. He hypothesized, therefore, that protons are essential components of all atomic nuclei.

Actually the weight of the same invariable object varies with the force of gravitational attraction. Thus a more precise term, *mass*, is used to refer to the invariable property that accounts for *weight*. It was not until 1932 that an uncharged particle that exceeds the mass of the proton only by that of an electron was found and named the *neutron*. All of the other many elementary particles now known were discovered after the Second World War! Some of them apparently play important parts in holding nuclei together, but their masses are small and their lifetimes outside the atomic nucleus are measured in minute fractions of a second.

The elementary particles that dominate the structure of matter thus are protons, neutrons, and electrons. Since free neutrons (outside the nucleus) break down to release an equal number of protons and electrons (half of them every thirteen minutes), protons and electrons can be thought of as breakdown products of neutrons and neutrons as the primary state of matter. The positively charged protons and uncharged neutrons make up the heavy nuclear center of atoms. Together they comprise all but a few tenths of a percent of the mass of our universe. The minuscule electrons, however, each with only about 1/1,800 the mass of a proton, perform the important function of neutralizing the atom.

The proton and the electron are essentially stable masses, as is the neutron as long as it remains within the nucleus—the normal state of affairs on Earth outside of cosmic radiation, nuclear reactors, or exploding nuclear devices. Free neutrons are the heart of nuclear transformations and atomic power. For neutrons in energetic motion, lacking an electrical charge, can penetrate to the center of the atom, whereas a charged particle would be repelled. In penetrating they can displace a proton, eject an electron, change the atomic structure, and generate energy. Under the conditions that generally prevail within our planetary system, however, neutrons just sit calmly in the nucleus, supplying units of elemental mass.

Thus arises the current picture of the once supposedly indivisible atom, clear in its main aspects, but, like an impressionistic painting, still fuzzy in detail. Atoms of all elements except hydrogen consist of a nucleus of protons and neutrons (together called *nucleons*), about which there moves in orderly procession a swarm of electrons, as the corps de ballet arrays itself around the prima ballerina, only moving millions of times faster. The simplest element, hydrogen, consists of a single proton, surrounded by a single orbiting electron—a "swarm" of one. Other elements have swarms of varying numbers of electrons, which in the elemental state are always exactly matched with the number of protons in the nucleus.

The atomic structure is very open, however. Imagine a greatly enlarged motion-stopping photograph of a large atom. The nucleus would resemble a cluster of cannonballs levitated at the center of an otherwise empty sphere the size of Yankee Stadium. The electrons, frozen in their orbits, would show up as occasional nearly weightless ping-pong balls, so spaced as to conform with the surfaces of a set of invisible concentric spheres or shells around the nucleus of cannonballs and within the open space of the great stadium-sized sphere. Such a structure, despite its openness, is capable of making elements as tough as iron and as heavy and durable as platinum, as well as the ethereal hydrogen and helium. The secret is that, although ordinary matter is mostly empty space, the shells of electrons orbiting around the nucleus are held in position by the tension between centrifugal force and the attraction of the matching and opposite electrical charge of the nuclear protons. In order not to be drawn into the nucleus, the electrons must speed about their orbital shells millions of times a second. In the process they exclude everything except highly energetic matter such as alpha particles and neutrons, just as

if they were everywhere at once. Like any good satellite, a well-behaved electron moves at just the right speed to stay in orbit, neither flying off into space nor crashing into the central body.

An element, then, can be precisely defined as a neutral substance, all of whose atoms have the same number of protons in the nucleus (figure 1). This number is the *atomic number*. Although the atomic number specifies the main properties of an element, however, two other properties and three other terms are involved in considering how atoms are put together to make the elemental substances and compounds of which all familiar matter consists.

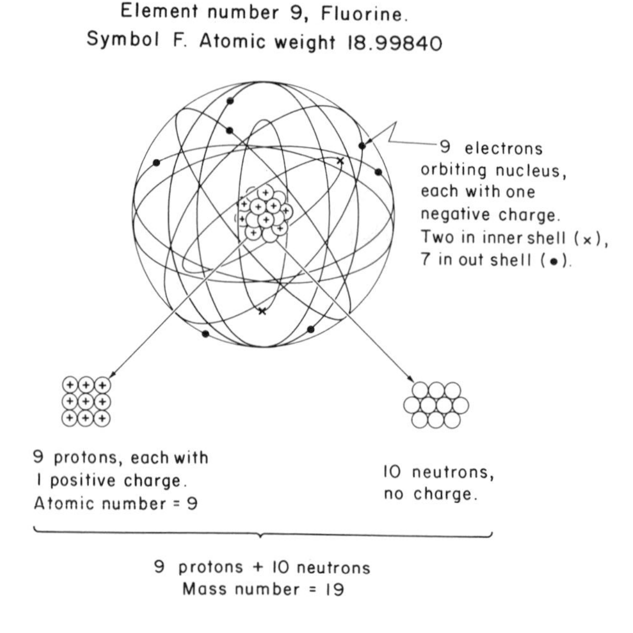

Figure 1. Structure of an atom.

Because the maintenance of atomic neutrality requires the number of orbital electrons and nuclear protons to be the same, the atomic number also specifies the number of orbital electrons. It tells us little about the number of neutrons in the nucleus, however. Yet the atomic nuclei of all elements except the common form of hydrogen contain neutrons, and there are even rare forms of hydrogen whose

nuclei contain one or two neutrons in addition to the single nuclear proton characteristic of hydrogen. Inasmuch as atoms of the same atomic number may have different numbers of neutrons, their mass also varies. Thus the number that specifies the sum of protons plus neutrons is called the *mass number*, and elemental substances that consist of atoms having the same atomic number but a different mass number from their common form are called *isotopes*. Isotopes are designated by the name of the element plus their mass number. Thus lead, with 82 protons, 82 electrons, and an atomic number of 82, has 11 known isotopes whose varying numbers of neutrons result in mass numbers that range from 203 to 214 (with 213 as yet unknown). As all of these isotopes except lead-204 are the end products of *radioactive breakdown or decay*, a process that proceeds at rates that are constant for any given decay system, they, their parent materials, and other radioactive decay systems are important in determining the age of the earth and of rocks formed at different stages in Earth history, as discussed in chapter 6. Finally, a more familiar but less useful property of an element is its *atomic weight*. As the reader may guess or already know, the atomic weight is the sum of the masses of all component parts (protons, neutrons, and electrons) of the atoms that make up an average sample of the element divided by the number of atoms in the sample. In contrast to the atomic number and the mass number, which are invariably whole numbers, the atomic weight is never a whole number.

The preceding paragraphs have been leading up to the idea that, in order to form a quantity of a given element or compound, atoms have to be stuck together in a particular way. And atoms, like the interdependent organisms of a particular habitat, are finicky about how and with what they are associated. The joining of atoms is called bonding, and bonding properties are determined by the arrangement of the electrons around the nucleus according to their energetic characteristics, as shown by quantum mechanics. *Chemical bonding* is achieved either by sharing or by exchanging electrons in the outermost (or only) shell of electrons that surrounds the nucleus (figures 2 and 3). Where a single shell consists of two or the outer shell of eight electrons, the atom is said to be stable. Elements that consist of such atoms, appropriately called the noble gases, combine only with par-

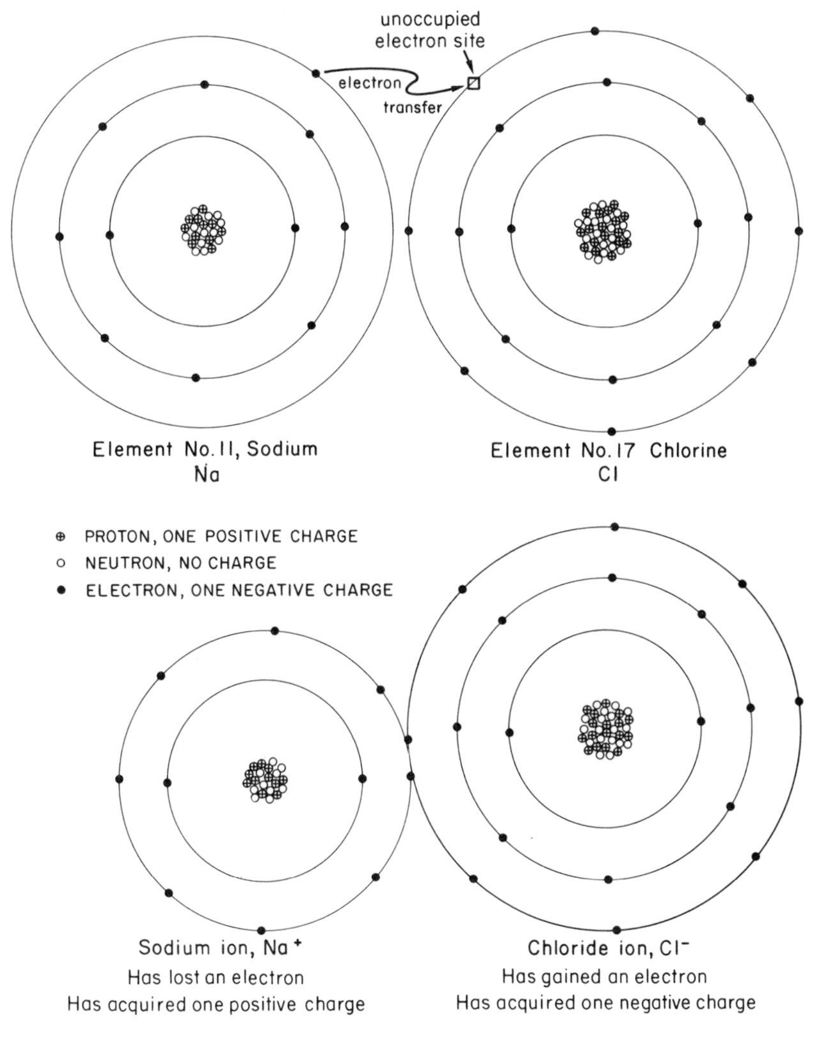

Figure 2. Ionic bonding. Transfer of an electron from sodium to chlorine (upper pair) yields a sodium ion with one positive charge and a chloride ion with one negative charge (lower pair) to form sodium chloride.

ticular common elements under very unusual circumstances. Where the outer shell consists of other numbers of electrons fewer than eight, atoms of the element join with themselves or others having complementary electrical properties.

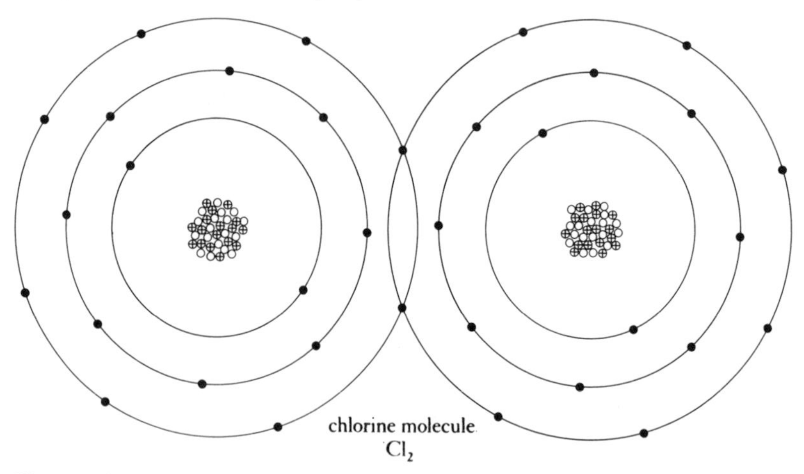

Figure 3. Covalent bonding. Two chlorine atoms, each with 7 electrons in outer shell, share electrons to gain the stable configuration of 8 and become a chlorine molecule, Cl_2.

Each atom of an element with fewer than two electrons in a single shell or eight in the outer shell, the usual case, plays a kind of chemical lonely-hearts game. It tries to find a partner for the giving, receiving, or sharing of electrons, so that each may achieve the magical status that comes with having a stable number of electrons in the outer shell (two or eight). That partner may be another atom of the same element or an atom of a different element. One hydrogen atom, for instance, may pair with another to gain a stable molecular configuration of two shared electrons, or two hydrogen atoms may lend one electron each to an oxygen atom to fill out the stable number of eight and form a molecule of water. The number of electrons that an atom can donate, receive, or share is referred to as its *valence;* the valence of an element is positive where electrons are given up, negative where received.

Valence determines combining ratios. Thus one sodium atom may join with one chlorine to make salt, two hydrogens to one oxygen to make water, two oxygens to one carbon to make carbon dioxide. The kinds of combinations achieved are illustrated in their simplest forms in figures 2 and 3.

When an atom or combination of atoms gains or loses electrons it acquires charge, because the negative electrons no longer balance the positive protons. Then it is no longer an ordinary element or compound. It is called an *ion*, which denotes the fact that it bears a charge and is ready to combine with suitable ions of opposite charge. In figure 2 sodium, with a positive valence of one, gives up an electron to chlorine, with a negative valence of one. The resulting sodium and chloride ions combine as the compound sodium chloride, common salt. This is called *ionic bonding*. In figure 3, two chlorine atoms share electrons to make a chlorine molecule. As electrons are not exchanged, ions are not involved. Such a joining is called *covalent bonding*. Most compounds in nature are much more complex than these simple examples, but all chemical bonding is one of the two types illustrated—either ionic or covalent.

It is by the formation of ions that substances go into solution. When salt is dissolved in a pan of water it does not form little particles of liquid salt suspended in water, but a solution of separate sodium and chloride ions. Where conditions are right, ions are prone to combine with others of opposite sign to form a precipitate. The way they function is well illustrated by the sedimentary rock called limestone, which consists of the compound calcium carbonate ($CaCO_3$). Rainwater combines with carbon dioxide in the atmosphere to form a weak acid that dissolves limestone. The limestone goes into solution as positive ions of calcium (with two missing electrons) plus negative ions of bicarbonate and carbonate (with one and two extra electrons respectively). And it stays in solution as long as the ionic concentration remains sufficiently low. If such solutions form in fresh water, the water is said to be "hard". When the concentration of calcium and carbonate ions reaches a certain level the solution is said to be "saturated". At higher concentrations, in a state of supersaturation, the dissolved ions may combine to form a calcium carbonate precipitate, which, on settling and solidifying, becomes limestone again. The principle is illustrated in the making of old-fashioned sugar candy or by dissolving sugar in water and then boiling the water away to get the sugar back.

The balance between the solution and precipitation of limestone in fact depends on all of the factors that affect the ionic concentration and chemical availability of either combining ion—carbon dioxide content of the water, temperature, agitation, other ions in solution, and so on. A similar balance has had a profound regulating effect on the chemistry of natural waters and the formation of many kinds of ore deposits throughout the ages. Because of it, sea water shows little variation in its level of acidity from place to place, remaining slightly alkaline the world over today and probably having varied little throughout the history of life.

Had the acidity of the sea varied significantly, it would have been inhabited by different kinds of organisms and the blood of land animals might have had a different composition. For it is a fact that the chemistry of blood is strikingly close to that of sea water, suggesting the emergence of ancestral animals in a sea whose salinity has remained pretty much the same since their origin. Such an interpretation is consistent with the idea that persistent natural regulating effects have maintained sea-water chemistry in a nearly steady state throughout much or all of geologic time.

The relations among the elements are of stunning regularity and possess great predictive power. Hydrogen is the lightest and commonest element in the universe. It takes the atomic number of one, from its single nuclear proton. It also has a mass number of one, for there are no neutrons in the nucleus of ordinary hydrogen. After hydrogen comes helium, with two protons and two neutrons in the nucleus, an atomic number of two from its two protons, and a mass number of four, signifying the sum of protons plus neutrons.

But there are also isotopes of hydrogen with mass numbers of two and three, called hydrogen-2 or deuterium (with one neutron) and hydrogen-3 or tritium (with two neutrons). And there is a rare helium isotope, with a mass number of three, produced during nuclear fusion as a result of the loss of an electron from one of the neutrons in the tritium nucleus: hydrogen-3, with one proton and two neutrons, changes into helium-3, with two protons and one neutron.

And so it goes through all 106 of the presently known elements. Their atomic numbers, from 1 to 106, designate the number of protons in the nucleus, equivalent to the number of orbital electrons. Their mass numbers, from 1 to 263, give the corresponding numbers of nucleons. As we saw in the case of hydrogen and helium, it is possible for isotopes of different elements to have the same mass number, but no two elements have the same atomic number. It is the number of protons and orbital electrons that determines the chemical properties of an element. The isotopes of an element differ only in mass and not in chemical properties. This distinction underlies the development of nuclear energy and of a time scale for planetary evolution.

One of the most fruitful scientific insights of all time was the recognition of this beautiful regularity of the elements. Although the discovery is conventionally attributed to the great Russian chemist, Dmitri I. Mendeleev and dated 1869, like many scientific insights it was in the winds of the times, and Lothar Meyer of Germany and J. A. R. Newlands of England were working on the same idea at about the same time. The genius of these three men was to see that all of the then known elements could be grouped in a single table according to their atomic weights and valence. Such relationships comprise the periodic system or table of the elements, or simply the *periodic table*. A slightly modified current Russian version of it is reproduced on the back endpapers. As is there shown, all elements are arranged in the order of their atomic numbers, in horizontal rows called periods, and, according to the number of electrons in the outer shell, in vertical columns called groups. This symmetry is really quite astonishing. When Mendeleev's table was first published, it included many vacant places, thus predicting that new elements having the properties indicated would be found. All places are now filled, including those occupied by the noble gases, unknown in Mendeleev's time. If the dynamic elegance of atomic structure can be compared with the flowing grace of the corps de ballet, the precision of the periodic table is reminiscent of a well-drilled marine platoon.

Moreover, the predictive power of this elegant table is still great, despite the filling of all places (except as new, very short-lived elements are added at the end by the powerful new atom-smashers of modern physics). This deceptively simple looking table can still tell us not only which elements are likely to combine with which and in what proportions, but also which are likely to be associated with which, although not chemically combined. Such clues bear importantly on the geological assessment of future commercial sources of certain unusual metals and other elements, for instance.

Nevertheless, there are also striking departures from symmetry in the distribution of the elements in nature. Ninety of the first ninety-two are known to occur naturally on earth, and they have a relative abundance that is, in general, inversely related to their atomic and mass numbers. In other words (and omitting isotopes), the heavier the element the rarer it tends to be, and the reverse. Hydrogen, with its atomic number and weight of one, is the most abundant element in the universe. All other elements (and isotopes), being, in effect, multiples of hydrogen with added neutrons, are correspondingly less common. More significantly, perhaps, they can all be thought of as multiples of helium, modified by neutron capture and exchange—and helium is formed in the stars by the "burning" of hydrogen.

Origin of the Elements

It is now the prevailing view about the nature of matter that all of it, living and otherwise, is descended from neutrons. A few would still have it that these neutrons are constantly appearing in empty space, there to evolve into hydrogen atoms, which in turn are driven together by the pressure of starlight and their own gravitational attraction to form new stars. A now more prevalent view—the so-called "big-bang" hypothesis of astronomer George Gamow—is that the entire visible universe once consisted of one or several balls of neutrons endowed with such energy and mass that it collapsed or blew up, flinging the bulk of its matter into space with such force that it has expanded continuously ever after to form the present universe.

Where the ball (or balls) of neutrons and the accompanying

radiation may have come from is a question that transcends the bounds of science. It is the ultimate or penultimate question of first causes, belonging to metaphysics and theology. Constraints can be placed on scientifically permissible answers in terms of a time framework and the nature of the starting materials, but, beyond that, there is as yet no apparent way to deal with this question scientifically. Perhaps the "big bang," as it is sometimes called, was simply an initiating event in the most recent of a continuing succession of pulsating expansions and contractions within an eternal universe. Perhaps it was a divine act. We have no way of knowing.

The importance of having an initial ball of neutrons is that everything else can be made from them. Given a sufficient mass of neutrons, they react back and forth, generating energy, plus a variety of products, among which protons, electrons, and hydrogen are prominent. Given a starting temperature upwards of 10 or 12 million degrees Kelvin (a scale that begins at absolute zero, and which is equivalent to degrees Celsius or centigrade +273), hydrogen can be converted to helium (4 hydrogen atoms to 1 helium atom). This transformation is accompanied by a release of energy equivalent to the mass lost times the speed of light squared, as stated in Einstein's famous equation. Our sun, by burning hydrogen to helium at a deep internal temperature of perhaps 20 million degrees Kelvin (cooling to 5,750°K at its surface) thus generates the radiant energy to warm our planet and sustain the delicate web of life that clings to its evolving surface.

Gravitational heating at the centers of giant stars, together with heat from radiation, produces the conditions under which elements heavier than helium can be made. When temperatures rise above 100 to 200 million degrees Kelvin at densities 1,000 times greater than that of water, helium itself burns and is compacted to form new elements whose mass numbers are small multiples of its own—carbon, oxygen, and neon. At almost unimaginably high temperatures above 4 to 5 billion degrees, elements with mass numbers up to 14 or more times those of helium are made—like iron and related elements. Most heavier elements now appear to be made in red giants or comparable stars by processes involving neutron capture and decay. Very heavy elements approaching the atomic weights of uranium, however, are formed only at very high rates of neutron flux such as are achieved when a star implodes as a nova or a supernova to form a neutron star and blast its remnants into space.

Such an implosion was recorded by a heavenly flare, brilliant even in broad daylight, seen by Chinese observers in the year 1054 A.D. Its brilliance faded rapidly as it evolved into what is now known to astronomers as the Crab Nebula (figure 4). That correlation is established by backward counting based on the steady rate of expansion of the Crab Nebula. The actual implosion that produced the flare that evolved into the Crab Nebula, however, took place some 3,000 years before the beginning of the Christian Era. Like most stars, it is so far away that it takes several or many millennia for its light to reach the earth. Perhaps six such supernovae have been recorded in our galaxy in the last thousand years, but they are so short-lived that others may have passed unnoticed. And no fewer than seventy-seven bright supernovae were observed between 1885 and the end of 1976 in those galaxies that lie within easy telescopic seeing distance of our own. Thus they are not particularly unusual.

Be that as it may, it requires a really big bang—a supernovation—to synthesize the heaviest elements. And such supernovations not only bring about the synthesis of the heaviest elements, they also fling them into space for the later building of other stars or planets. Each new generation of stars and planets builds from the ashes of the last. The abundance and range of the heavy elements and isotopes found on Earth implies that the matter of which our solar system is made has been through at least one supernovation. Our sun, therefore, is evidently a second- or third-generation star, and the evolution of the population of stars with which it is associated, our galaxy, preceded that of the solar system.

Thus the elements are created from neutrons, hydrogen, and helium in giant stellar furnaces, novations, and supernovations. The statistical likelihood of any element forming is related to such events, as well as to the relative abundance of the basic building

Figure 4. Remains of supernova of 1054 A.D. The Crab Nebula in the constellation Taurus. (Photograph by 500-centimeter Hale telescope on Mt. Palomar. © California Inst. Technology and Carnegie Inst. Washington. Reproduced with the permission of the Hale Observatories.)

stuff involved in its construction. This explains why the abundance of elements in the universe tends to be inversely proportional to their atomic weights. As atomic weights increase, each heavier element is built from decreasing quantities of immediately ancestral elements by increasingly unlikely events. The population of elements to be found in any given solar system such as ours is both a product and a record of its previous history.

For Further Reading

Alfvén, Hannes. 1969. *Atom, man, and the universe.* W. H. Freeman & Co. 110 pp.

Krauskopf, Konrad, and Beiser, Arthur. 1973. *The physical universe* (3d ed.). McGraw-Hill Book Co. 717 pp.

Peierls, R. E. 1955. *The laws of nature.* George Allen & Unwin. 284 pp. (Undated paperback issue by Charles Scribner's Sons.)

Shapley, Harlow. 1963. *The view from a distant star.* Basic Books. 212 pp.

Weinberg, Steven. 1977. *The first three minutes.* Basic Books. 188 pp.

3

OF STARS AND SPACE

Man has always sought to discern his beginnings. The most rudimentary civilizations have legends to explain the origin of Earth and its inhabitants. Astronomers and their predecessors the astrologers from time immemorial have looked out to the stars and sought to read in them the structure of the universe and the fates of princes. Despite the application of their observations in matters of navigation, magic, and state, however, general theories of the universe were long a-brewing. Although improvements made during the sixteenth and seventeenth centuries by Copernicus and Kepler demolished the time-honored geocentric model of the Graeco-Egyptian Ptolemy, they still limited the universe to a single cluster of stars fading into space beyond our sun-centered solar system.

Although it was long recognized too from their degrees of brightness that stars were at different distances from Earth, the measurement of those distances and the discovery of relative stellar motions were slow in coming. Distances that could be measured directly using Earth's equatorial and orbital diameters were limited to the nearest stars. The fact that the apparent brightness of stars varied as the square of their distance gave crude clues at best. But exciting things were in store for a young Harvard astronomer named Harlow Shapley and his contemporaries early in this century. They discovered, from measurements on stars that fluctuated regularly in brightness, the *Cepheid Variables*, that the brightness changed in a cyclical manner that could be related to distance. This was based on the relation between brightness, color, and distance. The key to the universe was in their hands. Soon they were measuring distances to stars that clearly lay beyond the boundaries of our own *galaxy* (Latin for "milky way," designating an aggregation of stars that is separated

lastose

Figure 5. Neighbor and look-alike to our Milky Way galaxy, the Great Nebula in Andromeda, 1.7 million light years away but nearest to ours in space. (Photograph by 120-centimeter Schmidt telescope on Mt. Palomar. © California Inst. Technology and Carnegie Inst. Washington. Reproduced with the permission of the Hale Observatories.)

in space from other such aggregations—for an example, see figure 5). Man had his first glimpse into remote space, and with it came the sobering realization that, far from being at the center of the universe, he lived on a minor planet of a very ordinary star far from the center even of his own galaxy.

As stellar distances grew, new methods were needed for estimating them. That employed for the greatest distances involves the phenomenon of the *red shift*, which had been giving speed of motion of stars since first measured by Sir William Huggins in 1868, but which was first calibrated against the Cepheid Variables for the estimation of distance by Edwin P. Hubble of the Mt. Palomar Observatory in the 1920s. Red shift is a displacement toward longer wavelengths in the measured light-spectra of stars or galaxies that move away from the observer. It is now determined by calibrating the absorption spectra of stars against a laboratory standard and then measuring the distances between distinctive lines observable in both, although Huggins did it by eye. Red shift is to light as the change toward deeper tones of a receding automobile horn is to sound. If the change in wavelength or pitch can be measured and related to distance from the observer, the speed of motion of its source and its direction of movement can be calculated. Hubble recognized that such an interpretation called for galaxies with large red shifts to be both very far away and receding from us at fantastic velocities. Indeed the galaxies that make up the visible universe are receding in all directions—much in the manner that spots on an inflating balloon (or, better, raisins in a loaf of baking bread) move in all directions away from one another and the center, only at incredibly high speeds. Continuing observation and measurement of red shifts led to the recognition that, although some stars in our own and nearby galaxies are actually approaching us, almost all galaxies everywhere in the visible heavens appear to be retreating in all directions from all others. Thus, omitting a few places where galaxies seem to be colliding, no matter where one is in the universe it always looks about the same, and one's own galaxy always appears to be at the center of it.

A century-old riddle had found an explanation. A German astronomer of the early nineteenth century, H. W. M. Olbers, had pointed out that the quantity of radiation from fixed stars should keep everything much too hot for life to exist. Now it was seen that planets could have been made habitable because the heating effects of stellar radiation were canceled by the expansion of the universe. And, thanks to the work of Hubble, astronomers can now correlate degree of red shift, speed of retreat, and distance, such that measured rates of retreat from red shift can be converted to distances for stars whose position in space is unknown from other evidence.

In this scheme of things the farthest stars and star clusters that can be observed in our optical and radio telescopes are at distances so unimaginable that they are expressed in billions of *light years*—a light year being the distance light travels in a year's time, or slightly less than 9.5 trillion kilometers. The farther away the object, the faster it seems to be fleeing. The most fantastic are the quasi-stellar bodies, or *quasars*, concerning whose interpretation there is vigorous disagreement. Quasars are visible radio sources that display unusually strong emission of radio energy and very large red shifts. If their red shifts are interpreted as a measure of velocity of recession, some of them would appear to be flying into space at rates approaching the speed of light and thus to be at astonishingly great distances for their degree of brilliance. Such stars are thought by some astronomers to reflect the very early history of the universe—to be phenomena as distant in time as they are in space, objects that may no longer even exist. Observing such distant stars thus could be a way of studying the history and former structure of the universe.

Other astronomers have noted that some quasars appear to be associated with galaxies of stars that have lesser red shifts and that some relatively closely associated quasars have dramatically different red shifts. It is not understood why this should be so, although ingenious hypotheses have been advanced to explain the greater red shifts. Some suggest that they result from the splitting of light particles *(photons)*. Others have proposed that quasars are not stars at all but strange events in the nuclei of galaxies. Still others claim that quasars are not even at cosmologically important distances.

Such competing hypotheses, challenging each other for preeminence are the sign of healthy science. But until this problem is resolved, quasar red shifts cannot be taken as conclusive evidence of distance or speed of retreat. In fact, even ordinary optical red shifts have periodically been questioned as indicators of speed of flight and distance. It was suggested in 1972 by the international team of J. C. Pecker (France), A. P. Roberts (Australia), and J. P. Vigier (France) that, if the light particle or photon had even the tiniest conceivable rest mass (less than one to the forty-sixth decimal place of a gram), collision of photons could cause a shift of frequency toward the red end of the spectrum. This modern version of the old idea of "tired light" visualizes photons on their way out of a light source colliding with other photons outside the source and thus being shifted toward the red. If this were true, the red shifts of very distant objects could be at least partially explained as results of light-scattering by the universal background radiation. And calculation of the temperature of the background radiation based on the assumption that it is the cause of the red shifts observed yields temperatures equivalent to those calculated from other observations.

The preferred view among astronomers, nevertheless, is that the optical red shifts of galaxies, despite local seemingly random motions, do support the interpretation of a universe that is expanding, and at staggering rates. The age of this universe since the primordial "big bang" (see preceding chapter), based on red shift, is estimated to be about 17 to 18 billion years. This age, and the time-distance significance of red shift, now seems to be confirmed by studies of the decay rates of the rare radioactive isotope rhenium-187 to stable osmium-187 and the concentrations of these isotopes in meteorites. Successful application of this method, first suggested by D. D. Clayton of Rice University in 1963, had to await refinements then unavailable. Now determinations independently carried out by cosmochronologists at Lawrence Livermore Laboratory and the University of Chicago indicate an age of 18 billion years for the beginning of the formation of the elements. We can round this off to an age of roughly 20 billion years for the visible universe—the *visible universe* being that region of space where everything is flying outward at less than the speed of light and thus can, in theory, be observed. Whether anything lies beyond such limits can be neither affirmed nor denied.

The restless mind of man does not like such answers, however. Thus we find it suggested by some that perhaps the edge of the

visible universe is simply a division between two antithetical but symmetrical parts—a universe of matter where everything travels at less than the speed of light, and a mirror-image universe of antimatter where everything goes faster than the speed of light. Each universe would thus be invisible and unknowable to the other. Another idea, advanced by the British cosmologist Fred Hoyle, is that the edge of our universe is a boundary between positive and negative time. Light does cross the boundary, he claims, in the form of so-called "universal black body radiation."

It takes an unusual kind of mind to think such thoughts. For most of us there is plenty to be learned within the limits of the universe we can see. One of the most interesting discoveries is that the history of stars, like that of mice and matter, is probably one of evolutionary progression, and that the likelihood of a star having habitable planets is related to its mass and its position in the evolutionary sequence. It is possible to represent this sequence, into which all stars can be fitted, in the form of color-brightness diagrams in which color, translated to surface temperature, is plotted against luminosity or brightness relative to the sun. A graph of this sort is called a Hertzsprung-Russell diagram, after the Danish astronomer Ejnar Hertzsprung, who first constructed one in 1910, and Henry Norris Russell of Princeton University, who improved the method.

In a Hertzsprung-Russell diagram such as shown in figure 6, surface temperatures of stars included range from about 3,000 to 15,000° K, while brightness varies from about $^{1}/_{10}$ to 100 times that of our sun. This sun, a yellow dwarf star with a surface temperature of 5,750° K, lies near the middle of the spectrum of the so-called *main sequence stars*, which range in color from red, through orange, and into yellow where they finally attain surface temperatures of 7,000 to 10,000° K. At that point, corresponding to the burning of about 10 percent of their hydrogen to helium, they expand and leave the main sequence. The expansion causes their surfaces to cool, while their interiors remain hot. After it leaves the main sequence, the star's further evolution depends critically on its mass. Some collapse steadily to form white dwarfs and eventually denser states of matter. Others implode in novae or supernovae.

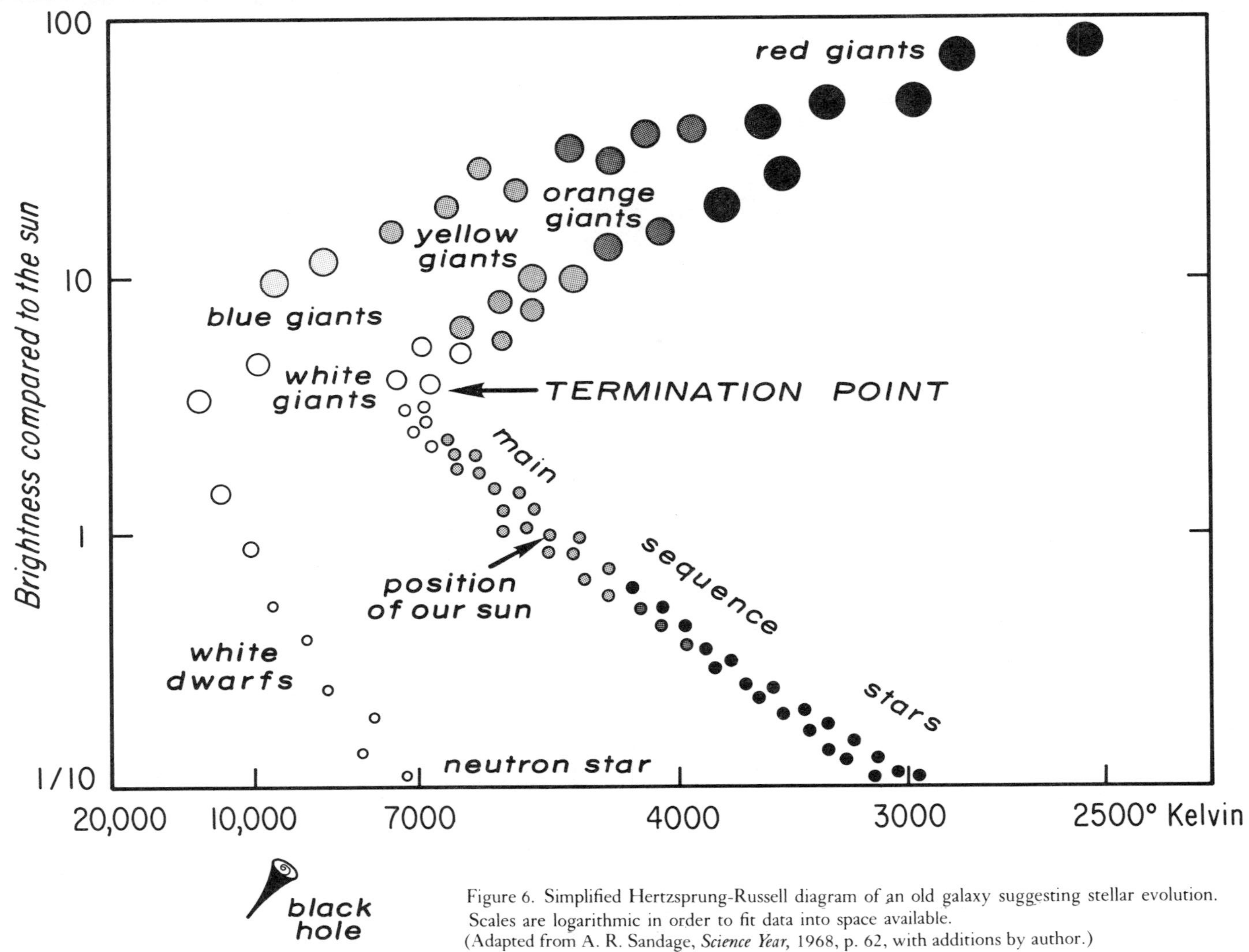

Figure 6. Simplified Hertzsprung-Russell diagram of an old galaxy suggesting stellar evolution. Scales are logarithmic in order to fit data into space available.
(Adapted from A. R. Sandage, *Science Year,* 1968, p. 62, with additions by author.)

White dwarfs are so compressed that a thimblefull of white-dwarf matter would weigh about ten thousand tons. Yet the white dwarf is the end stage of collapse only for stars whose mass is less than about 1.2 times that of our sun (1.2 solar masses). More massive stars are incapable of supporting their own weight at the density of a white dwarf and must collapse into even smaller volumes and greater densities. They can stop collapsing only when crushed to the incredible densities of neutron stars or black holes. During such an abrupt collapse the outer gaseous layers lag behind and the shock wave of the collapse blasts its quota of new elements outward in what appears as an explosion—the supernovation. A *neutron star* is characterized by the fact that most of its former orbital electrons are compressed into its protons to form neutrons. Its expected diameter would be no more than 10 to a few tens of kilometers, compared with perhaps some thousands or several tens of thousands of kilometers for a white dwarf and 1.4 million for the sun. A piece of such a star as big as the end of your index finger would weigh about a billion tons—100,000 times that of an equivalent amount of white dwarf matter. If it fell on Earth it would go right through it, like a bullet through soft butter. The force of gravity on a neutron star is so great as to defy imaginable simile. If you could survive a journey to such an Alice in Wonderland world you would arrive compressed to a size so small that an eminence as high as your thumb is wide would loom as a great mountain, requiring all the biological energy you could generate in a lifetime to ascend.

The conversion of an ordinary star into a neutron star also increases its speed of rotation as the mean distance of its mass from its axis of rotation becomes smaller—just as a pirouetting figure-skater turns faster and faster as she lowers her arms. Theory predicts that in the end such neutron stars would be rotating about one to ten times a second or faster, generating radio signals that pulsate with each rotation. Evidence that such stars exist is found in the presence of powerful local radio sources called *pulsars*, first discovered in 1967 by Anthony Hewish and his associates at Cambridge University, in England. Their first pulsar was observed to emit bursts of radio waves at precise intervals of

1.3373019 seconds. Other observers soon found more, including one in the Crab Nebula that pulsed thirty times a second. It is believed that these pulses are beams of neutron-star radiation, concentrated in the radio range, that sweep across Earth like the rays of a searchlight with each rotation of the neutron star, focused by a magnetic field millions of times stronger than any we have been able to generate on Earth.

These thoughts may seem like science fiction, but they are founded on a large body of reliable astronomical observation. And there is more to the story. As early as 1798 the French astronomer and mathematician Pierre Simon de Laplace suggested that, if a star were dense and massive enough, the velocity required for gravitational escape from its surface would be greater than the speed of light and that it would, therefore, be invisible. Even neutron stars cannot support a load greater than two or three solar masses. A more massive star must collapse until, consistent with relativity theory, light cannot escape and the condition visualized by Laplace is fulfilled. Such objects, assuming they exist, have been appropriately called *black holes.* A black hole may be thought of as a region in space of such fantastically high density that nothing ever leaves it. Volume at its center shrinks toward zero and density approaches infinity. Its ancestral star has essentially vanished from the visible universe—although, when quantum effects are taken into account, it appears that black holes actually do emit small amounts of radiation.

Astronomers are now engaged in an intensive search for black holes. Inasmuch as objects of this nature can be only a few kilometers in diameter and can scarcely be seen in any case, their existence can be established only by indirect evidence such as gravitational field and charge. As suggested by Caltech's Kip Thorne in 1968, such indirect evidence might be seen in the behavior of matter about to be swallowed by the tremendous gravitational attraction of a black hole. It has now been reported by several students of the black-hole problem that behavior of the sort hypothesized is manifested by X-ray sources associated with the double star system Cygnus X-1. Even more indirect, but also suggesting the existence of black holes according to some investigators, are peculiarities in the behavior of galactic gamma radiation. Indeed, thanks largely to the insights of Stephen H. Hawking of Cambridge University, the concept of the black hole as the final stage of stellar degeneration is rapidly becoming a part of the conventional wisdom of astronomy. Not only do black holes in fact appear to exist; it seems that they may be fairly common. Could it be that the structure and motion of spiral galaxies, our own or the Great Nebula in Andromeda (figure 5) for example, reflect a process of collapse into black holes at their centers?

Our sun, which entered the main sequence as a result of the collapse of dispersed matter (to form it and its planets) around 4.6 billion years ago, is happily in no immediate danger of collapsing into either white dwarf, neutron star, or black hole. Eventually, however, when about 10% of its hydrogen has burned to helium, it will expand, cool at the surface, and leave the main sequence to become a red giant, incinerating all life within its expanding radius. From the sun's dimensions, position in the main sequence, and rate of hydrogen burning, it can be calculated that this stage in solar history still lies some 4 to 5 billion years in the future.

Here it is appropriate to anticipate the existence of conditions that could give rise to life as we know it. The most restrictive conditions we can set call for life to be limited to planets about the size of Earth, at about the same distance from a central sun that has roughly the brightness of ours. The only evidence we have that such situations might exist beyond the earth is indirect and statistical. This was developed at length by the late Sir Harold Spencer Jones, then astronomer royal of Great Britain, and others. The estimated number of stars in the visible universe is equal to 10 followed by 21 zeros (shortened to 10^{22} in exponential form, where $10^1 = 10$ and $10^2 = 100$). Among them it seems statistically likely that some will have planetary systems, that some of those with planetary systems will possess suitable levels of brightness and radiation, and that some of their planets will be similar enough to the earth and at the right distance from their suns to support conditions favorable to the origin and persistence of life. If each of these three possibilites has as much as a one in a million chance of being true, there might be ten thousand or more inhabited planets in the universe. As there is small likelihood, however, that we will ever be able to visit such planets outside the solar system or communicate with whatever life forms may exist on them, such projections offer little prospect either for verification or disproof.

What will be the ultimate fate of the universe and whatever life exists within it? Is the universe open and infinite so that everything will continue to expand forever? Or will the energy needed for expansion eventually dwindle and be overridden by the intergalactic attraction of gravity so that the pieces will fall back together once again—a closed and finite universe? Or will it all reach, or does there now exist, a steady state condition wherein matter and energy throughout the universe are constantly being exchanged for one another? Would a view of the universe from any point in space or time always be the same? In recent years, the idea of a true steady-state universe has essentially collapsed under the weight of accumulating contrary evidence. In addition, all other experience tells us that it is a fundamental attribute of all systems that over the long term they evolve. Why should the universe be an exception?

Moreover, when we look out into space we also look back into time relative to ourselves. Some of the radiation from the outermost star clusters observed left them billions of years ago. What is particularly interesting about these very distant and therefore very primitive sources of radiation is that some of them appear to be more densely aggregated than those closer to us. As the visible light from the closer stars and galaxies has not been on its way toward us nearly as long as the radio waves from the distant ones, we are seeing them at a more recent stage in their history. Such differences in time and density of aggregation imply that stellar matter has progressed from a more to a less dense state over billions of years of cosmic time. This is strong evidence in favor of the interpretation that sees the universe as an evolving system. We will return to the question of whether it is open and infinite or closed and finite at the end of the next chapter.

For Further Reading

Coleman, James A. 1963. *Modern theories of the universe.* Signet Science Library, 211 pp.

Gamow, George. 1947. *One, two, three . . . infinity.* Mentor Books. 318 pp.

Laurie, John (ed.). 1973. *Cosmology now.* Published by the British Broadcasting Corp. 168 pp.

Ley, Willy. 1963. *Watchers of the skies.* Viking Press. 528 pp.

Sandage, Allan; Sandage, Mary; and Kristian, Jerome (eds.). 1975. *Galaxies and the universe.* University of Chicago Press. 818 pp.

4

WHY THE SKY DOESN'T FALL

What keeps the stars in the sky? Why don't the heavens collapse? Thomas Jefferson is reported to have reacted to a Yale geologist's account of a meteorite fall in 1807 near Weston, Connecticut, with a remark to the effect that it was easier to believe that a Yale professor would lie than that stone could fall from the heavens. Not so Chicken Little, who sought his king to warn him that the sky was falling.

Like most things children are curious about, the question of what keeps the sun, the moon, the stars, and the other planets from colliding with the earth is neither silly nor easy to answer. Indeed it is a way of asking about the laws of science and the dynamic principles that account for the structure of the universe, our galaxy, and our solar system.

It is now well known that Earth and its cosmic companions are not static objects. All parts of our solar system, galaxy, and universe are in constant motion. Planets and other satellites rotate on their axes and trace orbits about their parent bodies. Stars also rotate, follow prescribed courses through their galaxies, and send blasts of cosmic radiation streaming into space. Asteroids chase one another through a belt that follows a regular course around the sun, unless they break out of it to become stony or metallic meteoroids that go soaring through the seeming emptiness beyond until they collide with something solid to become meteorites. Comets of dirty ammonia-ice and other frozen substances volatile at Earth's surface temperatures pursue highly elliptical orbits about the sun—"tails" of volatilized matter blown behind or ahead in the solar wind, depending on whether they are moving toward or away from the sun, until they crash or vaporize. And galaxies flee into space at incredible speeds toward unknown destinations.

Why aren't there more collisions among all these flying objects? Why doesn't everything come together somewhere? What keeps Earth from falling into the sun or the sun from approaching and engulfing Earth? Why *doesn't* the sky fall?

It makes one wonder how Newton could be right about everything in the universe attracting everything else according to a law of gravitation, for taken at face value this seems to imply that everything should fall together. In fact, most or all of what we observe is explained by a combination of gravitational and other forces, the former attractive and the latter repellent. In the case of the solar system, the planets remain in orbit around the sun for basically the same reason that electrons remain in orbit around the nucleus—the force of attraction by the central mass is exactly balanced by the centrifugal force that would otherwise cause them to fly off in a straight line through space.

In fact all of the attractional and repellent forces between the objects within our solar system and beyond are so balanced that their effect is to maintain the motions of essentially all components in predictable patterns. This may be illustrated within our solar system by the ordered relationships between the distances of the various planets from the sun.

The average distances from the sun of the elliptical orbits of the planets known in 1772 were found by J. D. Titius, professor of mathematics and physics at Wittenberg, Germany, to have a regular relationship—later independently discovered by J. E. Bode of Berlin and formulated in what is now known as the Titius-Bode rule (sometimes wrongly given as Titus-Bode). If we take Earth's average distance from the sun of 150 million kilometers as one *astronomical unit* (A.U.), the average distances from the sun of other planets and the asteroid belt can be found by means of a simple empirical formula. This consists of taking the observed value of 0.4 A.U. for the distance of Mercury from the sun and adding to this 0.3, 0.6, 1.2, 2.4, 4.8, and 9.6 A.U. Thus we obtain the values 0.7 A.U. for Venus (0.4 + 0.3), 1.0 A.U. for Earth, 1.6 A.U. for Mars, 2.8 A.U. for the asteroid belt, 5.2 A.U. for Jupiter, and 10 A.U. for Saturn. The corresponding values for planets unknown in 1772 would predict the next three positions outward from the sun at 19.6, 38.8, and 77.2 astronomical units.

Enter William Herschel, a professional musician who fled from Germany to England to avoid military service and first took up astronomy as a hobby at age thirty-five. Eight years later, in 1781, on a telescope of his own making, he discovered the planet Uranus at an average solar distance of 19.2 A.U., within 2 percent of the "predicted" distance of 19.6 A.U.—a feat for which he was knighted by George III and set up with his own observatory and a lifetime pension.

When attempts were made to compute the orbit of Uranus, however, things went wrong. Older sightings that had not recognized its planetary nature were used, but no ellipse of past motion that could be found accurately predicted future motions. Additional observations only increased the discrepancy, until it began to seem that Newton's law of universal gravitation might not apply under some circumstances. That would have imposed fantastic difficulties in trying to unscramble the universe beyond our immediate solar system. Then it occurred to astronomers that the discrepancies in the orbit of Uranus might be caused by the gravitational attraction of another then unknown planet beyond, which would deflect Uranus slightly from the control of the sun. The probable location and properties of such a body were independently calculated in 1845 by John Adams, then a student at Cambridge University, and in 1846 by the French astronomer Urbain Leverrier, both agreeing that the object perturbing the orbit of Uranus should have a mass about twenty-five times that of Earth and be at an average distance of 36 A.U. from the sun.

When the object, Neptune, was found by astronomers at the Berlin Observatory in September 1846, it turned out to have a mass only eighteen times that of the earth and to be at an average distance of only 30 A.U. from the sun—a good bit closer to the sun and to Uranus than predicted by the Titius-Bode rule, by Adams, or by Leverrier. The discrepancy, however, is in no sense a violation of Newton's law. Neptune's closer-in position is balanced by its smaller mass. The combination of smaller mass and closer position produces approximately the same gravitational deviation as would have been produced by the predicted larger mass and distance. Newton's law was vindicated. A planet that did not fit the Titius-Bode rule had its mass compensated in the right direction to make up for it.

But even with the discovery of Neptune, small discrepancies continued to be found in the orbit of Uranus, as well as in that of Neptune itself. Past experience suggested another planet still farther away from the sun. The expected planet was so far away, however, that its discovery required special instrumentation, a painstaking search, and more than the usual amount of luck. The quest for it was begun in 1906 by Percival Lowell at his observatory in Flagstaff, Arizona, and came to an end there in 1930 when the then twenty-four-year-old Clyde Tombaugh spotted its giveaway planetary motion from differences in position relative to adjacent stars on photographs taken six days apart. Named Pluto, it has an average distance from the sun of 39.5 A.U.—barely more than half that predicted by the Titius-Bode rule. As in the case of Neptune, however, the nearer distance is compensated by much smaller mass than expected. The mass of Pluto is so small, in fact, that it has not yet been accurately measured, although the recent determination that it consists at least partially of methane ice implies a mass smaller than that of Earth's moon. It has been suggested that Pluto, instead of being a primary part of the solar system, may actually be an escaped satellite from another planet or a foreign object that wandered into our region of the universe at some date subsequent to the origin of the other planets, to be captured by the sun's gravitational field. The view current among planetary astronomers, however, is that Pluto is a remnant of solar system debris, like that of the asteroid belt between Mars and Jupiter. We see it today only because it is locked into place by a combination of centrifugal and gravitational forces.

Without the sun's gravity to constrain and curve their paths of flight, the planets would soar off into space along straight lines instead of following regular elliptical orbits about their central star. But each of the planets also exercises its own gravitational attraction on all of the other planets as well as on the sun itself. Could this have anything to do with the Titius-Bode rule, or is that just a numerical curiosity? What makes it seem unlikely to be a mere mathematical curiosity is the observation that the spacings of the inner satellites of Jupiter, Saturn, and Uranus display similarly regular relationships.

Such relationships have led some astronomers to conclude that the original orbits of the dust clouds from which the planets condensed may have been different from those now observed for the planets themselves. Early in the history of the solar system the several condensing dust clouds of the initially lumpy solar disc probably affected one another in such a way as to produce orbital changes. As a result, the several planets found those orbits that were least affected by external forces. This is known as the "theory of dynamical relaxation," and it has recently received strong support from computer modeling.

And that is why the sky doesn't fall, thanks to Newton, the balance between gravity and centrifugal force, and the theory of dynamical relaxation.

At least that's why it won't fall in our corner of the galaxy any time soon. But what would happen to the expanding universe were the expansive forces to weaken and gravitational collapse to take over? One group of cosmologists, to be sure, holds that the universe will expand forever. That view is based on the concept that we live in an "open" universe that does not contain enough matter to generate the gravitational attraction needed to pull it back together again. However, studies by Riccardo Giaccone and other X-ray astronomers at the Harvard-Smithsonian Center for Astrophysics, using the Uhuru satellite, have revealed quantities of gas in intergalactic space that had not been known before because it is mostly primordial hydrogen and helium that is too hot to show up in the visible spectrum. They believe that this newly discovered intergalactic gas "could represent a significant percentage of the so-called 'missing mass' needed to close the universe." If this is true the stars, galaxies, and star-clusters cannot rush away from one another indefinitely into an infinite void. Instead universal gravitation must eventually overcome the energy of flight of these objects so that they either slow down and start falling together again or reach a truly steady state of mutual balance between equal numbers of galaxies of matter and antimatter.

Estimates of the present age of the universe and where we are in the hypothesized cycle of expansion and contraction vary widely. From studies of cosmic black body radiation and relativity considerations, astrophysicists R. A. Alpher of the General Electric Company and Robert Herman of the General Motors Corporation proposed in 1975 that the half-life of a closed universe would be about 119 billion years—that is, 119 billion years expanding and a similar time collapsing. Since, as noted in the preceding chapter, estimates employing a variety of criteria converge on an age of 18 to 20 billion years for the initial "big bang" that gave rise to the present universe, it could be another 100 billion years before general contraction sets in. Even now, however, the work of astronomer Beatrice Tinsley of Yale University and others shows that the lesser galaxies of a cluster may fall into its brighter members from time to time.

Evidence that collapse on a broad scale has begun, if and when it comes, will consist of the arrival of starlight that shifts toward the ultraviolet rather than the red end of the spectrum (excepting the local ultraviolet shifts seen in our own and nearby galaxies as an effect of rotation or local motions). Then we will know, like Chicken Little, that the sky is really falling. But it will probably have been falling for billions of years and will likely still have billions of years to go when the first signal reaches us. If collapse is to be the fate of the universe, all life will have been incinerated or frozen long before it ends.

At or near the end of such a collapse, all matter may be condensed into a new ball of neutrons, similar to the one from which it started. But this would be a temporary state. It would either continue to collapse into a universal black hole, or the rapid generation of energy in the relatively gigantic neutron furnace would lead to a new big bang and a new dash into space. Neutrons now tied up in the molecules of which we are made would be a part of that bang. In such a model of the universe, the molecules of microbes, men, and minerals are made of the same fundamental stuff, consigned to the same end, and, if they do not disappear forever into the ultimate black hole, all are ultimately destined for an unknown reincarnation. In that sense they are all equally eternal. They cannot know this however, unless they are in people or similarly advanced forms of life on the planets of other stars.

For Further Reading

Abel, George. 1973. *Exploration of the universe* (updated brief edition). Holt, Rinehart & Winston. 483 pp.

Brandt, J. C. and Maran, S. P. 1972. *New horizons in astronomy*. W. H. Freeman & Co. 496 pp.

Gott, J. R. III; Gunn, J. E.; Schramm, D. N.; and Tinsley, B. M. 1976. Will the universe expand forever? *Scientific American*, vol. 234, no. 3, pp. 62-79.

Page, Thornton, and Page, L. W. (eds.). 1966. *The origin of the solar system*. Macmillan. 336 pp.

Wood, John A. 1968. *Meteorites and the origin of the planets*. McGraw-Hill Book Co. 117 pp.

5

ON THE BIRTH OF PLANETS

How and when did Earth and its sister planets begin? What is their relation to the sun and the other stars? Have they always existed in a solid state, with liquid water and a gaseous atmospheric envelope, or have they evolved from a once more primitive state? If the latter, what was that formerly more primitive state like?

Man has pondered such questions throughout the millennia of his conscious history. Observations by shepherds, sailors, soothsayers, and scientists over those same millennia led gradually to the recognition that to account for the earth is also to account for the solar system. This means that whatever explanation is proposed, it must be consistent with everything we now know or can find out, not only about the sun, moon, and Earth, but also about the eight other planets and their satellites, some 30,000 asteroids, and untold numbers of meteoroids and comets.

No society is without a story of creation. Philosophers of many lands and times have invented accounts of Earth's origin, some mythical, some scientific or quasi-scientific. A long dominant one, thought up by Laplace in 1796, called for simple condensation of a solar disc or nebula, leading to an increased rate of rotation as called for by the conservation of rotational force. As the proto-sun whirled faster and faster, contracting hot gases were thought to have condensed to form gaseous or molten proto-planets from which those we now know evolved. Another idea, suggested early in this century by two British scientists—the astronomer Sir James Jeans and the geophysicist Sir Harold Jeffreys—called for condensation of the planets from a jet of hot gas, drawn out from the sun by near-collision with a passing star. As long as such thoughts prevailed, it was generally supposed that Earth had started out as a molten mass that gradually cooled and shrank to its present size. In the process seas were imagined to have appeared as a result of condensation from an external shroud of water vapor around the originally molten but later cooling earth. Water from such condensation supposedly collected in low places underlain by heavier rocks, while continents stood high because underlain by lighter rocks. And mountains were thought to have originated by contraction incident to cooling, like the wrinkled skin of a baked apple. It was from such a model that the brilliant nineteenth-century English physicist William Thomson, later Lord Kelvin, calculated that the age of the earth was most probably no more than about 40 million years—based on the time required for an initially molten Earth to have cooled to its present state assuming no secondary sources of heat.

Such ideas can now be tested against some well-established observations and measurements. As we shall see in the next chapter, Earth's age is more than 100 times as great as Kelvin thought. In addition computations made in the early 1950s by the American geologist W. W. Rubey imply that only about 10 percent of the water now in the oceans could have existed as water vapor in contact with even a completely molten Earth, and that 90 percent of our hydrosphere, therefore, is not accounted for by the Laplacian or Jeans-Jeffreys hypotheses. Other critical measurements have to do with the sun's mass and rate of rotation, and with the orbital velocity and masses of the planets.

From relatively simple measurements we can calculate an important property of a rotating system such as the sun or the entire solar system. It is a measure of rotational force called *angular momentum.* Angular momentum is the product of the sum of the masses of all parts of a rotating system multiplied by their distances from the axis of rotation times their rotational velocities. It is characteristic of rotating systems that they conserve their angular momentum. The total amount must always remain the same. If there is more than one body in the system, angular momentum may be transferred by sufficiently canny bookkeeping from one account to another, for example from the sun to its planets or from Earth to its moon. It cannot leave the system as long as no external forces are involved. This means, as we have already seen in several connections, that if the average distance of a mass from its axis of rotation is increased, the rate of rotation must slow down. If that distance is reduced, the rate of rotation must speed up. When that pirouetting figure skater slows down or speeds up by extending or dropping her arms, her mass and angular momentum remain unchanged. She is simply extending and contracting her distribution of mass relative to her rotational axis. She eventually stops, of course, because of external friction, but no significant forces of this kind affect the solar system.

If either Laplace's nebular hypothesis or the so-called gaseous-tidal hypothesis of Jeans and Jeffreys were true, the conservation of angular momentum would require the sun to rotate much faster than its present rate of once every twenty-seven days, (at its equator; 31 days at its poles). Two more beautiful hypotheses killed by stubborn facts! Indeed the sun, with almost 99.9 percent of the mass of the solar system, has less than half of one percent of its angular momentum. Jupiter, on the other hand, with only a minute fraction of the total mass, but with great distance from the rotational axis of the solar system, great orbital velocity, and about 2.5 times the mass of all the other planets combined, represents about 99 percent of the angular momentum. This seemed a puzzle to match the legendary Gordian knot. To explain planetary origins it was believed necessary to move a large fraction of the angular momentum outward from the sun without transferring very much of its mass.

An old English madrigal sings of "pretty maidens more nimbler than eels." What was needed to get out of this dilemma were scientists with equally nimble wits. They were not long in coming up with alternative hypotheses. A number of modern theoreticians, including the German physicist C. F. von Weizsäker and the Americans Gerard Kuiper and Harold Urey, have contributed to a model that produces the planets by condensing them from dust clouds at varying distances from the center of a solar disc that once occupied the entire space defined by the present planetary orbits. This model also seemed to run into trouble with angular momentum, however, unless a variation was added. One suggestion that copes with this astronomical puzzle adds to the general idea of a condensing solar nebula a special mechanism for the outward transfer of a large share of its angular momentum without an equivalent transfer of mass.

When all of the clues are fitted together, the following plausible story emerges. An imploding star reaches temperatures of billions or tens of billions of degrees Kelvin and a high rate of neutron interaction. New and heavier elements are created from the fusion of helium atoms and from neutron exchange and decay to form protons and electrons. The star becomes a supernova, flinging its outer gaseous shells into space. But expansion is a cooling process, and soon the gaseous cloud or nebula so produced begins to condense again and to rotate. Our sun is thought to have emerged from the last ancestral supernovation as an enormous rotating disc of highly diffuse, hot gaseous matter. It extended beyond the reach of the outermost planets—beyond where Neptune now is, or even Pluto. It has been suggested, however, that the sun did not rotate faster as it condensed, as the law of the conservation of angular momentum would seem to predict. Only relatively small amounts of matter were transferred outward or left over after condensation, yet a lot of angular momentum resides in the planetary accounts, especially that of Jupiter.

One explanation for this apparent cosmological sleight of hand, suggested by Fred Hoyle, calls on the special properties of the hot ionized gases (or plasma) that would have made up most of the solar disc. Magnetic fields within such gases may have permitted angular momentum (and some matter) to spiral outward along lines of force with relative ease, whereas movement across them would be difficult. Matter moving outward from our sun, insofar as it was not lost to outer space, presumably condensed in the neighborhood of the present planets. The farther a given planet is from the rotational axis of the sun, which is also the orbital axis of the planets, the larger will be its share of the total angular momentum per unit mass and speed. Thus Jupiter, far from the sun, and with a relatively great mass and orbital velocity among the planets of the solar system, gets the lion's share. Neither this nor any other explanation, however, as yet resolves the difficulties with angular momentum to the satisfaction of all. Indeed it is now believed by some astronomers, such as Jeremiah Ostriker of Princeton University, that the problem doesn't exist.

They conclude that initial irregularities in the distribution of matter (and angular momentum) within the primordial solar disc could account for the apparent anomalies.

Just how did the condensation of the planets in our solar system take place then, and what was the nature of Earth when it first reached approximately its present mass? A clue to the basic process is found in an unexpected source. As comets approach the sun, their tails stream out behind them as good tails should; but as they move away from the sun the "tail" rushes ahead. The ionized particles of ammonia and methane that make up the cometary tails are "blown" outward by the sun's corpuscular radiation—the solar wind. In a similar way, the hot gaseous matter of the condensing solar nebula was probably also blown outward from the center, much of it being lost to interstellar space. The lighter elements traveled farther, while the heavier ones remained closer to the condensing sun. And some quantity of residual matter was present from the beginning in local clusters throughout the solar disc. Thus the outer or Jovian planets (Jupiter, Saturn, Uranus, Neptune, and Pluto) seem to consist mainly of hydrogen and helium, while the inner or terrestrial planets are made of varying amounts of heavier matter. Their initial envelopes of lighter gaseous materials were driven outward by the solar wind at then-existing temperatures, to coalesce with or bypass the outer planets. Support for such a view is found in the fact that there is a general decrease in the densities of the planets going outward from the sun.

Planet Earth, in short, seems to have grown by an initial aggregation from a cloud of dust particles and the subsequent infall of larger masses that collected around a center about 150 million kilometers from the sun. Once the solar nebula had cooled enough to permit the heavier atoms of this previously gaseous matter to condense into a solid state, such aggregation could proceed rapidly. Collisions within the dust cloud would lead to the snowball-like growth of larger masses, whose gravitational attraction for one another would escalate the process. The terminal stages of accumulation presumably would be marked by a very rapid infall of remaining particles. Theory and evidence to follow are in accord with the idea that this process of aggregation occurred rapidly in a geological sense in the terminal phases of the shrinking ancestral solar disc.

One more point before going more deeply into the question of ages. The relation between Earth and its moon, Luna, requires explanation. Lunar exploration has thrown much light on this problem and promises eventually to open a window on at least some parts of the presently obscure early history of the earth-moon system. Compared to her big sister, Luna is relatively poor in volatile components and rich in elements that withstand great heat. If the moon did originate by detachment from Earth, these differences show that it must have been detached at an early stage of the planet's evolution. The moon, therefore, could not have left a scar, such as the Pacific Ocean was once supposed to be. Other explanations for the earth-moon system propose separate condensation of Luna and Earth from the same dust cloud or from different but closely adjacent dust clouds, or formation of the moon elsewhere in the solar system, to be captured later by Earth's gravitational field. Continuing lunar studies have weakened hypotheses of either capture or detachment for lunar origin but have not eliminated them. All major hypotheses are still vigorously defended and are likely to be so for some time to come. Nevertheless we can place tighter constraints than ever before on the age, crustal differentiation, and early history of both Earth and its moon. The day can no longer be remote when we shall have a unified and well-dated history of these interesting but discrepant twin planets.

For Further Reading

Alfvén, Hannes, and Arrhenius, Gustaf. 1976. *Evolution of the solar system.* Natl. Aeronautics & Space Admin., NASA SP-345. 599 pp.

Henderson, Arthur Jr. and Grey, Jerry (editors), 1973, *Exploration of the solar system.* Natl. Aeronautics & Space Admin., NASA EP-122. 67 pp.

Hoyle, Fred. 1975. *Highlights in astronomy.* W. H. Freeman & Co. 179 pp.

Soffen, G. A., and others. 1976. Scientific results of the Mars Missions. *Science* 194: 1274-1353.

Whipple, Fred. 1964. The history of the solar system. *Proceedings of the National Academy of Sciences* (U.S.) 52: 565-94.

II

EARTH

6

ON THE RECKONING OF TIME

Where do the sun and its planets stand in the time scale of the universe? Can events in the development of the solar system be dated and placed in sequence relative to one another? Above all, how can we establish a succession of events in Earth history and calibrate that succession in years, including the origin and ascent of man?

In the 1650s Archbishop James Ussher of Ireland epitomized a then-common belief in uncommonly precise terms by announcing that the earth was created in 4004 B.C., on a particular day and at a specified time. Ussher's estimate is entirely consistent with then-prevalent assumptions as to the literal accuracy of biblical genealogies and not very different from that of contemporary scientists or some Old Testament fundamentalists today. But nineteenth- and early twentieth-century geologists, although often as devout as the good archbishop himself, were unable to accept so stingy an allotment of time. A regular, extensive, and—most important—predictable succession of fossils and rocks had been observed the world around. This succession was found to be consistent with other criteria for ordering events in a sequence of before, after, and contemporaneous with. The total known historical-sedimentary pile represented a thickness of hundreds of thousands of meters of stratified rock. Within this vast sequence the fossils changed, generally speaking, from relatively simple organisms of low diversity, in the lower parts, to more complex and more highly diverse organisms above. Considering observed rates of sedimentation and biological change, it was inconceivable that even this poorly known

Figure 7. Grand Canyon from the north rim. (Photograph by G. A. Grant used in preparing a two-cent stamp for the U.S. mails in the 1930s. Supplied by and reproduced with the permission of the U.S. Geological Survey.)

and incomplete geological succession of strata could be the product of a mere few thousand years of history.

By observing the succession of plants and animals, geologists early developed and tested a scale of "sequence-geochronology" (expressed by a succession of words rather than years) for those rocks that contain readily visible fossils. The idea behind such a geochronology and its extension is illustrated by figure 7, a famous view of the Grand Canyon which appeared on a two-cent postage stamp during the 1930s. In this the eye can trace the exposed edges of individual layers of the same nearly flat-lying rocks over the entire field of view. Now picture those rocks to be marked, as they are, by the same kinds of distinctive and easily recognizable fossils as appear in equivalent layers in distant canyons and faraway places. Even when crumpled and stood on end by mountain-building forces, these talismans of time, the lowly but distinctive seashells and primitive plant remains of ancient sedimentary deposits, tell the experienced paleontologist which stratum of a set is older or younger and what their sequence is. That kind of geochronology continues to prove useful in practical matters such as exploration for petroleum and ore deposits—often more useful in fact in the practical matters to which it applies than newer methods for obtaining ages in actual years. Other criteria now available allow conversion of the relative ages of such a sequence geochronology into a reliable numerical scale of years. But rates of accumulation of sediments and other changes of natural phenomena long ago yielded well-reasoned if crude estimates of duration. Such estimates tended to give around 100 to 400 million years as the time required to accumulate that part of the historical sequence of sedimentary rocks that is characterized by animal fossils—not that far from the 680 million years now obtained for the same historical interval, one which, as we now know, represents only the last 15 percent of Earth history.

During the latter half of the last century geologists engaged in making estimates of Earth's age became locked in heated and, as it seemed, losing debate with the earlier mentioned physicist Lord Kelvin. Kelvin, with impeccable logic and elegant mathematics, calculated how long it would have taken a body the size of Earth to cool from an initially molten state to its present temperature. There seemed to be no good way to extend his calculated age of 20 to 40 million years except by assuming a rate of cooling very much slower than available estimates would support.

That, however, was before Becquerel's discovery of radioactivity revealed the previously unknown, long-lasting internal heat source that the great University of Chicago geologist T. C. Chamberlin had earlier hypothesized might invalidate the assumptions behind Kelvin's mathematics.

Radioactivity provided light as well as heat. It proved to be the key to an independent method of calculating ages, based on the rate of transformation of uranium to lead. To obtain a fair approximation all one has to do is to measure the amounts of uranium and lead now present, add uranium and lead to approximate the uranium originally present, divide that by uranium remaining, and multiply by the decay constant. Simple! Almost like measuring time with an hourglass, but with one very important difference. The decay constant is a negative exponential. It is a quality opposite in its effect to the exponential growth of populations, the consumption of resources, and other growth curves that we worry about because of their very rapid rates of increase expressed as doubling times. *Instead of doubling, radioactive decay halves.* And it is only after about twenty halving times, or "half-lives" as they are called, that what is left of the original radioactive parent material has decayed to the point where it is essentially indetectable.

Suppose you wanted to build an "hourglass" to measure geologic time and took Kelvin's estimate of 40 million years as the outside limit of what you would need to measure. And suppose you were able to design a glass such that only one grain a day would pass from the upper to the lower compartment. Your glass would need a capacity of almost 15 billion grains of sand. If, however, you could design an hourglass in which half of the sand fell through to the lower compartment in the first 713 million years, half of the remaining sand (one-fourth of the original total) in another 713 million years, half of that remainder (one-eighth of the original) in a third 713 million years, and so on, at the end of 20 half-lives, more than 14 billion measurable years would have elapsed.

That is exactly what happens as the kind of uranium (the isotope

U-235) used in nuclear reactors decays to form lead (in this case the isotope Pb-207). Thus, although the quantity of sand that flows through the orifice of an hourglass in a given time is independent of the amount remaining above, the amount of lead formed from uranium in a given interval depends on the amount of uranium left (figure 8). One problem is that the half-life by means of which this process of exponential decay is measured must be independently determined for each radioactive isotope.

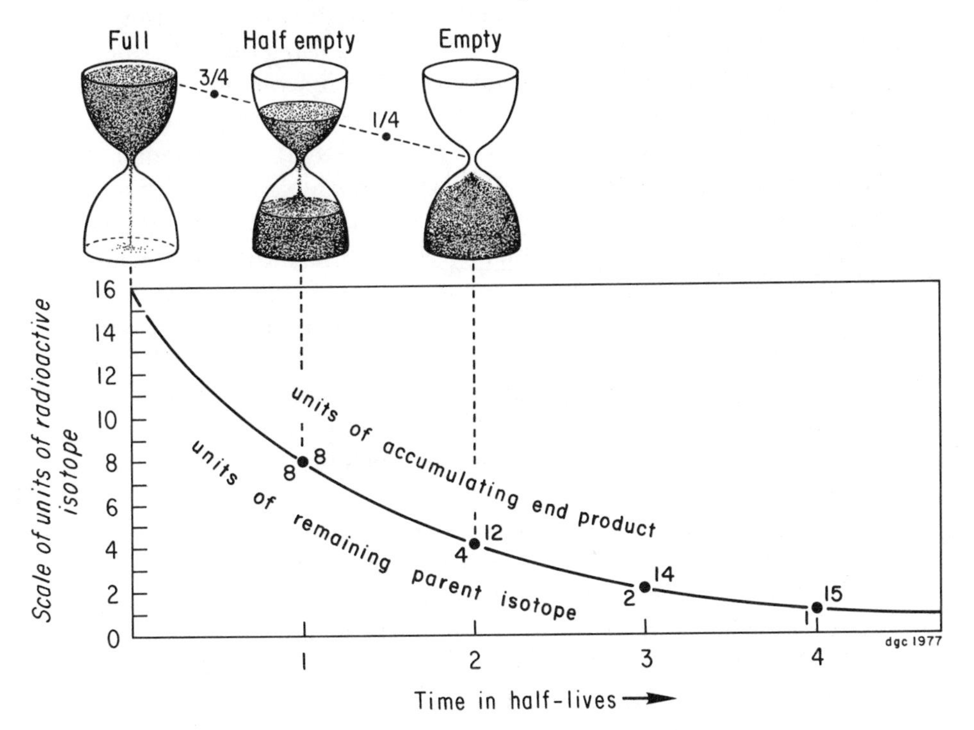

Figure 8. Timekeeping characteristics of radioactive decay compared with those of an hourglass.

Suppose you start with 8 grams of *parent material*, having a half-life of 4.5 billion years, which is about that of common uranium (U-238). At the end of 4.5 billion years only 4 grams will be left, the other 4 grams having converted to the *end product* (lead-206 in the case of uranium-238) by way of a sequence of intermediate *daughter prod-*

ucts. Another 4.5 billion years later, after a total of 9 billion years, the remaining 4 grams of parent material will have been halved, leaving only 2 grams of parent material and 6 grams of end product. And so on as indicated in figure 8. Such a process approaches the vanishing point of its parent material very gradually, so that its ability to measure age extends over a very long time, provided that our means of measurement are sufficiently refined.

Although science seeks simplicity, scientists have learned to be wary when things look too simple. Interesting results supporting the idea that the earth is of great age were obtained using the early simple uranium-lead method of geochronology. But it soon became apparent that there were problems. Lead consists of several different isotopes, not all of them produced by radioactivity (radiogenic). If common nonradiogenic lead-204 and the radiogenic leads with mass numbers of 206 and 207 are added together, excessive ages are obtained. A way out of this difficulty is to use only samples in which no nonradiogenic lead-204 is present, implying that all lead in the sample is radiogenic. Using only numbers based on such key samples, geologist Joseph Barrell of Yale University was able in 1917 to announce a time scale extending back about 1.8 billion years—a time scale in which the individual numbers are not greatly different from numbers given for the same samples by much more refined methods today.

Luck as well as good judgment was involved in Barrell's success. There were other problems, unknown to him. Another lead isotope, lead-208 was a product of the decay of thorium-232. In addition, some radiogenic lead was inherited by Earth at the time of its formation. No one then knew how much. Nature had labeled the samples for us, but the reading of those labels depended on the perfection of an instrument capable of separating and measuring the various isotopes of the parent and daughter elements involved. Also needed was a way of estimating the isotopic ratios of Earth's original heritage of lead—its *primordial lead.*

A difficulty in separating and measuring isotopes is that, as noted earlier, there are no differences in chemical characteristics or behavior between the isotopes of a given chemical element. The only differences are very small variations in mass: for example, an atom of lead-206 differs from an atom of lead-207 by the mass of one neutron—equal to the mass of one atom of hydrogen and less than half a percent of the mass of the lead atom. As in the case of the charged elementary particles mentioned in chapter 2, the atoms of isotopes with a difference that small can be separated and counted only by passing a volatilized mixture of them through a strong magnetic field. Since there is no difference in charge to deflect them, however, they must be separated by mass differences alone, detected by suitable collectors, and counted electronically. An instrument that can do this is called a "mass spectrometer." The prototype of the mass spectrometers now in use was perfected by the physicist A. O. C. Nier in the 1930s at the University of Minnesota, but instrumentation was greatly improved and diversified by Nier and others following World War II, especially under the impetus of space exploration and preparation for it.

It was the geochemist Clair Patterson of Caltech who in 1956 came up with the liberating insight that provided the key to the isotope ratios of primordial lead—building on microanalytic techniques for the study of radioisotope systems which he had earlier developed at the University of Chicago, along with George Tilton and Harrison Brown. Patterson reasoned that, because of the negatively exponential nature of radioactive decay, a meteorite with very little uranium and thorium left would have essentially the same ratios of lead isotopes as those inherited by Earth at the time of its aggregation from the solar dust. Using a mass spectrometer, he measured the lead isotopes in fragments of a meteorite of this kind from the famous Canyon Diablo meteor crater near Flagstaff, Arizona. The ratios obtained by Patterson are now used to correct apparent uranium-lead and thorium-lead ages, which, if uncorrected, would come out much larger.

A confirmation of ages obtained and of the constancy of decay rates over geologic time is provided by the different half-lives of the various parent materials. For instance uranium-235 decays to lead-207 more than six times as rapidly as uranium-238 decays to lead-206. Thus the fact that each of these two separate decay series gives the same age for rocks from which there has been no lead loss amounts to proof that their respective decay rates are constant, as predicted. Because all isotopes of lead and uranium exhibit comparable chemical behavior, a discrepancy in age is a sure sign that uranium, lead, or both have been removed from the rock by some subsequent event. That allows us to extract much more history from the rock than might otherwise be possible. In addition, unless the entire system has been reset or otherwise distorted by dramatic changes in temperature or pressure, and granted that we know the initial primordial ratios of the isotopes involved, the true age of the rock is given by the ratio of lead-207 to lead-206, which is unaffected by lead loss from chemical leaching and does not require separate analyses for uranium.

The existence of other decay series, each with a significantly different half-life, gives a self-checking quality to this system of radiometric clocks. In addition to the thorium-lead series, they include the conversion of rubidium-87 to strontium-87 and of potassium-40 to argon-40. When the ages given by these systems are in error, the error is almost invariably on the side of giving too young an age, except in the case of the absorption of argon-40 by certain minerals at high temperatures, which gives anomalously great ages. All such problems can be sorted out by comparing the results of one method against the results of another through discriminating techniques of geochemistry and geology.

An approximate age for the earth and the solar system was an unexpected bonus from Patterson's study of meteorite ages. When he plotted the ratios of nonradiogenic lead-204 to radiogenic lead-207 on a vertical scale and lead-204 to radiogenic lead-206 on a horizontal scale, all leads from metallic meteorites fell on a straight line, the slope of which gave the time of origin as 4.55 plus or minus 0.07 billion years ago. Random mixtures of Earth leads obtained from deep-sea sediments clustered around a point on a line of the same slope. This implied a similar age for Earth and meteorites. As a result, however, of subsequent refinements by George Tilton of the University of California at Santa Barbara, in collaboration with R. H. Steiger and others, the currently accepted age of the earth is taken as close to 4.57 billion years, which may be rounded off to 4.6.

Although the lead isotope ratios mentioned indicate an age of about 4.6 billion years for the earth, and although there is reason to believe that this number will not change much, no terrestrial rocks that old are yet known and it is unlikely that any will be found. The oldest unequivocally dated Earth rocks, found in southwest Greenland, have a radiometric age of only 3.76 billion years. However, a redeposited granite boulder in an ancient conglomerate in South Africa has now been dated at 4 billion years and rocks more than 4 billion years old have been reported in the news media from east Siberia and the Soviet sector of Antarctica. These ages, however, need to be independently confirmed, for they are very close to the oldest that any Earth rocks are likely to be. That is because the initial earth probably did reach melting temperatures, not for the reasons postulated in older hypotheses of origin, but as a result of combined heating from the gravitational collapse of Earth's primordial dust cloud plus early active radioactivity. It would probably have taken such a hot planet several hundred million years to cool enough to make solid crustal rocks. Thus, although the isotopic composition of our inheritance of primordial lead tells us the approximate time of aggregation of the earth, no rocks are likely to have solidified from the molten state until some later time, probably at most not much longer ago than about 4.2 billion years.

How long did it take the solar system to form? Is there any reason to think that all parts originated within a relatively short time, aside from theory and the concordance of the ages of oceanic leads and metallic meteorites?

An unexpected way of resolving this question emerged from the study of residues from nuclear explosions. An isotope of the rare gas xenon, xenon-129, is formed by the emission of a nuclear electron from iodine-129, converting one of the iodine neutrons to a proton and thus changing the atomic number from 53 to 54 (see the periodic table on the back endpapers). But the half-life of iodine-129 is very short. Although there is a reasonable amount of the essentially inert xenon-129 on Earth and in meteorites, no iodine-129 is found except as a product of nuclear explosion. Iodine-129, however, must have existed in the solar nebula. As early as 1947 the Caltech geochemist Harrison Brown had suggested that early fast-running but now extinct radiometric clocks might provide information about early solar system events. Thus if one could find meteorites with an excess of xenon-129, this would imply that the parent materials of those meteorites had remained in contact with the primitive solar disc after the solar dust that formed the planets had become detached.

In 1960, the Berkeley physicist John Reynolds announced the discovery of just such excesses of xenon-129 in a meteorite from Richardton, Minnesota, and in other meteorites. The abundances of xenon-129 in these meteorites, as well as xenon isotopes from other decay systems, strongly imply that the aggregation of the planets from the hot solar nebula took place during a time of planetary condensation that lasted no longer than 120 to 290 million years. Together with the above-mentioned lead isotope ratios, this indicates that planetary condensation was completed about 4.6 billion years ago.

Thus a spectrum of radiometric clocks has gradually emerged by means of which the events of Earth history can be correlated with a time scale of atomic years. A time-calibrated skeletal outline of those events, along with associated steps in biologic evolution and the sequence-geochronology derived from them, is shown on the front endpapers of this book.

Although the isotopic ratios of metallic meteorites and well-mixed leads from oceanic sediments indicate a time of origin for Earth and its sister planets of the solar system, the first 800 million years of Earth's history remains unrecorded by any Earth rocks confirmed as belonging to that interval of time. One of the reasons lunar exploration is so exciting to geologists is that lunar rocks, now giving ages approaching 4.6 billion years, may fill in part of this historical gap — the moon is seen as a window through which may be glimpsed the early history of the planet.

It is also known from the geochemical similarity of oceanic basalts throughout recorded Earth history that the gross concentric structure of Earth's interior evolved during that first 800 million unrecorded years— for geochemistry tells us that Earth's mantle, the source of such basalts, had already been differentiated from the deep internal core before they were formed. This would have required temperatures high enough to cause melting and settling of heavier elements such as iron and nickel into Earth's core, while generally lighter materials such as granite, basalt, and other silica-rich rock melts floated to the outer crust like slag to the surface of molten iron ore.

In addition to early melting or near-melting temperatures arising from the conversion of gravitational energy into heat and strong early radioactive heating, all planets and their satellites experienced intensive early meteoritic bombardment. Traces of this bombardment are now known to be abundant and well preserved on Earth's moon, Mercury, and Mars. They have been almost completely eliminated on Earth, however, as a result of intense, long-lasting atmospheric weathering and erosion by running water.

A few final words must be added about the reckoning of time. The geochemical methods described above apply especially to rocks formed from melts within Earth or extruded to its surface (igneous rocks) and to specific minerals that contain appropriate parent and daughter isotopes within such rocks. Sedimentary rocks can be dated using the same radiometric clocks under special circumstances, but the most trustworthy radiometric ages for sedimentary rocks are obtained from associated igneous rocks. Sedimentary rocks younger than about 680 million years can ordinarily be dated more conveniently by fossils than by radiometric methods, inasmuch as the relative ages given by the established sequence of fossils are now calibrated to the radiometric age scale.

In the case of very young rocks and sediments, of interest for dating the evolution of man and ice-age (Pleistocene) geology, only the potassium-argon method is helpful among those described so far. This method applies to events as recent as about a million years ago, and here the accuracy is greater than for older deposits since there has been insufficient time for distortion of age by escape of the gaseous end product, argon-40.

Carbon-14 dating applies routinely only to deposits younger than about 40,000 years (special techniques will take it back to 70,000 or perhaps 100,000) and then only to those that contain carbon and into which there has been no introduction either of more radioactive modern carbon or of old or "dead" carbon. Reliable results also depend on very precise determinations of radiation emitted, involving the exclusion of extraneous radiation. For carbon-14, with a half-life of only 5,730 years, is a product of the substitution of a neutron for a proton in the nucleus of nitrogen-14 as a result of cosmic-ray bombardment. Given no great variation in solar activity, carbon-14 forms from ordinary nitrogen at nearly invariable rates and is incorporated into biological carbon as a definite percentage of total carbon. Being radioactive, it undergoes subsequent loss of a nuclear electron—an event that results in the conversion of one neutron to a proton within the nucleus. For each such conversion one atom of carbon-14 reverts to nitrogen-14. These conversions occur spontaneously and with great regularity, providing the works of the *radiocarbon clock.* Unlike the various methods described above, radiocarbon dating involves the measurement of radioactivity itself and not of end products. In determining ages allowance must be made, however, for variations in the rate of formation of C-14 in the past and recent large additions of "dead" carbon to the atmosphere from the burning of fossil fuels.

An interesting method for estimating the ages of younger rocks, supplemental to C-14 and potassium-argon dating, is based on the fact that Earth's magnetic field undergoes abrupt episodic reversals. These reversals can be recognized by the polarity of magnetic materials in volcanic rocks and then can also be dated by the potassium-argon method. They show distinctive signatures in volcanic sequences that allow individual polarity episodes in undated rocks to be recognized and correlated with dated ones. The ingenuity of geochronologists, it seems, is matched only by the relevance of their results. All of modern anthropology, ethnology, prehistory, geology, planetology, even biblical history is done within a framework of radiometric numbers.

But what of our place in the time scale of the universe? We saw in chapter 3 that the red shift of fleeing stars implies an age for the universe of 12 to 17 billion years, while the decay of radioactive rhenium-187 to stable osmium-187, as measured in meteorites and corrected for the age of the solar system, indicates a passage of 18 billion years since the first massive formation of the elements. These ages are consistent with numbers of 16.6 to 18 billion years given by the relativistic energy equation assuming the now well documented relict black body radiation temperature of about 2.7° Kelvin. Yet the planets of the solar system are only 4.6 billion years old. What was happening in the meantime? And are there other ways of checking the apparently great age of the universe?

The answer to the last question is yes, within limits. A third way of estimating an age for the universe derives from the hydrogen-burning characteristics of main sequence stars—those that array themselves in a continuous succession from red, through orange, to yellow dwarf stars in the main sequence of the Hertzsprung-Russell diagram (figure 6). The astronomer A. R. Sandage has observed that a given star remains on the main sequence until it has burned 10 percent of its hydrogen to helium, at which time it heats up, expands, and leaves the main sequence. The mass of such a star can be computed from gravitational effects. From Einstein's equation we calculate the total energy generated by burning 10 percent of this mass. The brightness of the star, adjusted for distance, gives the rate at which energy has been radiated into space. Total energy radiated divided by rate of radiation gives the length of time the star has spent on the main sequence. Numbers so obtained give a range in age for different galaxies from 6 to 16 billion years. That implies an age for the evolving universe of 16 billion years or more—in good agreement, all variables considered, with ages computed from red shift and rhenium to osmium decay. Uncertainties involved are signified by rounding these numbers off to an even 20 billion, as suggested on the diagram in the front endpapers.

As for intermediate events, the age of the supernovation that created our heaviest elements and gave birth to the present sun has been estimated at 6.6 billion years by Geoffrey and Margaret Burbidge, Fred Hoyle, and William Fowler, working together at Caltech. They calculated from the properties of atomic nuclei that, if all the uranium in the primordial solar disc were made in one supernovation this would have produced 1.65 times as much uranium-235 as uranium-238. Measurements in various rocks on Earth show that the ratio today is only 0.00723. It would take 6.6 billion years for the decay rates of these isotopes of uranium to shift their ratios that much. Even if the estimated 1.65 initial ratio were as little as 1.0 or as much as 2.0, the age indicated would not vary greatly. On the other hand, there is no compelling reason to believe that all of this solar system uranium was made in one event, and other models give solar ages that range from 5 to 18 and average around 12 billion years. Still, I prefer an age of about 6.6 billion years as the most reasonable available for the ancestral supernovation at this time.

Other estimates of the ages of stars and galaxies based on their heavy elements give ages for the birth of these elements that range from 7 to 15 billion years. Given that the big bang that started it all took place about 18 to 20 billion years ago, it seems that star formation has probably been going on since the beginning and that our sun may be one of the more recent stars. Not long after its origin it became the proud parent of a whole family of planets, on one of which we try to imagine what the past was like and what the future may bring.

For Further Reading

Berry, W. B. N. 1968. *Growth of a prehistoric time scale.* W. H. Freeman & Co. 158 pp.

Cox, Allan; Dalrymple, G. B.; and Doell, R. R. 1967. Reversals of the earth's magnetic field. *Scientific American,* vol. 216, no. 2, pp. 44-54.

Eicher, D. L. 1968. *Geologic time.* Prentice-Hall. 149 pp.

Harper, C. T. 1973. *Introduction to geochronology: radiometric dating of rocks and minerals.* Dowden, Hutchinson & Ross. 469 pp.

Libby, W. F. 1961. Radiocarbon dating. *Science* 133: 621-29.

7

OUR RESTLESS EARTH

We think of Earth as made of rocks—not quite true but close enough. Earth, therefore, is "solid as a rock." But just how solid is a rock? Suppose you squeeze one very hard, deep within the earth or in a machine designed to duplicate Earth's stresses. The molecules and atoms that make it up will react to stress just as other natural systems do. The rock may change its physical state and mineral composition or alignments, it may bend, fold on itself, break, or become abruptly dislocated along a surface of rupture so as to transmit shock in the form of an earthquake. Our seemingly solid planet is constantly being stressed by the gravitational effects of the sun and moon, as well as by its own internal forces. These internal forces are generated by Earth's great interior heat machine, fueled by radioactivity and internal gravitational effects, and lubricated by the universal solvent, water.

The effect of all these forces is to keep Earth constantly in motion. Solid and stable though they may seem, the very rocks beneath our feet rise and fall, ever so slightly, in a twice daily tidal rhythm, responding to the gravitational pull of the sun and moon—breathing, as it were, like the ocean tides but on a lesser scale. Earth's rocks are heaved up by mountain-building forces and worn down by the erosive action of water, ice, and wind, combined with gravitational downslope movement in the form of soil creep, falling rock, avalanches, and landslides. The planet shudders under the occasional impact of large meteorites and whirls beneath our feet at 1,620 kilometers an hour. It speeds through space at around a million and a quarter kilometers a day on its gravitationally prescribed orbit around the sun. It vibrates, shakes, and rings like a bell from seismic shocks called earthquakes, which originate both within its outer rocky *crust* and in its *mantle* beneath the crust. Even the *fluid outer core* of the earth, enveloping its innermost *solid core* (see chapter 8 for definitions), rotates, dynamo-like, within the solid mantle, transforming the planet into a giant magnet whose field of influence reaches into space itself, trapping cosmic radiation in the Earth-encircling Van Allen belts and shielding Earth from the solar wind.

Of all these motions the one that evokes man's greatest awe is earthquakes. Suddenly, often without warning, by day or night, in winter or summer, the earth shudders, buildings collapse, chasms yawn, water pours out at the surface where no water had been before, wells run high or dry, "tidal" waves (seismic sea waves) are generated, dams fail, avalanches roar out of the mountains, people die, and the survivors are sore afraid. No wonder! The great New Madrid earthquake of 1811 changed the topography of the lower Mississippi Valley, rang church bells as far away as Boston, and cracked plaster in Virginia. At the time of the Lisbon earthquake of 1755 rivers in Germany rose and fell, and waves were generated in Loch Lomond in the west of Scotland. This great quake killed a quarter of Lisbon's 235,000 residents. Earthquakes in southern Italy in 1908 and in western China in 1920 and 1927 killed around 100,000 people each. Densely populated and primitively constructed China holds the record for casualties. Some 830,000 are reported to have perished in the Huasien quake of 1556 in the Huang Ho Plain of northern China. As if to emphasize that such numbers are not apocryphal, the seismic catastrophe of 28 July 1976 utterly demolished the industrial city of Tangshan, about 160 kilometers south of Peking, killing an estimated 750,000 of its 1 million residents. The Tangshan earthquake had a magnitude of 8.2 on a 1-to-10 logarithmic scale of magnitude, in which if 1 equals 10, 2 would equal 100, and 9 would equal a billion. Other regions that have suffered tremors of large magnitude recently include Alaska, Chile, Central America, Uzbekistan, Turkey, and the Balkans. Even areas of the world that are not normally earthquake-prone may suffer occasional violent tremors, such as the great New Madrid earthquake mentioned above, and the less intense but more traumatic earthquake of 1886 centering on Charleston, South Carolina. Modern engineering and building codes have greatly reduced damages in earthquake-prone areas of developed countries in recent years. Yet it is still a sobering thought to reflect on what could happen in the San Francisco Bay area or the great cities of the Mississippi Valley today were earthquakes of the magnitude of those of 1906 or 1811 to be repeated in these now thickly populated regions. Depending on the time of day and the degree to which structures withstand stress, the United States Office of Emergency Preparedness in the San Francisco Bay area estimated that between 14,000 and 100,000 people would be killed or injured in a recurrence of the 1906 earthquake.

Curiously enough, the earthquake waves that do the most damage are not the ones by which the most energy is released. A very long, rapidly traveling wave like a seismic sea wave does little damage and, in fact, goes unnoticed until it runs into a resisting obstacle. But should it strike, say, a shore line, the sea may rush with literally irresistible force to heights of tens of meters above sea level. It is the chop and heave of relatively low-energy waves and swells that do the most damage and cause the most alarm at sea. Similarly, the long, fast-traveling seismic waves that are the distinctive and energy-intensive features of earthquakes do little damage of themselves. That is because they keep right on rolling through and bouncing around within the earth, disturbing little except seismometers until they play themselves out. When these waves encounter discontinuities, however, at places where there are changes in the density or rigidity of different layers of rock within the earth or at the boundaries of rock with air or water above, they generate new families of waves. Among these are *surface waves* that travel horizontally through the outer layer of rocks or sediments. Such waves travel much more slowly than the main wave types and are more complex. The disturbances they produce at Earth's surface are what topple buildings, open chasms, ring bells, and generate avalanches and seismic sea waves. It is as if one set out a toy landscape on a rug and then gave the rug a hard snap. Everything would topple over from the long waves going through the rug. On the other hand, if you struck the floor with an ax the light fixture overhead might sway but the damage to the play landscape would be focussed at the point of impact.

Earthquakes nevertheless bring such interesting messages about the otherwise unknowable deep internal structure of Earth that a group of geophysicists who call themselves seismologists spend most of their time studying both natural and man-made quakes. An earthquake starts in a limited area of energy release, the earthquake *focus*, where adjacent moving blocks or plates of Earth's solid outer shell or *lithosphere* first stick and then abruptly come unstuck. Or it may start, from less well understood causes, at depths up to several hundreds of kilometers. The point on Earth's surface that lies directly above the focus is the *epicenter* (figure 9). The energetic waves move away from the focus, changing course and bouncing back at internal concentric surfaces that separate shells of different density and rigidity within the earth—as rays of light are both bent and reflected when they pass from air into water. Under the same principles that govern the behavior of light in lenses, seismic signals from these energetic waves are used to trace a picture of Earth's internal structure (explained more fully in the next chapter). Seismological studies have also proved to be of great practical use in differentiating between natural earthquakes and those produced by nuclear explosions, thus monitoring nuclear testing in regions not accessible to inspection. They may, in addition, eventually provide mankind with its first means of predicting and perhaps even to some extent controlling the fearful effects of natural earth tremors.

Why does Earth quake? It is easy to understand why this happens around erupting volcanoes, and it is a fact that volcanoes and earthquake epicenters tend to follow similar trends at Earth's surface. But many earthquakes strike far from the nearest volcanoes. And even the energy released by the most violent volcanic eruptions is trivial compared to that of a major earthquake, which may move thousands of cubic kilometers of rock for many meters along deeply penetrating earth-fractures called *faults*. Thus, over the last few tens of millions of years, sidewise movement resulting from a succession of earthquakes and related earth slippage along the San Andreas Fault has apparently caused the western edge of North America from north of San Francisco Bay to the Gulf of California to move several hundred kilometers northward with respect to the rest of the continent.

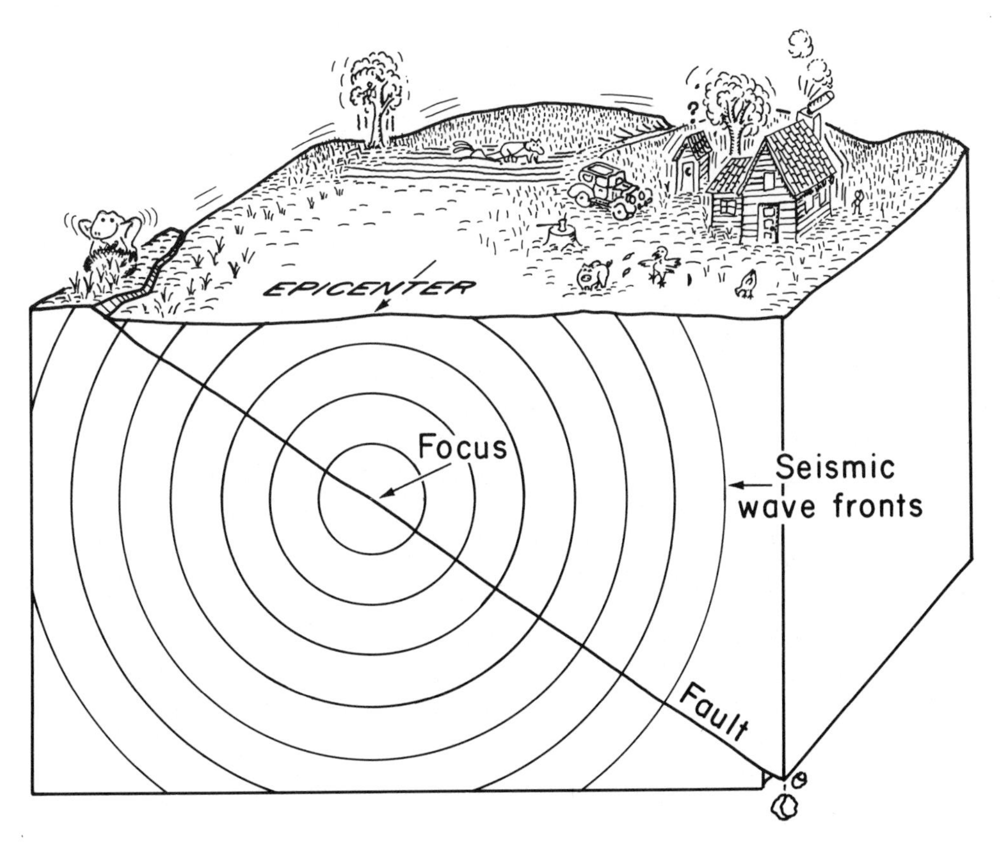

Figure 9. Focus and epicenter of an earthquake. (Adapted from an unpublished sketch by J. Gilluly.)

We now know that, although each individual movement is sudden, the energy released accumulates slowly as stresses build up in adjacent blocks of Earth's crust between earthquakes. As the rocks west of the San Andreas Fault, for instance, creep slowly northward with respect to those to the east (and the reverse) they stick together here and there along the fault until stresses accumulate to the breaking point. Then the rocks suddenly snap apart, offsetting streams, fence lines, rows of trees, highways, bridges, and buildings. The stress is released, the strain is visible, the damage is done. After such a shock the rocks may continue to glide slowly past one another, or they may lock together again until the accumulated stresses within

Earth once more exceed its elastic strength, producing abrupt offset and another earthquake. That, in essence, is the elastic rebound theory of earthquake genesis, as first elaborated by the late Johns Hopkins geophysicist H. F. Reid. Think of it as comparable to the force that impels an arrow as the bow string is drawn back and then abruptly released.

But what makes the western sliver of California drift northward? The key to that question, remarkably enough, is found far from California, in fact far from land, in the middle of the oceans. The fiery birth of the new volcanic island of Surtsey, near Iceland, in November 1963, sent out a loud signal to the effect that Earth's crust is splitting open like an overripe melon along the 65,000-kilometer ridge that divides the world's oceans, mostly at mid-width. Basaltic lava wells up along this crack to make new crustal rocks, new sea floor, and sometimes new islands like Surtsey—or like Iceland itself some 15 million years earlier. Earth's crust, however, cannot grow in mid-ocean without increasing the diameter of the earth unless crustal shortening is taking place elsewhere—by processes such as the folding of mountains and the local overriding or descent of crustal rocks above or beneath one another. The 1,200-kilometer-long San Andreas Fault, it turns out, is one of the more trivial consequences of that growth of sea floors and its related forcing apart of continents and buckling of their edges into folded mountain chains.

From this emerges the modern concepts of large-scale continental drift and *plate tectonics*, which are revolutionizing and integrating the earth sciences. Tectonics is the geologist's word for the architecture of the earth, referring to such matters as the folding of mountains, the uplift of plateaus, and, on a larger scale, the differentiation of continents and ocean basins and the internal structure of the earth. Tectonics has a strong dynamic connotation, and plate tectonics refers to large-scale motions of the upper 100 kilometers or so of Earth's outer skin—motions that create and wipe out ocean basins and raft the continents about like masses of scum on a river plunge pool.

My son Kevin asked me once, after studying his first globe for a while, if maybe North and South America had been torn away from Europe and Africa. It looked to his then very young and unprejudiced eye as if they'd fit almost perfectly if pushed back together again. You may have wondered the same thing, as have many observant and imaginative people ever since the first accurate maps were made. In fact, the striking geographical fit of the opposite coasts of the Atlantic was for many years the main evidence for the hypothesis of continental drift, popularized by the German meteorologist Alfred Wegener, beginning in 1915.

For fifty years after the publication of Wegener's first book advocating drift, the question as to whether continents were fixed in position or mobile remained unresolved, with geologists, geophysicists, paleontologists, biologists, oceanographers, and others taking sides depending on what they knew or didn't know, what subject they had trained in, or where they had gone to school. Although much evidence was accumulated and discussed, both for and against continental drift, that evidence remained equivocal as long as it came only from dry land and a nodding acquaintance with sea-floor topography. The balance shifted toward the side of drift soon after World War II, with the appearance of the concept of seafloor spreading as a result of submarine exploration employing sophisticated sampling gear and a battery of remote-measuring devices. The leading spirit in this shift was the geologist Harry Hess of Princeton University. Hess collected important new physical evidence with his automatic depth-reading instruments while on duty as a troopship commander in the Pacific during World War II. He also had the insight to integrate this new physical evidence with age determinations based on studies of fossils by others, arriving at the conclusion that islands and banks along present-day mid-ocean ridges were younger than former islands and banks, now seen as submerged seamounts far landward from these ridges. Indeed the patterns observed imply strongly that the ocean floors get progressively older from medial ridge to continental coast and that no seafloor anywhere is much older than perhaps 200 million years at the most. Compared with continental rocks almost 4 billion years old, the present ocean basins thus seemed young indeed.

In fact too young. Besides there are inconsistencies in the pattern. Although the oldest truly oceanic floor yet known is about 165 million years, in the northwest Pacific rocks up to more than 800 million years old are reported locally near the crests of oceanic ridges. And the data on seafloor spreading away from these ridges suggest spreading rates that are locally greater than needed to explain the present widths of the oceans, the crustal shortening related to the growth of fold mountains, or the observed thickening of sediments in ancient oceanic trenches.

The clinching argument for the great majority of informed Earth scientists, who have now accepted seafloor spreading and continental drift, comes from seafloor geophysics. This involves the measurement of variations in the magnetism of the basaltic volcanic rocks that make up the seafloor beyond the continental slopes and beneath a thin cover of tiny sedimentary particles and microorganisms that have rained down from suspension in the overlying waters. Lateral variations of seafloor magnetism form linear patterns parallel to oceanic ridges, the most persistent of which correspond closely to the mid-widths of the oceans, away from which the seafloor slopes in both directions. These linear variations in magnetism, parallel to the mid-ocean ridges, can be correlated with the sequence of observed and radiometrically dated reversals of magnetic polarity in vertical successions of volcanic rocks on land (polarity reversal referring to changes in compass orientation whereby the currently north-seeking needle points south, and the reverse). Such correlation with known intervals of normal and reversed polarity implies similar times of origin for distinctive magnetic patterns at the seafloor. Sequences of magnetic variations on opposite sides of the mid-ocean ridges match one another like the fingers of the right and left hands when they are placed together palm to palm. And the distance between dated and matching anomalies on opposite sides of the mid-ocean ridges is evidence for the conclusion that the seafloor is spreading and for how fast it is spreading at any given place. This is illustrated in figure 10.

The kinds of measurements needed to arrive at such conclusions could be made only after the invention of a deceptively simple instrument called a proton magnetometer and its perfection for use at sea. A spinning magnetic bottle towed behind a ship records the magnetic variations in rocks at the seafloor below on a strip-chart aboard ship. The variations are seen as small peaks and depressions on the

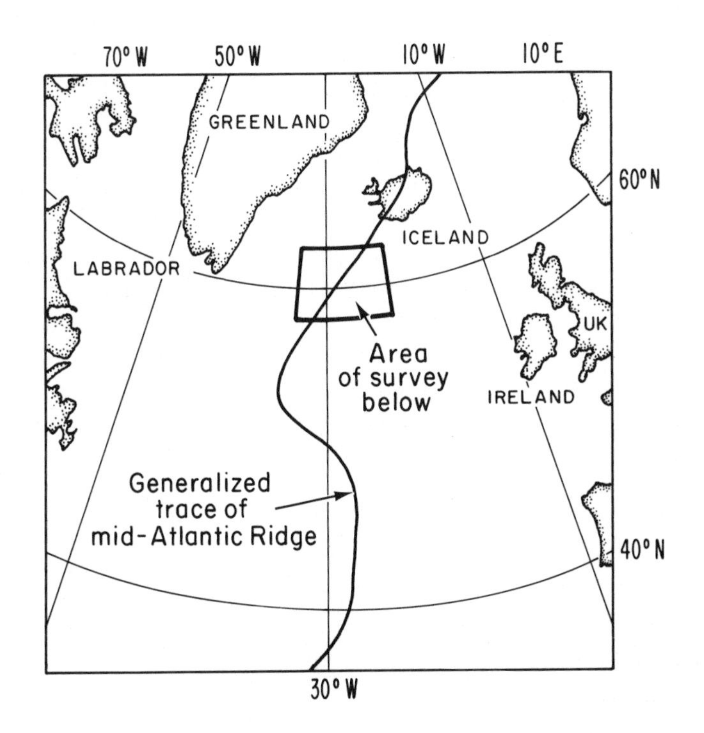

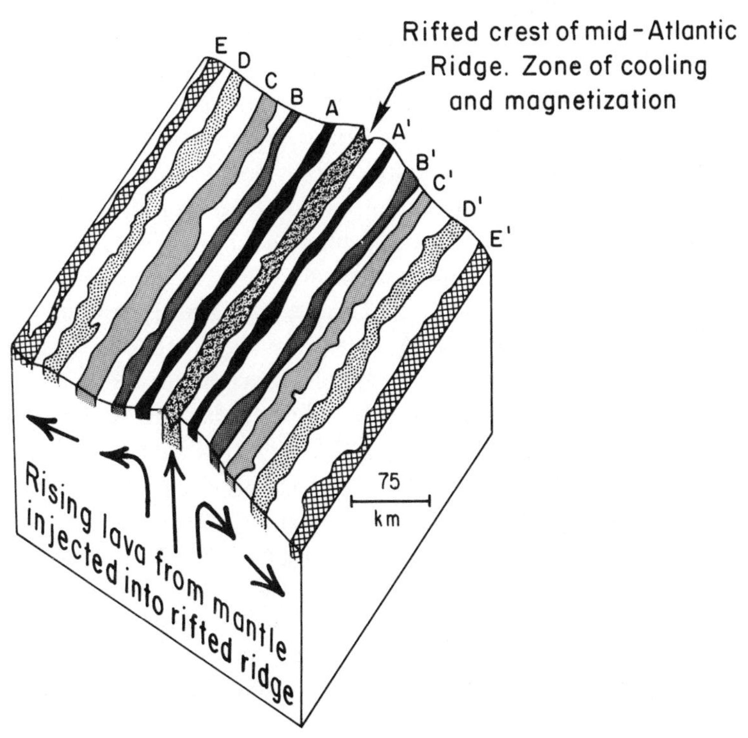

Figure 10. Seafloor spreading. (Slightly modified from published adaptations of original illustrations by F. J. Vine, *Science* 154, 1966, p. 1407.)

strip-chart. Strip-charts from a number of surface traverses that run at right angles to the mid-ocean ridge and along courses parallel to each other are converted into maps that show the linear regional magnetic variations of the oceanic basalts beneath their cover of pelagic muds. The mirror-image symmetry of these magnetic anomalies across the mid-ocean ridges correlates in detail with the known pattern of magnetic reversal. Highly contrived models are required to explain this symmetry by any mechanism other than movement of contemporaneously magnetized rocks to opposite sides of the same spreading ridges at essentially identical rates.

Thus the patterns observed are attributed to the periodic injection and subsequent rifting and drifting of basaltic lavas. These lavas ascend to the surface from Earth's underlying mantle along the axes of the mid-ocean ridges, register the contemporaneous magnetic polarity as they cool, divide at the center, solidify, and move landward away from the ridges. The spreading motion is probably a simple downslope gravitational glide of still cooling and contracting rock, which is, as it were, pulled downslope away from the *spreading ridges* as a function of its increasing density. The lava is fluid to begin with because ascent of hot mantle rock from beneath the spreading ridges brings it into lower pressure regimes where, for matter of its composition and temperature, the stable state is fluid. Such convective motions are seen as a way of ventilating the radioactively heated interior of the earth, which cannot cool fast enough by simple radiation.

Emplaced by fluid injection along the spreading ridge, the lava then cools below the temperature at which magnetic orientation is fixed (the Curie point). Tiny crystals of magnetite and other minerals in the congealing crystal mush swing around so as to orient their magnetic axes along lines of force in the direction of Earth's North and South Poles, like iron filings on a piece of paper above a magnet. Continued spreading while this lava is still warm, plastic, and weak leads to rifting down the center of the previously emplaced lava and the upward injection into this space of another surge of molten matter along the ruptured spreading ridge. Repeated over and over again, and associated with lava flows and other aspects of surface volcanism, this can be visualized as the basic process of seafloor spreading. The cumulative effect of this process registers paleomagnetically as the paired magnetic anomalies revealed by the magnetometer on opposite sides of the spreading ridge, as illustrated in figure 10.

The rates of spreading obtained by matching these magnetic anomalies on either side of the ridges range from 1 to 6 centimeters a year. This implies total spreading rates of 2 to 12 centimeters a year. At such rates it would have taken the present Atlantic Ocean between 36 and 288 million years to attain its present width. This is consistent with the Late Triassic, or roughly 200 million year, age estimated for the beginning of the breakup of the ancestral supercontinent of Pangaea and the drifting apart of its separate segments to arrange themselves as the world's present continents (figure 11). Thus it seems that the opening of oceans along spreading ridges is related to the growth of seafloor, the drift of continents, and the local adjustment of boundaries separating these moving segments of Earth's exterior along lines of horizontal slippage like the San Andreas Fault.

Another interesting thing about the (mostly mid-ocean) spreading ridges is that they are sites of both volcanic and seismic activity. Iceland, for instance, with its great array of volcanic and thermal centers, sits right at the crest of the mid-Atlantic ridge (figure 10). Indeed, a section of the ridge south of Iceland was the place at which the mirror-image symmetry of magnetic anomalies across the ridge was first demonstrated by F. J. Vine, then working at Princeton University with United States Coast and Geodetic Survey Data under the guidance of Harry Hess.

If you plot on a globe or world map the epicenters of all known earthquakes whose foci are at great or intermediate depths you find an extraordinary thing. With rare exceptions these foci cluster along well-defined trends (figure 11). These trends correspond with the mid-Atlantic ridge, the east Pacific ridge (or rise), the mid-Indian Ocean ridge, connecting oceanic ridges (or rises) that encircle Antarctica, the rim of the Pacific, the deep-sea trenches and island arcs, and the line of uplift defined by the Himalaya, Hindu Kush, and Zagros mountains, extending westward from northern Vietnam through Turkey to the northern Mediterranean. These more earth-

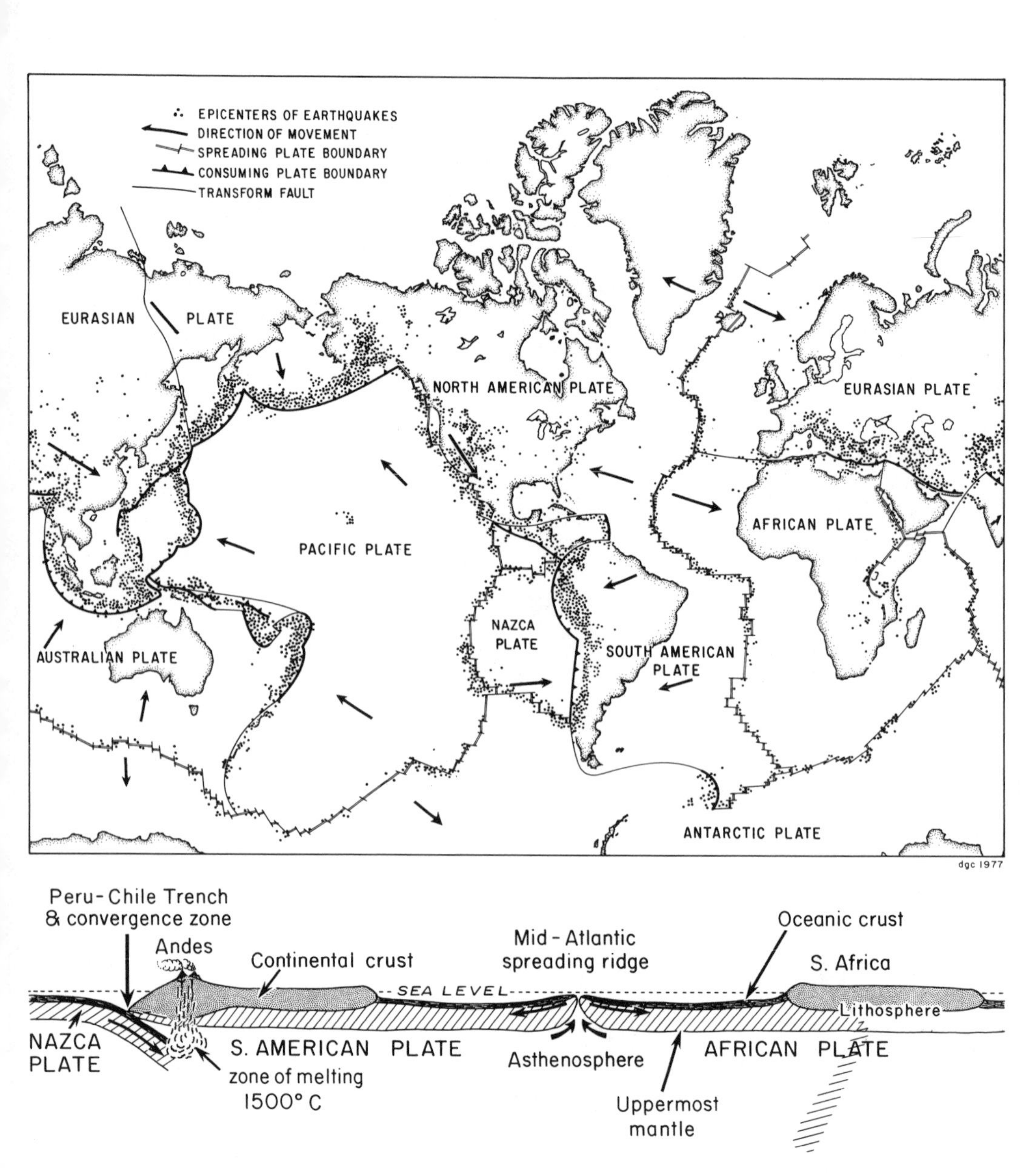

Figure 11. Principal lithospheric plates and their motions. *Above:* Map, slightly modified from a print by the Geophysics Research Board, National Research Council, shows relation between plate boundaries, earthquake epicenters, and plate motions. Magma wells up from the mantle along spreading ridges to form mid-ocean volcanoes and new oceanic crust, crust is consumed and more silicic sorts of volcanoes created where ocean floor plunges beneath converging plate boundaries, folded mountains arise where plates collide, and plates slide past one another along transform faults. *Below:* Schematic profile across the South American and parts of the African and Nazca plates.

quake-prone regions of the world are commonly but not invariably associated with volcanoes. The depths of earthquake foci show that great earthquakes are for the most part initiated along certain well-defined surfaces that descend steeply into Earth's rocky mantle, retaining their character to depths of up to 600 or 700 kilometers, where their seismically distinctive properties fade. Such surfaces and their related earthquakes denote boundaries along which the plates of Earth's crust are converging on and overriding one another, or where they are gliding sidewise along *transform faults* like the San Andreas. The local occurrence of strong earthquakes within the continental segments of plates and far from plate boundaries is not yet well understood. The earthquakes that occur along the mid-ocean and other spreading ridges are mostly shallow- to intermediate-focus earthquakes of relatively small magnitude, for here the rocks are too hot and plastic to accumulate much stress. Where great earthquakes appear to be related to spreading centers they are usually associated with one of the many short transform faults that offset them.

The locations of earthquake foci and the directions of motion along their generating surfaces are determined from study of the arrival times of earthquake waves. Over the years a body of knowledge has arisen from the seismological interpretation of such arrival patterns which yields much interesting detail about Earth motions. We now know that the surfaces along which earthquake foci line up define the boundaries of independently moving blocks or *plates* of Earth's solid exterior. These are usually thousands of kilometers across but only about 100 to at most 200 kilometers thick. They are laterally extensive slabs of rock that normally extend downward only to where temperature, increasing with depth, gives rise to a plastic state—a zone of weakness called the *asthenosphere* (Greek *asthenes*, weak), along which the overlying *lithospheric plates* glide.

Thus Earth's present surface can be divided into about eight major plates and a dozen lesser ones (figure 11), excluding the areas between incipient spreading centers which extend into the main plates, as they do along the African rift valleys. All of the crustal plates appear to be moving relative to one another, except for Africa, which has apparently been relatively stable in position for several tens of millions of years. Within some plates are lines of volcanoes, such as the Hawaiian chain and the seamounts beyond it, that are believed to originate over fixed *hot spots* beneath the moving plates and which therefore give directions and rates of motion.

From the idea of such motions over hot spots arises the charming and probably true story told by biologist Archie Carr and geologist P. G. Coleman of the green sea turtle that travels some 2,000 kilometers eastward from Brazil into the rising sun and against surface currents to breed at Ascension Island in the central South Atlantic. It is believed that this migration is but the continuation of an inherited behavior pattern that became fixed among its ancestors some 80 million years ago with the beginning of the South Atlantic spreading ridge. The distance of migration, small at first, increased stepwise along a chain of former volcanic islands (now seamounts) that arose successively over the same hot spot responsible for the present-day Ascension Island, the islands being successively submerged as seafloor spreading continued and new islands arose.

Surprisingly perhaps, considering the observed differences in rock type and geophysical characteristics between continental and oceanic crust, the boundaries of the moving plates do not consistently or even generally follow continental-oceanic boundaries. The Pacific plate is entirely oceanic except for the western sliver of California and its continuation in Mexico's Baja California west of the San Andreas Fault. The North and South American plates, however, include both continent and seafloor. They are bounded by the Pacific margin on the west and the mid-Atlantic spreading ridge on the east, and they are separated from one another by a variety of Antillean structures. The African, Indian, and Antarctic plates likewise include big swatches of both continental and oceanic rocks. The Eurasian plate, really two ancient plates annealed along the Ural Mountains, includes substantial land areas as well as much of the North Atlantic and Arctic oceans.

Plate margins are of several sorts. Oceanic spreading ridges are *accretional margins.* New basaltic oceanic crust is made here, as in the mid-Atlantic. The linear belts where such dense oceanic rock converges on and dives beneath lighter continental or thinner oceanic crust, as in the deep-sea trenches, are called *consuming margins.* At such places the downgoing plate is in effect consumed. It is warmed, melted, and loses the rigidity that gives it seismic identity at depth. The melting, in turn, gives rise to ascending volcanic plumes above the leading edge of the descending plate.

One effect of this is that consuming plate margins are marked by arcuate deep-sea trenches associated with matching volcanic alignments called island-arcs, where moving seafloor of normal thickness plunges under thinner seafloor adjacent to continents. Such island-arcs are prominent around the northern and western Pacific and in Indonesia. Where moving continent overrides seafloor, however, the deep-sea trenches are straight, and linear chains of volcanoes rise through the continents, as in the Andes. Not only is ocean floor consumed at such places, along with scraped off islands and seamounts, but chunks of adjacent continent founder into the trenches and are carried beneath the advancing continent along with truly oceanic rocks, as if dumped into a giant rock crusher. Such downgoing rocks eventually melt at great depths, mix with other molten matter of continental origin, and rise again to the surface as melts of intermediate composition (figure 11). Such mixed melts are important sources of certain types of ore deposit that are associated with continental-margin volcanism, probably because they sweat out and concentrate metals that were relatively dispersed in the downcarried deep-sea sediments and overlying continental crust.

Where oceans are drastically narrowed or wiped out by overriding and convergent continents that then bump against one another along *collisional margins*, the relatively light, high-standing continental crust is compressed, folded, shortened, and thickened to produce linear folded mountain ranges like the Himalayas, the Alps, or the Appalachians—the last formed as the ancestral Atlantic closed, beginning about 320 million years ago, before its current opening.

Transform margins are the last main kind. They are found at places where plates slide sideways past each other, as along the San Andreas Fault, offsetting or transforming one type of plate margin into another along a nearly vertical line of horizontal sidewise motion.

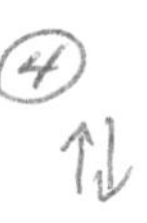

Movement along these different lines of contact between lithospheric plates explains a lot of things—the different kinds of deep-sea trenches, island arcs, fold mountains, volcanoes and earthquakes, the youth of ocean basins and the antiquity of continents. Although the seafloor is locally expanding, Earth's surface is not. For Earth's surface could not expand without its diameter also increasing, and that would change Earth's moment of inertia and affect Earth-moon relations in a way that finds no support in the evidence provided by Earth and planetary science. Thus, as the several plates of Earth's crust try to move past one another, there is no place to go but up, or down, or sideways. Instead of flowing smoothly past one another, however, the plates grind together and stick. When they finally come unstuck they do so with a jolt. The earth quakes. And man trembles.

Thus we have the model of a global framework of rigid surface plates that glide over the plastic asthenosphere beneath and grind against one another along relatively straight, seismically active boundaries to shake and deform Earth's crust. That is plate tectonics. It seems well on its way to becoming to geology what relativity has been to physics, what atomic theory has been to chemistry, and what molecular biology has been to genetics—a general field theory within which a great number of seemingly unrelated phenomena can be coordinated and explained. Geologists are now in the process of checking the concept of plate tectonics against the historical record and trying to work out the shifting patterns of paleogeography, paleobiogeography, and paleoclimatology on a geographic base that has itself been in constant motion—a global chess game in which the parts of the spherical board are shifting relative to one another, where pieces are lost down subduction zones at plate boundaries and the rules have not yet been clearly drawn up. It's a terribly exciting time to be a geologist.

What could cause the spreading ridges to develop and the plates to move? One idea that is gaining favor proposes that convective loss of heat from Earth's interior takes the form of rolls or plumes of rock or molten matter that rise from deep within the earth to form hot trends and hot spots at the surface. Such hot trends and spots may line up to create zones of weakness in Earth's crust, and, where they do, the hot, upwelling material causes elevation and rifting of the surface along spreading ridges. As proposed earlier, and as originally suggested in 1969 by geophysicist A. N. Hales, the solid oceanic lithosphere probably then slides downhill over the fluid asthenosphere away from the spreading ridges toward the lower parts of the ocean basins, cooling as it goes. This cooled oceanic lithosphere, being very dense, sinks beneath the overriding continents or thinner, convergent ocean floor (as in the Philippine Sea), descending into and beyond the deep-sea trenches to depths of around 600 to 700 kilometers in the mantle, according to seismic data. This is interpreted as the depth at which the downgoing plates approach their melting points closely enough to flow. Commonly the sinking solid plates also bulldoze wedges of overlying seafloor and continental margin sediments against the opposing plate margin (as in western California during Mesozoic time or Indonesia now). In the process they may also entrain large chunks of foundering continental material that is thereupon carried to depths and pressures such that it melts and is extruded upward again or plastered on to the base of the overlying continental rocks, leading to elevation of the land surface, plateau-building, and erosion.

Where continental and oceanic crust are both on the same plate, as in the North and South American plates, they move as a unit. Deep-sea trenches are not formed at the trailing contact of continent with ocean within such plates. Where continents (or limited areas of thin oceanic crust) override oceanic plates or oceanic plates dive beneath continents, as they do around much of the Pacific plate, trenches are ordinarily formed at the boundaries—with island arcs where there are oceanic rocks between trench and continent. North America, however, in contrast to South America, has ridden westward over the northern Pacific plate faster than new Pacific floor is being made along the East Pacific rise. The East Pacific rise, which disappears under North America at the north end of the Gulf of California, is a former mid-ocean ridge that has been partly overridden by this westward drift of the Americas, so that North America rests discordantly on older Pacific floor without a boundary trench. Indeed, it is likely that the geothermal regions and mineral deposits of the Great Basin region in the western United States are in some way related to ancestral North America's having overridden a succession of perhaps incipient spreading ridges during the last 60 to 120 million years of geologic time.

Finally, evidence is now coming to light in the geologic record to suggest that plate tectonic mechanisms were in effect long before the last major episode of plate movements began some 200 million years or less ago. Such evidence supports the conjecture that plate tectonic mechanisms may have obtained for perhaps the last 2.6 billion years. Before then Earth's crust was probably thinner and hotter, continents and ocean basins were only beginning to be differentiated, and crustal motions appear to have been mainly vertical.

All of which raises a host of interesting questions: Will the southwestern margin of North America (western California and Baja California) become a separate island microcontinent like Madagascar? Will the Atlantic continue to widen and the Pacific to become narrower? Will there be more devastating earthquakes along the San Andreas and other faults? The answer to all of these questions is yes, or very probably. What we don't know is how soon, or how fast, or when the motions involved will be halted, shifted, or reversed, as ultimately they will be. Except for the matter of earthquakes, these questions involve events so far in the future that they are of present interest to no one except Earth scientists. Even if estimated rates of motion are sustained, it will be another 10 million years before Los Angeles merges with that part of San Francisco east of the San Andreas Fault and another 50 million before it is carried beneath the Chukchi Sea by way of the Aleutian trench. That will allow plenty of time to think of a new name for California's unequivocally largest city and to prepare for the trip to Alaska.

Ruinous earthquakes, however, can come at any time. The last truly large motions along the San Andreas Fault were in the vicinity of San Francisco in 1906 and in the Imperial Valley of southeastern California in 1940. Another big quake is overdue. It will make more of an impression if the epicenter is in a populated region, just as the San Francisco earthquake registered as a major tragedy, while the Imperial Valley tremor of about the same magnitude was seen as just another earthquake.

What can be done to foresee or ameliorate the effects of major earthquakes? Unlikely as those eventualities seemed only a few years ago, study and information have now reached the point where prediction seems feasible and a degree of control possible. Prospects for such goals began to brighten in the mid-sixties when David Evans, a Denver consulting geologist, drew attention to the correlation between earthquakes in the Denver region and the high-pressure injection of fluid wastes from the Rocky Mountain Arsenal into faulted ground along the Rocky Mountain Front Range. A world network of seismometers, tilt- and creep-meters across active faults, and other sensitive instruments is beginning to have a cumulative effect. A large global corps of qualified operators and perceptive scientists, especially in the United States, the USSR, Japan, and China, is fast providing a framework of data and ideas for a predictive model. A list of contributors to this framework would include many of today's leading seismologists.

A variety of effects premonitory to earthquakes have been observed. These include changes in fluid pressure and well levels, variations in local electrical and magnetic fields, variations in the rate of emission of the rare noble gas radon, changes in the relative velocity of deep seismic waves, frequency of occurrence of small earthquakes, and changes in the ratio of smaller to larger shocks in the seismic zone—all related to small but measurable changes in volume of rocks in seismically active areas. C. H. Scholz, L. R. Sykes, and Y. P. Aggarwal of Columbia University, in a 1973 article in the magazine *Science*, note that "precursor effects occur before many, and perhaps all, shallow earthquakes and . . . have a common physical basis." These premonitory effects, moreover, "occur at a characteristic time before earthquakes that increases with the earthquake's magnitude and . . . every one of the premonitory effects found in field observations has been reported in rock mechanics studies in the laboratory." These scientists conclude that volume changes as registered by such related variations signify a state of instability that is invariably followed by an earthquake. Volume changes and fluid pressure are seen as the triggers that release the accumulating stresses. The lag time between the detection of such symptoms and the release of the earthquake is inversely proportional to the magnitude of the quake. Thus, on the one hand, warning times are longer the greater the impending shock; while, on the other, the magnitude of the quake seems to be a function of the time available to store up energy after stress begins to accumulate. Such variables make it easier to predict the probability than the exact time of an expected earthquake, yet data on stress fields does allow estimates for time of onset within broad limits.

Suppose it were practicable to set up earthquake warning systems comparable to weather warning systems. Should we do so? Imagine that you were to read in the morning paper that there was a 70 percent chance of a major earthquake near your residence within the next month. What would you do? Suppose that you sold your home at a large loss to migrate to a presumably stable area and no earthquake came? Or that the area to which you moved was struck instead of the one you left? Imagine what Los Angeles or San Francisco traffic would be like in case of a general panic. And, if a general exodus were to be controlled, how would it be decided who went and who stayed, or in what order people were permitted to leave. How about looting, or fires from severed fuel lines? Would police and fire department employees be permitted to leave too? Whether it came true or not, any prediction of an imminent damaging earthquake that came from usually authoritative sources would be fraught with shattering social, economic, legal, and political hazards.

The relation between the magnitude of earthquakes and the time available for the accumulation of unreleased energy suggests the possibility of a kind of earthquake control in which accumulating energy is released by many small steps instead of a few large ones. The idea involves a trade off of rare large and damaging earthquakes for many small and relatively harmless ones. The process usually visualized is basically a deliberate application of the Rocky Mountain Arsenal procedures mentioned above. Fluids would be injected at high pressure in deep bore-holes along the places where the opposite sides of a fault were sticking together, lubricating the fault and causing it to slip and produce an earthquake—with luck a small one.

Again the problem is not simple. Take a fault such as the San Andreas, along which there has been much creep but (as of the beginning of 1978) no major earthquake since 1940. Suppose that pressurized fluids are injected along it with the result that major movement occurs in a thickly populated region? Unfortunately a large movement on this fault is long overdue in middle California (especially south of the Temblor Range) and a lot more information than is now available is needed to evaluate the risks of tampering with it. There is always that one in a billion chance that the San Andreas won't generate another major earthquake of its own accord, and no responsible person or agency is going to take a chance on provoking a destructive earthquake any more than a riot.

After the next big quake along the San Andreas the situation may be different. If instruments along the fault at that time suggest that it is not then locking up dangerously anywhere near a major settlement, regular deep-well lubrication might be tried at points of sticking, provided we have by then learned where they are. In this way rocks on opposite sides of the fault might be kept moving past one another by bits instead of sticking and allowing stresses to accumulate that can only be released in major future shocks.

It remains to be seen when and if we will dare to try some such control method and whether it will work. Meanwhile the seismologists of the world continue to read the instruments they have, to try to get more of them installed, and to puzzle over the meaning of their data.

For Further Reading

Drake, C. L. (Chairman). 1973. U.S. Program for the Geodynamics Project. National Academy of Sciences. 235 pp.

McElhinny, M. W. 1973. *Paleomagnetism and plate tectonics.* Cambridge University Press. 358 pp.

Marvin, Ursala B. 1973. *Continental drift—the evolution of a concept.* Smithsonian Institution Press. 239 pp.

Matthews, S. W. 1973. This changing earth. *National Geographic Magazine*, vol. 143, no. 1 (Jan. 1973), pp. 1-37.

Wyllie, P. J. 1971. *The dynamic earth.* John Wiley & Sons. 416 pp.

8

THE ARCHITECTURAL EVOLUTION OF THE EARTH

In days of yore when a youth went off to make his fortune so as to return home a man of substance and experience, it was not unusual for him to "sail the seven seas." Different people had different seas in mind, to be sure, and the names had a tendency to change. On a 1783 French map of the Americas that hangs by my fireplace, the whole of the Atlantic Ocean is called the North Sea, while the Pacific is the South Sea. Today only a small part of the northeastern Atlantic retains the name North Sea, while the term Southern Ocean is restricted to the waters encircling Antarctica.

Examine a modern globe, however, and you will see that seven seas and seven lands do stand out. The oceans are the North and South Atlantic, the North and South Pacific, the Indian, the Arctic, and the great Southern Ocean that surrounds Antarctica. All of these flow together to make the global sea, with its many embayments and subordinate seas. The seven obvious lands are North America (including Greenland), South America, Eurasia, India, Africa, Australia, and Antarctica. These continental landmasses arise within and are surrounded by the world ocean, whose waters cover nearly three-quarters of the surface of our globe, locally even spilling over on to the continents themselves to produce continental shelves and epicontinental seas such as Hudson Bay, the South China Sea, the Gulf of Maracaibo, and the Baltic.

Why are these lands and seas located just where they are? Have they always looked the same or have they varied in the past? What causes the low places on the earth where ocean basins are found and the high places of the continents? The truly fundamental differences are geological. Amazing as it may seem, most of the numerous and diverse kinds of rocks on Earth can be grouped into two main types—continental and oceanic. Rich in variety, the continental types of rocks, whether sedimentary, igneous, or altered from one of these (metamorphic), tend to be characterized by high proportions of silica and alumina. The much less varied rocks that characterize the ocean floor, mainly silica-rich basaltic lavas, are distinguished especially by their relative abundances of magnesium and iron. I will refer to these two broad rock groupings simply as *continental* and *oceanic*, regardless of present location relative to shorelines or variations in composition. They have different mineralogical, chemical, and physical properties, which point to different origins. These cause the continental rocks to be lighter and the oceanic rocks to be heavier, account for the differences in elevation between ocean basins and continents, and lead to different velocities of propagation of earthquake waves. By measuring such seismic velocities on three or more seismometers at widely separated points not on a straight line, it is possible to map the patterns of distribution and the thicknesses of these seismically distinctive rock types even where we cannot directly observe them.

As a result of such seismic mapping combined with more conventional geological observations we know the distribution at and beneath Earth's surface of continental and oceanic types of rocks. It turns out that, while all the great oceans and some of the smaller seas are underlain by oceanic rocks, a good many of the lesser seas are not. Some of these, the epicontinental seas mentioned above, are simply places where excess ocean waters spill over the low-lying parts of the continents—where continental crust is relatively thin and low standing. Such seas, like the waters on the continental shelves, are underlain by continental types of rocks. Still others, like the Philippine Sea, rest on a thinner than usual layer of oceanic or oceanic-like rocks. A few, like the Mediterranean, appear to be true small oceans, underlain by oceanic rocks but trapped between opposing continents. The present geological boundary between the continental masses and the ocean basins occurs at an average depth in the sea of about 200 meters. Like forgotten sinks, the ocean basins are overfull.

How fortunate it is for air-breathing animals like us that we have these two major rock types. Without such a differentiation there would be neither high-lying lands nor low-lying ocean basins, but only a nearly even rock surface overlain by an outer shell of water and ice that would cover the entire surface of our planet to a depth of around 2,400 meters. If Earth originated, as suggested earlier, from a random aggregation of solar dust, how is one to account for the segregation of such initially unsegregated matter into continental and oceanic types of rocks? And, even more puzzling, why didn't all the lands wear down to sea level billions of years ago, filling the deeper basins and evening out the ocean floor in the process?

If you were to guess that this had something to do with dynamic processes within the earth you would be right. Although Earth's internal sources of energy are minuscule compared with those of the sun, they are enough, together with indirect uses of our share of the solar energy budget in weathering and erosion by rain and running water, to account for Earth's present configuration and the history of its surface features. The architecture of the earth, as I use the term here, includes both the planet's broad surface configurations, and the internal structure, properties, and motions that give rise to them.

We know from the records of historical geology that the continents continue to stand high even as they continue to be worn down and the ocean basins continue to lie low even as they continue to receive sediments from the eroding lands. The land is carried to the sea yet still the sea is full; still the land is dry. This could be true only if there were persistent processes of renewal. One of these processes is plate tectonics. The other involves the fact that Earth, beneath its solid crust, behaves over the long term like a fluid, deforming beneath the load of its crust somewhat like the surface of the sea beneath differently loaded ships, or like water in a pan with floating blocks of different densities—say wood, ice, and cork. That Earth's lands do indeed float in buoyant balance on a yielding interior has been amply verified by many geophysical measurements in different parts of the world. Major elevations of Earth's surface, therefore, are seen, not as loads supported by a rigid crust, but as lighter crustal segments that rise to different heights and put down roots to corresponding depths on a dense but plastically deforming interior. As they are worn down, the removal of load causes them to be heaved back up. In fact, erosional processes that cut deep valleys into mountain crests, removing load without significantly reducing the general elevation of the crest line, may actually cause mountain peaks to increase in elevation.

That helps in grasping the basic distinction between continents and ocean basins. But what explains how the density differences arose in the first place? And what other processes may be involved in maintaining or renewing the distinctions on an actively eroding and deforming earth? The processes that are of central interest are all connected with the interior of the earth, its thermal structure and history, and the convective circulation by means of which heat is lost from Earth's interior. Those processes, already touched upon in the preceding chapter, will be briefly reviewed following a short discussion of the planetary interior.

Earth's Interior

The deep structure of Earth is worked out using earthquakes for probes and an array of suitably spaced seismometers as recording devices. Times of arrival of distinctive seismic waves as recorded on calibrated strip-charts called seismograms tell us fascinating things about the internal structure and rigidity of our planet.

If you were standing on the bank of a pool and wanted to spear a fish in the water, you would have to compensate for the light's bending as it passed from water to air, conveying to your eye a false impression of where the fish really is. Seismic waves behave like light waves, passing successively through layers of different density like air, water, and glass. The speed and direction of travel of seismic waves changes, like light, with the density of the material penetrated, following mathematical laws worked out for the optical profession by the Dutch mathematician Willebrord Snellius early in the seventeenth century (Snell's Laws). The different arrival times of trains of seismic waves at different seismometers, like intersecting surveyor's lines, show the location of the earthquake focus and epicenter (figure

9). In addition, every time a wave-front crosses a density boundary within the earth, part of it bounces off this surface and part changes course as it enters a layer of different density beneath, generating new wave trains.

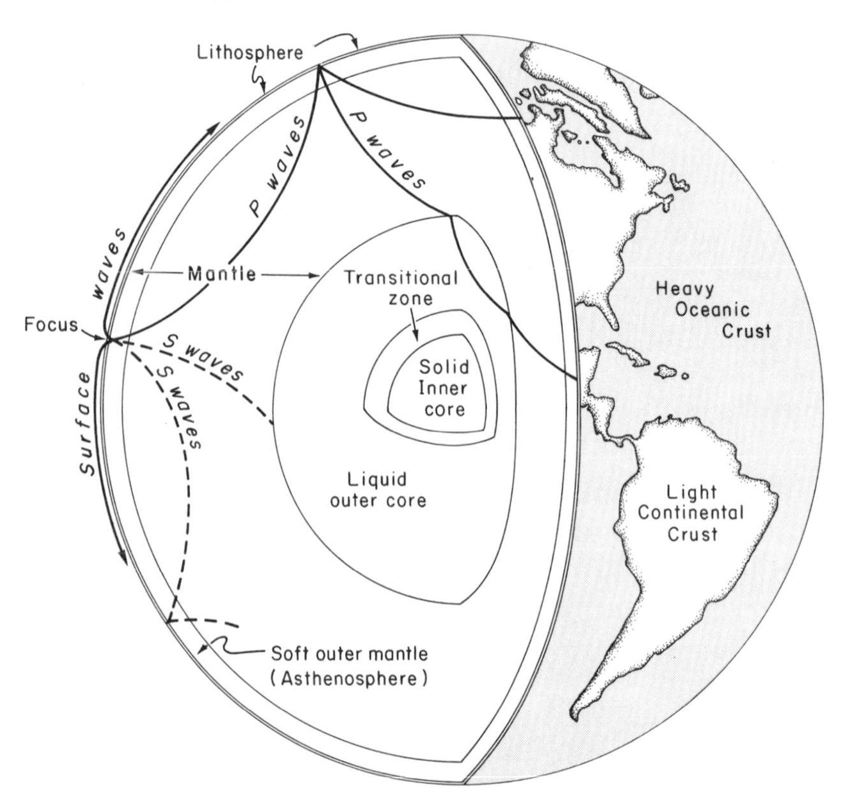

Figure 12. Earth's internal structure. Note that both *P*- and *S*-waves propagate in all directions from the focus; their separation here is only to avoid clutter. (Adapted from B. A. Bolt, 1973, *Scientific American* 228, no. 3, p. 25.)

These seismic waves are vibrations transmitted through the solid earth as a result of an earthquake. As indicated at the left of figure 12, a family of seismic waves generated by any given earthquake consists of several types. *Compressional or P-waves* involve alternating contraction and expansion of the matter through which they pass and these vibrations move in the same direction as that of wave propagation.

The vibrations of a *shear or S-wave* are at right angles to the direction of propagation. The letters *P* and *S* denote primary and secondary: *P*-waves travel faster and thus are the first to arrive at seismic stations. In addition to these two main types of seismic wave, relatively slow-moving waves of shorter wavelength, the surface waves, propagate along and parallel to Earth's surface. As noted earlier they are the ones that do the damage. The *P*- and *S*-waves are more interesting for what they have to tell us about the internal structure of the earth as a result of the way they bounce around within it, start new trains of waves at surfaces between layers of different density, or fail to be transmitted through some materials. *P*-waves are transmitted by either rigid, plastic, or fluid material, but because of their transverse motion *S*-waves are unable to propagate through anything having the properties of a fluid.

Using such evidence, the labors of generations of seismologists at observatories around the world have shown unequivocally that Earth has a concentric structure as idealized in figure 12—something like an onion but with fewer layers. The several concentric layers that surround Earth's solid central core differ in density, rigidity, and thickness. Three main divisions are: (1) the *core*, with a radius of about 3,500 kilometers and a density at the center indicated by seismic travel times to be about 12.5 to 18 times as heavy as water; (2) the *mantle*, about 2,900 kilometers thick and varying in density from 3.3 times the density of water at the top to 5.5 at the bottom; and (3) the *crust*, 10 to 40 kilometers thick at most places, up to 70 kilometers thick under the eastern Alps, and with an average density near 2.8. It is convenient in discussing such matters as plate tectonics to have a special word for the crust plus the upper part of the mantle above the zone of weakness on which the great plates glide. That combination of crust plus uppermost mantle is the *lithosphere*, and the plates are lithospheric plates. The simple term crust, however, is more useful in referring to the familiar outer skin of rocks that makes up our everyday earth.

Although crustal thickness relative to the rest of the earth is proportionally no greater than the skin of an apple, that crust is where most of the obvious action takes place. Variations in its density account for many features of the earth: the lighter parts become continents, the heavier parts become ocean basins, mountain chains arise at recent or ancient zones of collision between continents and ocean basins, and other types of mountain chains form where there have been continent-to-continent collisions. The base of the crust and its variations in thickness are clearly defined by an abrupt increase in seismic velocity as earthquake waves pass downward from the less dense crust above to more dense mantle rock beneath.

The deep interior of the earth is also more complex than is implied by a simple division into core and mantle (figure 12). The core consists of solid inner and fluid outer parts. The mantle is divided into lower and upper parts of different density with different earthquake transmitting characteristics. Comparison of these characteristics with those of experimental materials suggests that the inner core is solid nickel-iron or something like it. Only *P*-waves pass through the outer core. Although highly incompressible, this part of the core behaves like a fluid. It will not transmit shearing forces like *S*-waves, or does so only weakly. Thus, taking density into account, the outer core is thought to consist of chemically uncombined iron mixed with silicate materials in a molten or mushy state. The upper and lower mantle (or the more numerous mantle layers of seismologists) are apparently different density-phases of otherwise similar, solid silicate rocks that are rich in iron and magnesium.

How did Earth's concentric structure originate? Most but not all Earth scientists find it easier to explain this layering by calling on secondary rather than primary processes. In chapter 5 I describe how Earth probably originated as a mixture of unsorted particles that fell together rapidly once the central nucleating mass was big enough to generate a strong gravitational pull. It seems likely that conversion of the gravitational energy of such infalling particles to heat would have raised Earth's temperature substantially by the end of this initial aggregation. A second source of early heating is the decay of short half-lived radioactive isotopes. Half of all Earth's primordial uranium-235 decayed to lead during the first 713 million years of Earth history, with the energetic emission of eight heat-releasing alpha particles for each atom of lead formed. No wonder we don't find Earth rocks older than about 3.8 to perhaps 4 billion years. The initial gravitational and radioactive heating probably melted the early earth, leading to gravitational segregation of its component materials and development of the observed concentric layering. The heavier stuff settled to the center, the lighter rose to the surface, and matter of intermediate density took intermediate positions. The processes visualized are more complicated in detail but that gives the general idea. According to this view, the gases that make up the atmosphere and the water that condensed out of it to form the oceans and other waters of the hydrosphere arose in part from later episodes of melting and volcanic emission of gases—a process called *outgassing*. The crust has evolved and diversified as a consequence of its interactions with the evolving atmosphere, the water masses of the earth, and the products of igneous intrusion and volcanism.

A strong minority view, however, holds it to be more likely that core, mantle, crust, atmosphere, and hydrosphere were all formed early on, in roughly that sequence and in roughly their present volumes, as a consequence of three main processes: First is visualized a segregation of solid materials in space; second, a gravitational segregation of infalling materials in order of decreasing density; third, a heat front that moved upward through the accumulating materials, sweating out contained gases, including water vapor. Such processes, as championed by Gustaf Arrhenius and Hannes Alfven of the University of California at San Diego, suppose an early catastrophic accumulation of nickel and iron rich particles about 4.6 billion years ago, culminating in what is now the fluid outer core. The mantle and crust, they believe, formed during a following interval of slower accumulation, in which each region of the growing mantle was repeatedly remelted. The initial crust would simply consist of whatever scum floated to the surface. Further, Arrhenius and Alfven believe, most of the gases brought to the growing surface of the initial earth by infalling solid particles would have been transferred to the atmosphere on impact. They also judge that the energy of the infalling particles was such that Earth's surface temperature remained on an average low enough (under 100° C) for water vapor to condense and the hydrosphere to develop. According to such a view, the internal structure of the earth, its crust, the hydrosphere and the atmosphere are all products of Earth's initial aggregation and not of subsequent processes.

Both models of the origin of Earth's initial concentric structure envisage the present atmosphere, oceans, and crust as substantially modified from their original states, whatever they were, by persistent interaction with one another, and by photochemical and biological processes.

The Development of the Major Planetary Surface Features

What was the nature of those interactions and their effects on the evolving earth? How did the crust become divided into continental and oceanic parts? What has been the relation over time between the volume of the ocean basins and that of ocean water?

Whatever the history of Earth's interior, it is a fair conjecture that the volume and diversity of crustal materials has increased over geologic time. The processes involved can be observed because they are still going on. A record of similar processes extends billions of years back into the dim recesses of the geologic past.

As briefly noted in chapter 7, an interconnected chain of rifted spreading ridges defines the centers of the Atlantic and Indian oceans, runs part way along the eastern margin of the Pacific, and extends into the Arctic Ocean and the Red Sea. The crust of the earth is splitting open along rifts at the crest of this ridge. Mantle material here wells up toward the surface, melts as it moves into a realm of lower pressure, heats and distends the overlying lithosphere upward, and issues at the seafloor to become oceanic basalt. The rate of formation of new ocean floor is recorded by the spacing of magnetic reversals of known age on either side of oceanic spreading ridges.

This is the process of seafloor spreading, the driving mechanism of plate tectonics, by means of which the seven seas were created and the seven lands dispersed among them (figures 10 and 11).

How can heavy mantle material rise through the crust to cool and crystallize as seafloor basalt? If Earth's crust is splitting along the oceanic ridges, something must fill the space created, and, except for water, outer mantle material is all that is available. The upward journey of this material is made easier by the fact that it melts and becomes lighter as it approaches the surface. Yet why is the crust splitting? Expansion of Earth's circumference cannot account for this, even though a superficial examination might suggest such an explanation. If Earth were expanding, its radius would be increasing and this, as we have seen, would cause it to rotate at a slower rate. That would mean longer days and fewer of them in the year, and such a consequence can be checked against astronomical records.

Sure enough, the rate of Earth's rotation *is* slowing down, but not as much as would be the case were this attributable to increases in circumference (and thus radius) caused by the observed rates of sea-floor spreading. In fact Earth's rotation is slowing down at exactly the rate predicted from forces of tidal friction alone. And it has been slowing at close to that rate for the past 370 million years or more, if the record of daily, monthly, and annual growth rings in fossil corals, so elegantly worked out by the Cornell paleontologist John W. Wells and others, is to be believed. That means that the formation of new oceanic crust at spreading centers must be almost exactly compensated for by the consumption of old oceanic crust in subduction zones and crustal shortening in folded mountain chains.

This introduces the interesting story of the geological record of Earth's rotational rate, the length of the lunar month, and the conclusions to be drawn from both—providing a revealing glimpse into the synergistic workings of the collective scientific mind. Planetary physicists knew from historical records and very precise measurements that Earth's day was getting longer, which could only be because its rate of rotation was slowing down. Such a slowing down might occur because of expansion of Earth's diameter or because of tidal friction. The forces of tidal friction were about right to account for the retardation observed and there was no compelling evidence of diametral expansion. But if Earth's rate of rotation was slowing without change of mass or distance from its rotational axis, that meant it had to be losing angular momentum, and we have seen that angular momentum must be conserved. But even though Earth's apparent angular momentum was decreasing, it could be conserved within the Earth-moon system by transferring the missing fraction to the moon. That would imply either a speeding up of the orbital velocity of the moon or an increase in the distance between Earth and the moon, thus lengthening the moon's distance from its orbital axis. An increase in orbital velocity would mean a shortening of the lunar month.

It was not possible to test these alternative hypotheses until Wells, in 1963, saw significance in the fact that the surfaces of the limy skeletons of solitary corals (not the colonial reef-building kinds) are marked by low annular swellings or bands, which in turn consist of numerous thin rings. Wells hypothesized that the wider bands might be annual and the thin rings daily growth lines, so he started counting them. Sure enough recent corals had about 365 thin rings within a single wide growth band. But fossil corals about 370 million years old had an average of 400 of the thin rings to one coarse band, and fossils of intermediate age gave intermediate counts. From such data Wells concluded that the length of the day 370 million years ago was about 22 hours, which was close to geophysical projections from historical observations and theory. An equally remarkable discovery, also based on roughly 370-million-year-old fossil corals, was made in 1965 by the British geologist C. T. Scrutton. He found monthly growth bands which contained an average of 30.6 of the thin daily growth lines, very close to the present number of days in a month. This implies that the relation between day and month has remained the same over the 370 million years during which the day had lengthened by about two hours.

The choice between hypotheses is now clear. The lunar orbit cannot be quickening. Instead the distance between moon and Earth must be increasing. Charles Darwin's son George had much earlier judged that this should be the case, arguing that a backward projection in time of lunar retreat would bring Earth and its moon together early in Earth history. He thus arrived at the once favored conclusion that the moon had somehow become detached from Earth, leaving the Pacific Ocean as a birthmark. That would certainly have been enough to split the Earth. Indeed, as geophysics now tells us, it is unlikely that either Earth or moon could have survived such an event.

Large though the implications of a closer moon may be, however, for tidal effects and sedimentation on the early earth, this of itself does not tell us how the seafloor can continue to spread without an equivalent expansion of Earth's circumference.

A clue to that puzzle is given by the measured rate of loss of heat from the slowly cooling interior of the earth, producing heat flow at

Earth's surface. Thousands of measurements taken by sensors in shallow probes and bore holes all over the earth bear witness to the fact that the rate of heat flow is fairly uniform except in two linear regions. Heat flow is high over the oceanic spreading ridges where molten mantle rock wells up from Earth's hot interior, and it is low at deep-sea trenches. This is consistent with the previously expressed view that hot new oceanic crust generated at the spreading ridges many millions of years earlier has moved trenchward, cooled, contracted, and is being consumed in the deep-sea trenches at about the same rate as new crust accretes and moves away from the mid-ocean ridges. Although the descending oceanic slab remains relatively cool to depths of several hundred kilometers, it does eventually melt at around 600 kilometers down (figure 11). Lighter components of the melt can then gravitate upward to join and mix with continental rocks above, while heavier components reassimilate with basal crust or mantle. The probable effect of all this, besides compensating for seafloor spreading, is to increase the thickness of continental crust from beneath, helping the continents to maintain "freeboard" above the oceans. Such processes of melting and floating upward of lighter crustal components have been going on throughout Earth's history but were active on a much larger scale in the earliest days of crustal evolution when there was a higher level of internal radioactive heating by the shorter half-lived isotopes.

The ocean basins themselves are both constantly growing and constantly disappearing as seafloor descends beneath the continents. Thus, while the water in the oceans is old (though often recycled) the basins that contain it are geologically young. Even though oceans have existed as long as continents, no part of the present ocean floor is known certainly to be older than about 160 million years, and probably no part is anywhere older than about 200 million years, as compared to Earth's age of 4.6 billion years and continental ages approaching 4 billion years.

Additions of new gases, including water vapor, are believed by many geologists to have occurred during intervals of extensive volcanism throughout Earth's history so that the total volume of water in the oceans would have increased from time to time, although probably at decreasing rates and never on so massive a scale as in the first great outgassing. But the volumes of ocean basins and continents have also changed, and therefore the relation between land and sea has fluctuated. Often sea water has exceeded the storage capacity of the contemporary ocean basins to spill over the continental margins in the form of shallow seas. At times such seas have barely transgressed the continental margins to form narrow shelf seas and embayments like those of today. At other times—for example, during Middle Ordovician (about 465 to 475 million years ago) and Late Cretaceous history (about 62 to 100 million years ago)—the lower lying parts of the continents were very extensively flooded.

Such flooding can also be seen as a likely consequence of plate tectonics. For the rise of new spreading ridges would tend to decrease the total volume of the world's ocean basins, while associated volcanism may have increased the total volume of water to be accommodated by them. Both processes may be forcing factors in the growth of the shallow epicontinental seas whose marine sediments record so much of what we know today about the history of Earth's evolving surface. From that record, supplemented by the less common records of sediments deposited about sea level, and from volcanic and other igneous events, we can read and integrate the stories of the growth and erosion of continents, the invasions and retreats of the sea, the amelioration of climate with flooding and its worsening with drainage of the lands, the building and destruction of mountains, and the many other events that make up historical geology. Figure 13 shows in the most simplified way the breakup and repositioning of the continents as a result of seafloor spreading over the past 200 million years. In it, in our mind's eye, we can visualize all the other consequences referred to (and partly illustrated on the front endpapers) as mountains arose at colliding plate margins, seas overflooded the continents, and volcanism and erosion altered the face of the earth.

Thus it is that the seven seas, the continental masses between them, and all the minor seas and continental islands have an evolutionary origin that depends on processes and events within the earth. They are maintained, added to, and rearranged by internal melting, gravitational segregation, and sedimentary processes that account for the different rock types that give distinctive and contrasting geo-

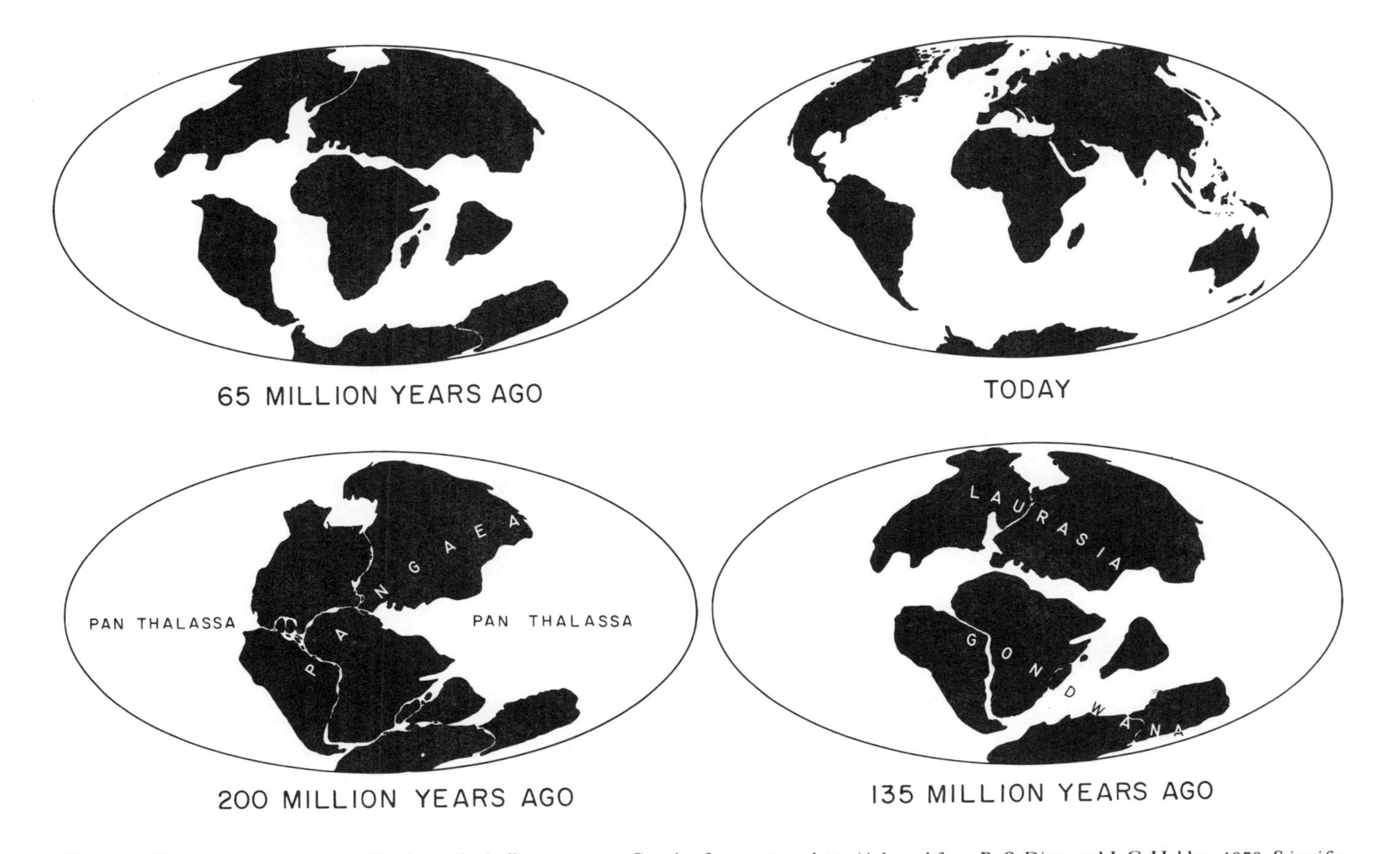

Figure 13. World geography today and in the geologically recent past. See also figures 11 and 23. (Adapted from R. S. Dietz and J. C. Holden, 1970, *Scientific American* 223, no. 4.)

logical characteristics to the continents and ocean basins and explain their differences in elevation. Growth and erosion of lands are in such good balance that the continents, although changing position and configuration with time (figure 13), are essentially permanent features. Although constantly worn down, they continue to stand above the surface of the global sea. The ocean waters too are essentially permanent, although they are probably added to with time and new volcanism. They are recycled through the atmosphere and fresh water systems about once every 2 million years. And they are certainly to some extent recycled in different form along with downgoing oceanic slabs in zones of crustal convergence. In contrast to the essential permanence of the water contained in them, the ocean basins themselves are transient, being created along the spreading ridges and disappearing beneath the continents on a cycle of perhaps 200 to 500 million years duration.

It is the task of historical geology to reconstruct that history and to relate it to other evolutionary cycles. I do not propose here to regale the reader with the details of historical geology but only to sketch in some of the highlights that are of special interest for the question of how we and the surface of our planet came to be as we are today. I will do that in succeeding chapters, but first let me arrange the furniture on the global stage of action which sets the scene for the great drama of life.

A good deal of historical geology has to do with the shifting patterns of land and sea, the elevation and erosion of mountains, volcanism, changing climate, and the effect of all this on life. The record of life is mostly one of sea life because the marine sedimentary rocks that contain most of the relevant data are thicker, display more complete sequences, and are less likely to be eroded away by later events than those deposited on dry land. The seas in which the best and most complete records are found are the shallow seas that spill over the continents and provide a diversity of habitats for a variety of the shelly marine faunas that loom so large in the records of the last 600 million years or so of geologic history. Deep-sea faunas contain few organisms that are well designed for preservation in the geologic record. In addition, as we have seen, the ocean basins proper are also transient features whose sediments are continuously being carried beneath the continents at the convergent boundaries of the giant conveyer-belt system we call plate tectonics.

As the continents are moved around by this system of plates they also come into different climatic zones and fall under different architectural influences. They become glaciated when they cross over or cluster around the poles. Folded mountain chains develop along continent to continent collisions. Volcanoes and remelted materials of mixed continental and oceanic origin rise above downgoing oceanic plates at consuming plate margins to produce another type of mountains, like the Andes or the Sierra Nevada. And the less elevated continental surfaces are successively flooded and drained as the volume of the ocean basins is reduced by the upward bulging of new spreading ridges and increased by the growth of new deep-sea trenches.

Finally, let us consider very briefly some of the motions that are writ large in the records of the past. The history of the earth before about 2.5 billion years ago was one of a high rate of heat loss and mainly vertical motions. Earth was attempting to cool itself convectively as it does today at the spreading ridges, but it had to lose more heat faster because of its greater radioactivity. The oldest rocks proclaim that this was a time of many hot spots and numerous mini-oceans and microcontinents but apparently of little horizontal motion. That phase was wiped out by a worldwide building of maxicontinents and consolidation of oceans between about 3 and 2.5 billion years ago. Plate tectonic mechanisms involving substantial lateral movement thereafter became effective.

By 680 to 600 million years ago (or perhaps much earlier), continent building seems to have been essentially complete, although the geographical arrangement of land and sea at that time is not clear. Thereafter the ancient geography is easier to follow. A bit less than 500 million years ago glaciers were active in what is now the Sahara Desert, while it lay at the South Pole. Africa and North America were then about 10,000 kilometers apart.

Paleo-Atlantic and paleo-Siberian oceans probably existed for some time in the interval before about 300 million years ago. By about 250 (or more) million years ago, however, both oceans were gone. Northern Asia (Siberia) had collided with eastern Europe to form the Ural Mountains and western Africa had closed the earlier 10,000-kilometer gap to join Europe in colliding with eastern North America, creating the Appalachian mountains.

Two hundred million years ago all of the present continents were nestled together to form a single great continent called Pangaea, meaning "all lands," surrounded by a single world ocean, Panthalassa, meaning "all seas." Pangaea began to break up and disperse its parts into Panthalassa about 200 million years ago, during later Triassic history. Figure 13 gives a highly generalized picture of the changing continental positions that have come about during the still continuing process of continental drift. Of all the dramatic motions that are revealed in this figure, the most remarkable is that of India. Wedged free from its position between southeastern Africa and northern Antarctica about 150 million years ago, it has since then traveled some 8,000 kilometers northward to collide with southern Asia and elevate the world's highest mountains.

It is exhausting merely to think of such massive forces and motions. Yet they are the underlying causes that have brought about climatic, geographic, and ecologic change and through those changes have influenced the development of life on Earth.

For Further Reading

Cox, Allan (ed.). 1973. *Plate tectonics and geomagnetic reversals.* W. H. Freeman & Co. 702 pp.

Press, Frank, and Siever, Raymond. 1974. *Earth.* W. H. Freeman & Co. 945 pp.

Takeuchi, H., Uyeda, S., and Kanamori, H. 1967. *Debate about the earth.* Freeman, Cooper, & Co. 253 pp.

Wilson, J. T. (ed.). 1972. *Continents adrift.* W. H. Freeman and Co. (Readings from the *Scientific American*.) 172 pp.

9

OF AIR AND WATER

The chain of evidence and ideas so far related has dealt mainly with the solid earth, its historical antecedents, and its large-scale motions and architecture. It is time to ask more explicitly where the air and water came from. For without these there could have been neither life nor sedimentary rocks to preserve its fossil remains for study. The long record of biological evolution leading to man would not exist. Nor would there be anyone to care about it.

An answer to this question about the air and water is suggested by the relative abundances of the noble gases. For studies of the absorption spectra of stars tell us that the rare and chemically inactive gases such as neon, krypton, and xenon occur in very much smaller quantities in our atmosphere than in the sun or other stars. Thus they could not be residual from some primary atmosphere that accumulated contemporaneously with the accreting earth. Either Earth accumulated without an atmosphere or any primary atmosphere was largely lost during an early high-temperature regime in which the motions of hot gaseous molecules reached velocities such that even the heaviest of the noble gases could escape its gravitational attraction. Because of this the atmosphere from which our present one has evolved is considered to be of secondary origin. That is also the case for the water that condensed out of it to form the oceans, fresh water, and ice masses of the world—the hydrosphere. The initial atmosphere and hydrosphere are almost certainly the products of volcanic exhalation or sweating out of gaseous components from the materials that originally fell together to form the planet, as we drive steam from a baked apple—a process referred to as outgassing and supported by our knowledge of volcanism and the composition of gases found in certain classes of meteorites.

Given that conclusion, a related question follows. Did all the water vapor (and other gases) emerge from Earth more or less at once, was it sweated out episodically during widely separated times of major volcanism, or has it accumulated gradually over a long interval of time as a result of the more or less continuous emission of water-releasing lavas and thermal waters? These three possibilities have been wryly but mnemonically called the big burp, the many burp, and the steady burp hypotheses. It is easier, however, to name them than it is to evaluate them.

Although the accumulation of Earth's air and water as a consequence of outgassing was first proposed by the Swedish geologist A. G. Hogbom in 1894, it has been most effectively articulated by the late W. W. Rubey, long of the United States Geological Survey. Rubey visualized that Earth's water had accumulated at a more or less constant rate during billions of years. Although he granted that such accumulation was probably faster during the initial phases of heating and volcanism, he supposed that it continued through geologic time. In effect he was a continuous burper or a modified many burper. Rubey's detailed analysis seemed to show not only that the amount of water had increased systematically through geologic time, but also that the water in the world oceans has varied little over geologic time in its content of dissolved salts. This contradicted the (until then) accepted view that the water accumulated first and only gradually became salty upon continued leaching of the lands by rain and transport of dissolved minerals to the sea in streams. Rubey's conclusion, however, seemed to be strongly supported by the salt content of waters thought to be newly arrived at the surface of the earth (juvenile waters) in volcanoes and hot springs. The concentrations observed, taken literally, imply a near-constant composition of dissolved salts in the oceans over geologic time.

This was all so beautifully consistent that Rubey's model properly stood as a ruling hypothesis until isotopic techniques were devised for testing how much of the volcanic and hot springs water is in fact juvenile and how much recycled. Measurements obtained from volcanoes and thermal waters in Iceland indicate that less than 1 percent of these volcanic waters are truly juvenile in Rubey's sense. Rather the great bulk of the Icelandic hot waters and vapors are recycled from the present ocean and other parts of the hydrosphere and are *not* newly arrived at Earth's surface. If Icelandic volcanoes, hot springs, and geysers are typical, it would follow that nearly all of the waters and gases that are associated with *modern* volcanoes and hot springs are the recycled products of earlier outgassing and thus not a proper basis for estimating primary compositions or rates of growth of the total hydrosphere.

A position nearly opposite to Rubey's was proposed in 1971 by F. P. Fanale, then of Caltech. Fanale argues for a single major outgassing as a consequence of a kind of heat wave or succession of heat waves that swept upward through the accumulating materials of the initial earth, sweating out most or all of the juvenile water vapor and other gases that were ever to reach the surface of our evolving planet. As mentioned in a different context in chapter 8, a similar view is expressed by Arrhenius and Alfven.

Earth could not, of course, have been completely outgassed by such a mechanism; otherwise it would be a dead planet. For without internal water to lubricate its deep motions, nothing much would happen—no seafloor spreading, no volcanism, no mountain-building, and no further evolution except what took place as a consequence of meteoritic bombardment or as part of solar-system and galactic history.

Because Earth was clearly not completely outgassed in the beginning, it seems likely that *some* juvenile water and gases have arrived at its surface as a result of later volcanism. Initial heating, as previously suggested, was presumably accentuated, both by much more intensive radioactivity than now during early planetary evolution, and by the conversion of the gravitational energy of Earth's collapsing parental dust cloud to heat. Thus one would expect unusually voluminous early outgassing, including quantities of juvenile water. Similarly it seems unlikely that the great outpourings of lava that periodically inundated the later evolving earth brought no great quantities of new water and gases to its surface.

Thus I favor the middle course, the many burp hypothesis, starting with a gigantic burp of long duration and evolving through a succession of subsequent outgassings in which the proportions of juvenile water decrease as we move toward the present.

The picture emerges of a world ocean which was of substantial volume to begin with but nevertheless grew in pulses through time as surface-flooding volcanism waxed and waned. Intervals of quiescence were punctuated by major episodes of active continental drift and volcanism during which most of the new additions of water and associated gases probably took place, along with much recycling of older water. In addition, it seems that the evolving Earth always preserved a rough balance between the quantity of water at its surface and the volume of its main storage basins, the oceans.

Even today, however, a part of the ocean spills over the continental shelves. In the past, spillovers on the continents created great inland seas that moderated world climates and created a diversity of habitats for marine life. Such events are recorded in fossil-bearing sediments on all continents. They tell us, among other things, about the climatic evolution of the earth. The sea, a great reservoir of heat (and cold), is like a giant global thermostat. Past epochs saw reduction of climatic variation where continents were extensively invaded by the sea and more variable climates where marine invasion was slight. The advance of inland seas, carrying an unimpeded flow of heat-transporting and cooling waters over the continents as mid-ocean ridges grew, reducing the volume of ocean basins, led to mild climatic conditions such as those during which the great coal forests and the dinosaurs flourished. Restriction of inland seas, high continental relief, and barriers to free communication between world oceans, also functions of plate tectonics and changing continental orientation, led to more severe climatic conditions that often seem to have stimulated biological adaptation and change. These things bespeak an intimate relationship between plate tectonics, global geography, climatic variation, and biological evolution.

As with the evolution of the hydrosphere, the details of the evolution of the atmosphere are still disputed. One thing that is agreed upon, however, is that there was essentially no chemically uncombined or "free" oxygen in the primitive atmosphere—neither ordinary molecular atmospheric oxygen (O_2), nor the rarer atomic oxygen (O), nor ozone (O_3). That may seem odd if you happen to be an organism for which oxygen is essential, but there are three compelling reasons for this agreement. (1) No likely hypothesis about the nature of the primitive atmosphere includes a plausible source of free gaseous oxygen. (2) Any plausible model of the early atmosphere includes large quantities of gases that would rapidly combine with and remove any free oxygen produced. And (3), as explained below, free oxygen in the early atmosphere would have inhibited the origin of life, which, of course, is known to exist on a large scale today, and which also existed billions of years ago at a time when geochemical evidence implies a lack of atmospheric oxygen.

There are also three reasons why oxygen is unfavorable to the origin of life by natural processes. (1) Chemical evolution of the molecular building blocks from which living cells could be constructed depends on an adequate source of energy, of which the most likely is high-energy ultraviolet radiation. Yet if as much as 1 percent of oxygen is present in the atmosphere, ozone forms in its upper layers where it absorbs ultraviolet radiation of the intensity needed. (2) Any free oxygen present would rapidly degrade any large organic molecules that might have been produced by prior chemical evolution, causing them to be "burned"—in the sense that they would combine with oxygen and become unavailable for further chemical evolution toward living forms. (3) If life did somehow manage to evolve despite such hazards, it would find the free oxygen in its atmosphere a lethal poison. The most fundamental life processes take place in special regions of the cell from which uncombined oxygen is excluded. These processes can tolerate neither free oxygen nor the peroxides and superoxides it produces. As will be discussed in the next chapter, nature has had to be very clever in finding ways to carry out the oxidations essential to higher forms of life in indirect ways that exclude free oxygen, and in evolving special enzymes that keep any free oxygen or corrosive oxygen products from attacking the vital centers.

Scientists can advance only informed guesses as to the composition of the primitive atmosphere and hydrosphere, how rapidly they accumulated, and what the earliest interactions in which they were involved were like. Good evidence, however, confirms that substantial amounts of air and water were already present at the time the oldest well-dated rocks were formed on Earth.

That evidence consists of the metamorphosed equivalents of common types of sedimentary rocks from southwestern Greenland, rocks that have been dated as being 3.76 billion years old by geochronologists from Oxford University. The initial rocks were igneous, formed by solidification of molten materials at or near Earth's surface. In order for sedimentary rocks to be produced from these, there had to be atmospheric gases like carbon dioxide and water vapor to disaggregate and chemically alter them by weathering processes so as to produce fragmental particles, clays, and dissolved matter. The sedimentary associations and characteristics observed are evidence that, in addition, there had to be running and standing water to transport the fragmental and dissolved materials to a site of accumulation, impose the sedimentary characteristics observed, and provide an aquatic burial place. There are much thicker, more diverse, and better preserved sediments about 3.4 billion years old, in eastern South Africa and Swaziland, which tell a similar story even more clearly. From then onward an extensive sedimentary record yields clues to the further progress of atmospheric and hydrospheric evolution.

Under what kind of atmosphere were the oldest known sediments deposited? One hypothesis holds that the ancestral atmosphere consisted mainly of ammonia and methane, with subordinate carbon monoxide and, of course, water vapor. Another holds that carbon monoxide, carbon dioxide, nitrogen, and water vapor were the main components. Considered on a global scale, however, older sedimentary rocks, up to as recently as about 2.6 billion years ago, include relatively little limestone but are rich in chemically precipitated silica. Had ammonia been abundantly present at that time, its effects on water chemistry should have favored the extensive formation of ordinary limestone or its magnesium-rich equivalent, dolomite, while severely limiting the formation of chemically precipitated silica. It therefore seems probable that, if there was an initial methane-ammonia atmosphere, it had evolved into one in which those components were superseded by carbon dioxide and nitrogen by the time the oldest known sediments were formed, around 3.8 billion years ago.

These oldest sediments bring three other interesting messages from the past. Because the minor amount of dolomite they do contain is rich in unoxidized iron compounds, they support the position that the atmosphere of the time contained little or no free oxygen. The presence of any dolomite at all also supports Rubey's argument that the primeval sea was as salty as it is now—for dolomite is, above all, a product of fully marine or even hypersaline waters, or of contact zones between salt and fresh waters. Certainly the sea was fully salty by somewhat more than two billion years ago to judge from the extensive formation of dolomite of that age in southern Africa. And it is not much later that a potash-rich sedimentary mineral called glauconite, distinctive of modern marine habitats, becomes a normal component of some ancient sediments.

The third message is accessory to the evidence these and younger sediments convey about the continuous presence of liquid water on our planet from 3.8 billion years ago until now. That means planetary temperatures generally above the freezing and below the boiling point of water. And that means either that the early sun was not as faint and cool as some astrophysicists hypothesize, or, if it was, that the early earth basked in relatively benign temperatures because of some process such as the heat-retaining effects of an early carbon dioxide-rich atmosphere, or a then high rate of loss of heat from its interior, or both.

It appears, in any case, that Earth's air and water are products of internal melting and the release of gases. Man, with his peculiar metabolic requirements, would have had a low opinion of the purity of this primordial atmosphere. Smog would seem preferable. Although the early oxygen-free atmosphere was well suited to the chemical evolution of organic molecules and the origin of life, it would have been a lethal mixture to those forms of life we now like to think of as advanced. Biologic evolution, beyond the origin of life itself, depended on the further evolution of the atmosphere. That took place as a result both of chemical changes and of interacting biological processes. Students of the evolving earth now realize that there is a very complex interaction between the separate but related evolutions of air, water, life, and rocks.

For Further Reading

Brancacio, P. J., and Cameron, A. G. W. (eds.). 1964. *The origin and evolution of atmospheres and oceans.* John Wiley & Sons. 314 pp. (contains papers by W. W. Rubey and others).

Cloud, Preston. 1974. *Encyclopaedia Brittanica,* 15th ed., s.v. "Atmosphere, development of."

Fanale, F. P. 1971. A case for catastrophic early degassing of the earth. *Chemical Geology* 8:79–105.

Kuiper, G. P. (ed.). 1952. *The atmospheres of the earth and planets* (2d ed.). University of Chicago Press. 434 pp.

10

HOW THE AIR BECAME BREATHABLE

It seems that early intensive heating of the earth and subsequent volcanism sweated out the previously entrapped or chemically bound gases that went to make up the initial atmosphere. Among these gases was the water vapor that later condensed to form the first minioceans. The precise composition of the earliest atmosphere cannot be known, but the earliest to which we have some sedimentary clues probably included carbon monoxide, carbon dioxide, water vapor, nitrogen, and hydrogen. It may also have included some quantity of gases rich in nitrogen and hydrogen, such as ammonia and methane, perhaps residual from an earlier atmosphere in which such gases dominated. Not a very salubrious atmosphere from our point of view.

Advanced forms of life, which are dependent on a high rate of energy utilization and thus on respiratory metabolism, could not originate or survive in such an oxygen-free atmosphere or in the anoxygenous water-bodies beneath it. Processes were already beginning to operate, however, that would eventually result in the origin of life, the production and accumulation of free oxygen, the onset of respiratory metabolism, the development of advanced forms of life, and the preservation of a sedimentary and biological record of Earth history.

Ample carbonic acid, derived from rain falling through the gaseous atmosphere, loosened the rock-forming minerals of Earth's early crust, partially dissolving them and partially converting them to sedimentary particles. Flowing water, aided by ice, wind, and gravity, moved such dissolved matter and sediments down slopes and along stream courses to lakes and seas, where the particles came to rest. Here also, given an appropriate chemical setting, dissolved matter could precipitate to form chemical sediments. Weathering and sedimentation began. Deposits accumulated in subsiding rock-floored basins, there to be bound together into sedimentary rocks by chemical cements precipitated from water entrapped between the sedimentary grains and seeping around them. With the passage of time, the pile of sedimentary rocks became thicker and thicker. A sedimentary record was accumulating, from whose talismans we now read the history of our planet, the evolution of its air and water, the record of its life, and the interactions among these things.

The oldest preserved parts of this sedimentary record, however, contain minerals that imply an absence of free oxygen, for which there is indeed no known primary source. These facts are of great interest for the hypothesis that life originated by natural processes. Many well-controlled laboratory experiments show that, *given the absence of free oxygen*, chemical evolution, impelled by high-energy ultraviolet radiation, could have produced abundant large organic molecules under a variety of plausible early atmospheric compositions. Were such molecules to have been concentrated in pools or on catalytic surfaces such as clays, there to be further energized by external sources, this could have led to the evolution and chemical linkage of still more complicated molecules, and eventually to the origin of the first living cell. Such a sequence of chemical events, as briefly noted in the preceding chapter and discussed more fully in the following one, depends on the absence of free oxygen, which might burn up either the molecular building blocks of life or primitive life itself. Yet the emergence of advanced types of cells, oxidative metabolism, and the splendid panorama of animal evolution all depended in turn on the presence of atmospheric oxygen in ample quantities.

Where could this oxygen have come from, when it is not observed in volcanic gases or in an uncombined state within meteorites or Earth's primeval rocks? How and when did oxygen build up to present atmospheric concentrations? What variations, if any, have affected our oxygen budget in the past, and how might these have been linked with biologic evolution?

An atmosphere like ours, of roughly 78 percent nitrogen, 21 percent oxygen, and 1 percent argon, is certainly a most improbable mixture of gases. What happened to all the other products of Earth's outgassing? If the 4.6 billion years of Earth's evolutionary history could be compressed to the equivalent of a day in length, no one would have been around to observe the relevant events until the last second. But nature has kept its own account of the processes involved in the form of sedimentary rocks, which register the geochemical consequences of unobserved events. And these can be decoded with the aid of the Rosetta stone of now-observable processes and well-founded geochemical theory.

As soon as free oxygen began to appear in our atmosphere or hydrosphere, from whatever source, it would promptly combine with any ammonia or methane then present to release free hydrogen and nitrogen and to make more carbon dioxide and water vapor. Now atomic motions regularly cause certain amounts of hydrogen and helium to escape the pull of Earth's gravity field even today; but the process is slow, and free oxygen tends to combine with any remaining hydrogen to form water, as well as with carbon monoxide to form carbon dioxide. Indeed it is possible that all the carbon dioxide that ever passed through Earth's atmosphere was simply the "ashes" of combusted carbon monoxide of volcanic origin.

The carbon dioxide, as noted above, combines readily with water to form carbonic acid, which, as rain, is a great weathering agent of rocks, especially of limestone and its metamorphosed equivalent, marble. That is why, in humid climates the world over, the inscriptions on the marble headstones in graveyards become dim with age, as the carbonic acid in rain water neutralizes itself by reaction with the rock. Only in the parched regions of the planet do marble pleasure domes still gleam. Where carbonic acid accumulates in pools, or where carbon dioxide is added to lakes and oceans by the fermentation, respiration, or decay of aquatic organisms, something else happens. The carbonic acid loses hydrogen ion ("dissociates," a chemist would say) to form bicarbonate ion, which in turn changes to carbonate ion as some of the dissolved carbon dioxide leaves or is removed from a system with bicarbonate ion in it. The carbonate ion (CO_3 with two negative charges, written as CO_3^{--}) then combines with calcium or magnesium ion (Ca^{++} or Mg^{++}) in solution to precipitate as an insoluble neutral "salt." Such sediments harden by other processes to become limestone and dolomite, thereby locking up vast amounts of carbon dioxide and oxygen in the form of sedimentary rocks.

By far the greater part of the carbon monoxide and carbon dioxide that has ever passed through Earth's atmosphere is thus removed from it—a very good thing for people and other animals.

The processes described explain why carbon monoxide, carbon dioxide, and hydrogen are rare gases in our atmosphere. The reason why nitrogen is so abundant is just the opposite. At atmospheric temperatures and pressures nitrogen does not combine readily with other substances. It is an almost inert gas. Any nitrogen that came into the early atmosphere would have tended to remain there. Whereas ammonia in the atmosphere would have combined rapidly with any available oxygen to form nitrogen and water ($4NH_3 + 6O_2 \rightarrow 2N_2 + 6H_2O$), nitrogen could accumulate from several processes and thus increase relative to other primary gases, serving as a kind of inert carrier-gas for reactive free oxygen.

None of these processes accounts for the 21 percent of oxygen in our present atmosphere, however. We have seen many processes for removing oxygen from the atmosphere but none for putting it in. And oxygen is what most of us are likely to have in mind when we think of the atmosphere as a life-supporting system.

Although there is no primary source of free oxygen, there is plenty of bound oxygen in water (including water vapor), which has ample primary sources. Two important ways of releasing free oxygen from water are known, both dependent on solar energy. One is the breakdown of water vapor in the outer atmosphere as a consequence of *ultraviolet* radiation. The second is the splitting of the water molecule by photosynthesis, linked with assimilation of carbon dioxide to form carbohydrates plus free oxygen (in simplest form $CO_2 + H_2O \xrightarrow{\text{solar energy}} CH_2O + O_2$). Both processes, however, are easily reversed, yielding stored energy. Hydrogen and oxygen recombine to form water vapor in the outer atmosphere, while carbohydrates or their carbon residues recombine with oxygen to revert to carbon dioxide and water ($CH_2O + O_2 \rightarrow CO_2 + H_2O +$ released energy). The only way oxygen can accumulate, therefore, is for hydrogen to escape Earth's gravity field or for carbonaceous matter to be buried in sediments where it is unavailable to recombine with oxygen. For every molecule of oxygen that has accumulated in our atmosphere or combined with other molecules in oxygen reservoirs over geologic time, an equivalent amount of hydrogen must have escaped into outer space or carbon been buried in sediments.

There can be no doubt that both processes have contributed some oxygen. The question is how much, and which, if either, process was dominant. Although most scientists who have considered the problem think that photosynthesis was probably the main source of free oxygen, an origin from decomposition of water by the purely physical action of solar radiation has strong supporters. R. T. Brinkmann, for example, has argued that this process could have produced free oxygen equivalent to the amount now tied up in Earth's atmosphere, hydrosphere, and biosphere, plus the amount that combined with reduced iron compounds over all of geologic time. There are, however, other oxygen reservoirs or sinks to be accounted for—besides iron oxides—that are of equal or greater importance. These include the sulfate ion which is present in sea water and tied up in the minerals gypsum and anhydrite, the organic matter in sedimentary rocks, and the carbonate fraction of limestone and dolomite, not to mention all the oxygen that may have been involved at some stage in the oxidation of ammonia, methane, and carbon monoxide.

When geologists want to know whether a process was geochemically significant or not, and at what scale, they customarily prepare a geochemical balance sheet from the rock record. We can do this for the oxygen- and carbon-rich end products of photosynthesis, but not for the end products of the ultraviolet decomposition of water vapor. In order to write a geochemical balance sheet for that, we would need to have budgets for both oxygen and hydrogen. An estimate for oxygen is feasible, but we are thwarted in trying to write a hydrogen budget by the fact that there is no satisfactory independent way of estimating either how much hydrogen has escaped Earth's gravity field over geologic time or how much has been added by the fluctuating solar wind over the same interval. The solar wind, made a household word by the Apollo space flights, is the blast of ionizing radiation, rich in hydrogen, that spreads in all directions from the sun, and from which Earth is largely shielded by its magnetic field, continuously regenerated by the dynamo action of its rotating outer fluid core.

Although we cannot properly reckon the amount of hydrogen, we can write a balance of sorts for carbon by adding all the carbon found in atmosphere and hydrosphere to what is buried in sedimentary rocks. We cannot account for carbon that may have been carried beneath the continents by descending slabs of crustal materials during episodes of plate tectonic motion; but, since the oxygen budget is subject to the same sources of error on a comparable scale, and because both carbon and oxygen are presumably recycled by plate tectonic processes, the important thing to know is how closely oxygen and carbon balance one another geochemically in the observable parts of Earth's crust. That balance is important even though the oxygen released by photosynthesis actually results from the splitting of the water molecule as a way to get at hydrogen as a source of electrons for biologic energy. The *products* of photosynthesis, however, are just as closely associated with assimilation of carbon dioxide in some form, while the water itself becomes linked with carbon in carbohydrates and eventually recycled. Thus the ultimate balance we seek is in terms of a reconstitution of carbon dioxide.

If, then, we add all the oxygen and carbon in various geochemical sinks to the quantities of carbon and oxygen in circulation in either gaseous or dissolved states, we arrive at the sums indicated in figure 14. These sums are so large that they are written in units of 10^{20} grams, equivalent to 100 trillion metric tons. The chemical combining equivalents are obtained by dividing the estimated weights in grams by the molecular weights of the combining substances. Since in CO_2, two oxygens combine with one carbon, it will be clear upon consulting atomic weights in the periodic table of the elements on the back endpapers that the total weights for oxygen and carbon must be divided by 32 and 12 respectively to get combining equivalents. When this is done we find, in figure 14, that the combining equivalents of carbon and oxygen are almost the same, considering all of the

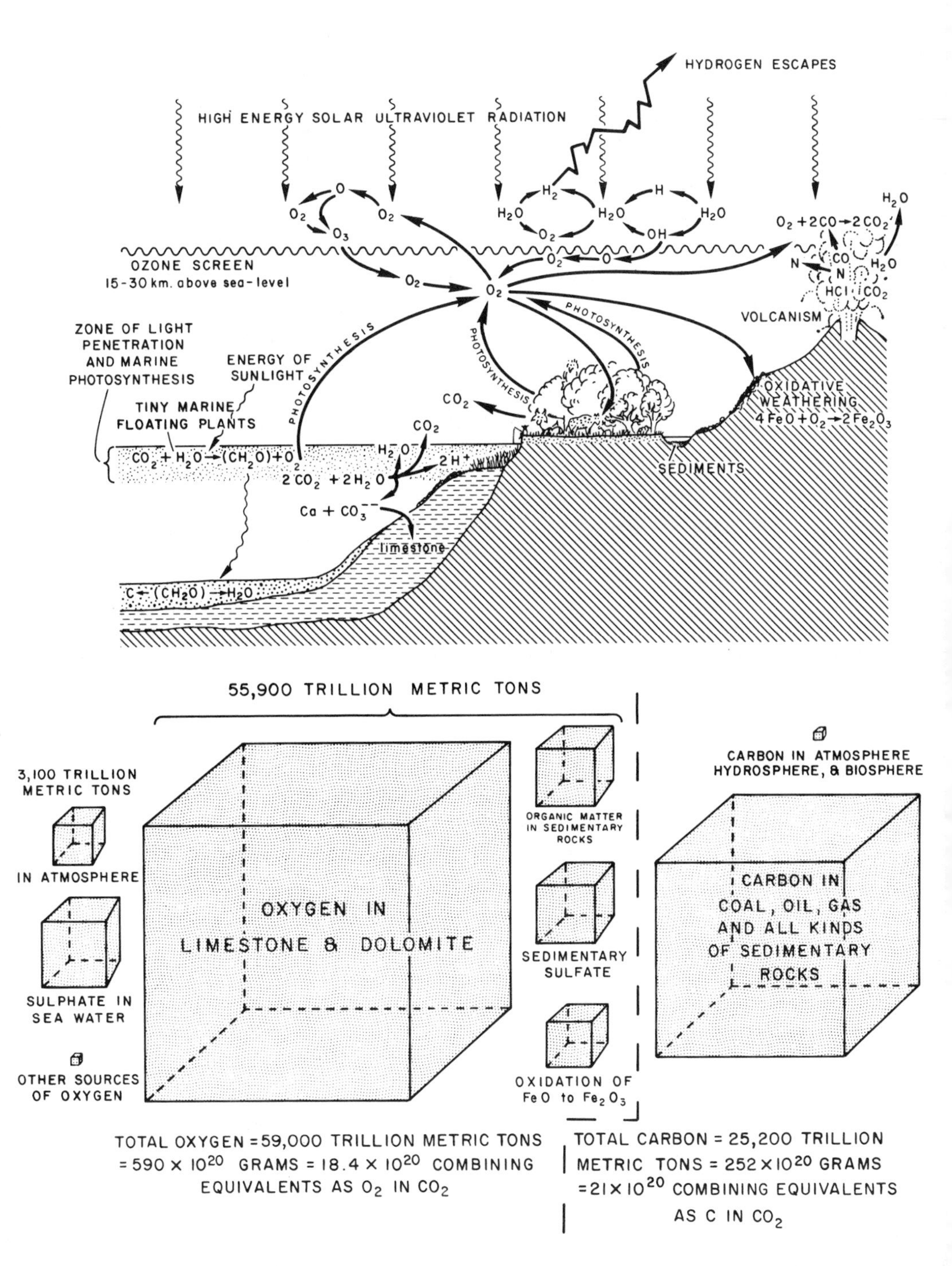

Figure 14. Origin of oxygen and its apparent geochemical balances. (Adapted from sketches by P. Cloud and A. Gibor, *Scientific American* 223, no. 3, 1970.)

uncertainties involved. This rough equivalence surely warrants some confidence in the conclusion that photosynthesis has been a major source of free oxygen over geologic time, if not its main source.

The excess carbon indicated in figure 14 indicates unaccounted-for oxygen that could be accounted for by the conversion of some carbon monoxide to carbon dioxide over geologic time. If, however, all of the carbon dioxide originated as carbon monoxide (which is thought to be unlikely), a sizable number of additional combining equivalents of oxygen must have come from some source, most likely from solar breakdown of water vapor. Such oxygen could have been important during earliest atmospheric evolution in changing the composition of the initial atmosphere toward a more oxidized state (that is to say, CO to CO_2, and so forth) over the billions of years that preceded the appearance of free oxygen as more than a fleeting component of the atmosphere. In addition to carbon monoxide, some geochemists think that methane and ammonia were important among these early gases. If so, we would require even larger sources of "excess" oxygen from breakdown of water vapor by purely physical processes of solar radiation to account for the invisible oxygen that became a component of water and carbon dioxide during the oxidation of this ammonia and methane. Unfortunately we have no good way of calculating these quantities.

Thus we see that both photosynthesis and the breakdown of water vapor by solar radiation have probably contributed to the oxygen budget over geologic time. Photosynthesis, however, seems to have become the dominant process after the origin of green plants and the establishment of an ozone screen. But man's curiosity is unbounded. Like children who want to know why the sky doesn't fall, we immediately think of other questions. How and when did photosynthesis begin? How and when did all the oxygen sinks get filled up so that oxygen could begin to accumulate in the atmosphere and generate a life-shielding ozone screen (by UV dissociation and recombination of O_2 to O_3)? What processes have worked to prevent the recombining of our atmospheric oxygen with the carbonaceous by-products of photosynthesis so that our air could become and remain breathable—a major miracle of atmospheric balance, or, one might say, imbalance?

I remarked earlier that the first living things on our planet were probably minute microbes, intolerant of oxygen and dependent on external sources of food. Although it is very likely that oxygen was being generated from the nonbiological breakdown of water vapor in the outer atmosphere even then, it could not last long in the presence of all the unoxidized substances in the atmosphere and at the surface of the earth—a fortunate circumstance in view of oxygen's lethal potentiality for living things. Life, however, could not continue, diversify, and evolve to higher levels on a photochemical food supply alone. The way out of this predicament was the emergence of a kind of microorganism that could manufacture its own foodstuffs, thus providing a nutrient base for the other creatures that would prosper on its initiative.

The first such organisms were probably bacteria (or bacteria-like things) that utilized the hydrogen from water as a source of energy for their life processes, while combining the potentially poisonous nascent oxygen with reduced materials such as sulfides or alcohols. This was a kind of photosynthesis, but not the kind that results in free oxygen. That came later, with the emergence of so-called green-plant photosynthesis in a creature which in its main aspects was a bacterium, but which utilized a different kind of photosynthetic pigment from its bacterial antecedents. That was the beginning of the blue-green algae (or blue-green bacteria, as some prefer), which, along with orthodox bacteria, comprise the simplest forms of cellular organization. Such primitive cellular organization is said to be *procaryotic*, referring to the primitive structure and unbounded nature of its nuclear materials. The procaryotes of modern times include many prime indicators of polluted environments. Despite the fact that their photosynthetic activities produce free oxygen, many blue-green algae even today tolerate or prefer relatively low oxygen levels. Some are so sensitive to oxygen that they live only in environments rich in hydrogen sulfide, which serves as a local oxygen sink to keep the oxygen depressed to tolerable levels.

Considering the lethal effects of unshielded free oxygen on living systems, the first microorganism to take up splitting the water molecule as a source of energy-rich electrons, releasing free oxygen in the

Figure 15. Oldest known sedimentary structures of probable biological origin compared with modern counterparts. *Below:* concentrically laminated domes of probable algal origin, on eroded bedding surfaces of 3-billion-year-old limestone in northern Zululand (photograph by Tom Mason, University of Natal, Durban, South Africa; reproduced with his permission). *Above:* similar unlithified blue-green algal structures in a brackish pond of northern Andros Island, Bahamas.

process, would have been courting trouble. It would have required some system for controlling or removing the oxygen. Whereas even the most primitive photosynthesis-related pigments would probably have offered some protection against intracellular oxygen, the first oxygen-producing microbes probably had very low tolerances to oxygen in the surrounding waters. And until such time as there was enough atmospheric oxygen to produce a shield of ozone that would effectively exclude damaging or lethal wavelengths of ultraviolet light, no organisms could exist on the land, except perhaps in limited areas of deep shade or within sediments. For even though high-energy ultraviolet radiation is seen as the source of energy for chemical evolution leading to the first organisms, that same radiation has a disruptive effect on the genetic mechanism. The cell, therefore, must either possess special genetic repair systems or be shielded from the disruptive wavelengths of ultraviolet radiation in some way. Repair mechanisms are found in some bacteria and blue-green algae, reinforcing the idea that their ancestors were in fact exposed to high levels of background radiation. Other (and all advanced) forms of life must be permanently shielded from radiation.

Such considerations, and the geological circumstances of their occurrence, lead me to think of the early oxygen-releasing photosynthesizers as occupying a shallow aquatic environment, one within reach of suitable wavelengths of light for photosynthesis but shielded from disruptive ultraviolet radiation by a thin layer of water or sediment, much in the way we attempt to shield ourselves from atomic radiation. The oldest record yet known of sedimentary structures believed to represent such a shielding mechanism is shown in the lower part of figure 15. The early photosynthesizers that inhabited and built such structures were also presumably surrounded by sufficient quantities of reduced substances to soak up any unrecombined photosynthetic oxygen almost as fast as it was produced.

Such a perception has interesting geologic consequences. The substance or substances that functioned as a sink for excess biogenic oxygen would, depending on their nature, either become oxidized sediments or be converted to gaseous components that, in turn should have other feedbacks to the sedimentary record. Because of its ultimate derivation from CO_2, each molecule of O_2 thus accounted

Figure 16. Oldest known sedimentary rock on Earth. Edge-on view of deformed layers of siliceous banded iron formation 3.76 billion years old from southwestern Greenland. Dark bands are iron-rich, light bands iron-poor.

for should be represented by an equivalent quantity of carbon somewhere in the associated sedimentary complex.

So we turn to the pile of ancient sedimentary rocks in search of records—signals that can tell us of events long past. Our attention is quickly riveted on a most unusual rock (figure 16). It is a beautiful rock consisting of red, gray, blue, black, and white bands of alternating iron-rich and iron-poor silica (or, less commonly, calcareous rock)—abundant among and characteristic of sedimentary rocks deposited between 3.8 and 2 billion years ago but uncommon in more recent strata. The iron is rich both in *oxidized* (oxygen-rich, or, in the case of iron, ferric) and *reduced* (oxygen deprived, in this case ferrous) components and the total amount on Earth is on the order of trillions of tons, if we include rocks of relatively low iron content as well as mineable ores. This rock has long been a major geochemical puzzle. Geologists agree that it is a chemical deposit from former open-water bodies of substantial size. It taxes the imagination, however, to visualize what kind of a chemical system might explain the observed continuity of individual iron-rich laminae, often only a fraction of a millimeter thick, for distances of hundreds of kilometers within certain prominent sedimentary basins that were being filled about 2 to 2.2 billion years ago (and, less prominently, in older ones). No convincing modern counterpart is known. On the one hand it is hard to visualize how such a wide dispersal of dissolved iron could occur except in solution in the reduced state. On the other hand, how could reduced iron have been abruptly converted to the insoluble oxidized state and simultaneously precipitated in laminae of nearly uniform thickness over very large areas *except* in the presence of oxygen?

This is the riddle of the geologically well-known banded iron formations, or BIFs, as geologists call them for short (figure 16). By combining this riddle with the earlier discussed riddle of biological sensitivity to free oxygen, we resolve both. The BIFs can be seen as oxygen sinks that kept early oxygen levels depressed, allowing the first photosynthesizers to operate. The iron could become widely dispersed in solution in an initially reduced state because the contemporary atmosphere and hydrosphere were, in fact, deprived of oxygen. This iron, in turn, is thought to have precipitated in the

Figure 17. Microorganisms from the 2-billion-year-old Gunflint Iron Formation west of Schreiber Beach, Ontario. Greatly enlarged views of specimens visible in thin rock slices, showing variety of forms. All bar scales are 1/100 millimeter long.

oxidized state because it subsequently combined, molecule by molecule, with biogenic oxygen from a primitive but widely dispersed, floating blue-green algal or related source—a primitive *phytoplankton* (floating microscopic plants). Thus it seems that, for some hundreds of millions of years, an approximate balance was preserved between limited populations of oxygen-sensitive microorganisms and their surrounding sink of oxygen-accepting reduced iron. To produce the lamination observed what is called for is the episodic introduction of iron, or the episodic flowering of phytoplankton, or both, at sites of essentially continuous silica precipitation. A prevalence of contemporaneous silica-saturated waters is assured by the fact that silica secreting and precipitating organisms had not yet evolved (all such organisms are eucaryotic). In this way one may visualize the origin of the alternating bands and microbands of iron-rich and iron-poor silica that comprise the BIF.

Much has been written about this process and the significance of different kinds of BIF for ancient ecosystems. Discussion centers on sources for iron and associated silica, the causes of the rhythmic banding and lamination, and the nature of the environment of precipitation. It is enough here to note that the story is more complicated than I have told it, but that the basic idea of a kind of mutually dependent relation between the early BIF sediments and primitive biogenic sources of oxygen is now widely accepted. The rarity of carbon in the BIFs needs to be explained, however. Two possible explanations come to mind, and each helps. First, carbon-rich sediments are found in the central parts of some of the old sedimentary basins, suggesting that a goodly fraction of the dead microbial carbon may simply have floated out there before settling. Second, carbon in the BIFs may have recombined with a fraction of the abundant initial mineral hematite, the most highly oxidized state of iron in the BIF, to form carbon dioxide plus magnetite, a less oxidized mineral that is a conspicuous component of BIFs today. Some of the kinds of microorganisms known to be associated with BIFs are illustrated from a two-billion-year-old rock in figure 17. It seems that all organisms of that age or older were of the simple procaryotic type.

Thus was the way prepared for the evolution of efficient biological oxygen-mediating systems that could lead to a final filling up of the oxygen sinks and a gradual buildup of atmospheric oxygen to its present level.

That, to be sure, did not happen all at once. Even with efficient biological oxygen-mediating systems, oxygen could accumulate no faster than carbon (or hydrogen) could be segregated and potential new oxygen sinks, such as volcanic carbon monoxide and sulfides, neutralized. What signs in the geologic record might tell when free oxygen began to appear in the atmosphere in perceptible quantities, thus assuring that the major oxygen sinks had previously been neutralized? It bears on this question that, when the buildup of atmospheric oxygen first started, there would have been little or no ozone (O_3) in the atmosphere to shield Earth's surface from high-energy ultraviolet radiation. Such radiation would have converted some of the oxygen to ozone, and some to atomic oxygen (O), both highly reactive substances. Even very small quantities of free atmospheric oxygen, therefore, would produce strong oxidizing effects on Earth's surface materials.

Instead of reduced iron being readily removed in solution from Earth's weathered surface, as had previously been the case, it would tend, under such circumstances, to accumulate there as insoluble oxides. Iron that did go into solution could not travel far in open-water bodies because it would immediately precipitate on contact with oxygen-saturated waters. Grains of sand coated with oxidized iron would accumulate in favorable continental and marginal marine environments to form the red sandstone deposits geologists call "red beds." True red beds known to be produced by processes not involving redeposition of BIF or oxidation related to younger weathering surfaces are unknown in rocks deposited more than about two billion years ago but are common from then onward. That combines with the essential disappearance of BIF from the sedimentary record at about the same time to suggest something important about the evolution of life, of the atmosphere, and of the interaction between them. This may denote the time of perfection of advanced biological oxygen-mediating systems and the beginning of evasion of free oxygen from the global sea to the atmosphere on a substantial scale. There is evidence in the record of ancient sedimentary rocks to suggest that these events may have been triggered by a vast, perhaps climatically induced, transfer of ferrous iron in solution from storage in the deep oceanic basins to the relatively shallow shelves and shelf-basins where conditions were favorable for blue-green algal photosynthesis and thus for the sedimentation of BIF.

The large-scale oxygen-depressor capacity thus afforded may have stimulated substantial increases in the number of photosynthetic, oxygen-producing procaryotes. In such organisms individual change by mutation is very nearly the sole source of evolutionary variability, not ordinarily assisted by sexual recombination as in advanced organisms. Thus the pace of procaryotic evolution becomes a function of the number of potentially mutant individuals. And because the normal manner of reproduction among such organisms is by simple vegetative division or some modification of it, one viable mutant is all that is needed to give rise to a new strain. Biochemical evolution of an advanced oxygen-mediating system may well have been a product of such a flowering in numbers of microbial ancestors, as I have hypothesized above. Once the problem of biological oxygen was solved, its photosynthetic production and accumulation could proceed as fast as carbon could be buried in the sediments. And once all the ferrous iron in solution was swept out as banded iron formation, perhaps, in the final stages, as a product mainly of seasonal upwelling, no further significant quantities could form, except in special local circumstances. Thereafter biologically generated oxygen would build up in the seas and begin escaping to the atmosphere. That event may be signaled by the worldwide appearance of the oldest true red beds about two billion years ago, or a little less.

The cumulative efforts of countless trillions of minute, photosynthesizing microbes, assisted by physical processes, transformed the primitive oxygen-free atmosphere into one where new kinds of life processes could prosper, new levels of energy utilization could arise, and new and more complicated kinds of organisms could evolve.

Another 1,300 million years of atmospheric evolution seems to have followed, however, before oxygen reached levels capable of sustaining multicellular animal life of even the most primitive types.

Despite efficient biological systems for coping with it, free oxygen could still build up no faster than carbon was buried and new oxygen sinks neutralized. Clues to levels of oxygen buildup are few, but there are some.

Somewhere between 1.3 and 2 billion years ago a new type of cell seems to have appeared. It contrasted sharply with the previous, exclusively procaryotic types of cells, in which the nuclear material is not clearly packaged or membrane-bound, intracellular structures are generally simple, and the single long circular chromosome (if there is one) is intricately folded on itself. This new, truly nucleate type of cell, the *eucaryotic cell*, has all its nuclear matter gathered within a double membrane in a single well-defined nucleus that contains a number of short linear chromosomes. There are many other differences as well, most of which can be detected only at submicroscopic levels. These contrasting cell types are illustrated in figure 18, where the broad resemblance between the entire blue-green algal cell and the so-called chloroplast of the eucaryotic plant cell invites notice.

One prominent difference between procaryotic and eucaryotic levels of life is that it is only among organisms made up of eucaryotic cells that sexual reproduction is regularly practiced.

The distinction between the procaryotic and the eucaryotic cell, in fact, is the most important one we know of among living organisms, far more important than the rather fuzzy distinction between animals and plants. Among other things, the first appearance of the eucaryotic cell denotes the prior or contemporaneous evolution of special enzymes (oxygenases) needed for the oxidation of certain components of the cell membrane (steroids) that are apparently distinctive of eucaryotes. As these enzymes do not function at oxygen pressures below 0.001 percent of the present level of atmospheric oxygen, the existence of probable eucaryotes as far back in time as 1,300 to 1,400 million years ago implies that atmospheric oxygen pressures were then already at that level or higher. It is more than likely that oxygen had by then already reached concentrations exceeding 1 percent of its present level, which is the concentration, first noted by Louis Pasteur, at which procaryotic forms capable of either anaerobic or aerobic metabolism switch over from one to the other.

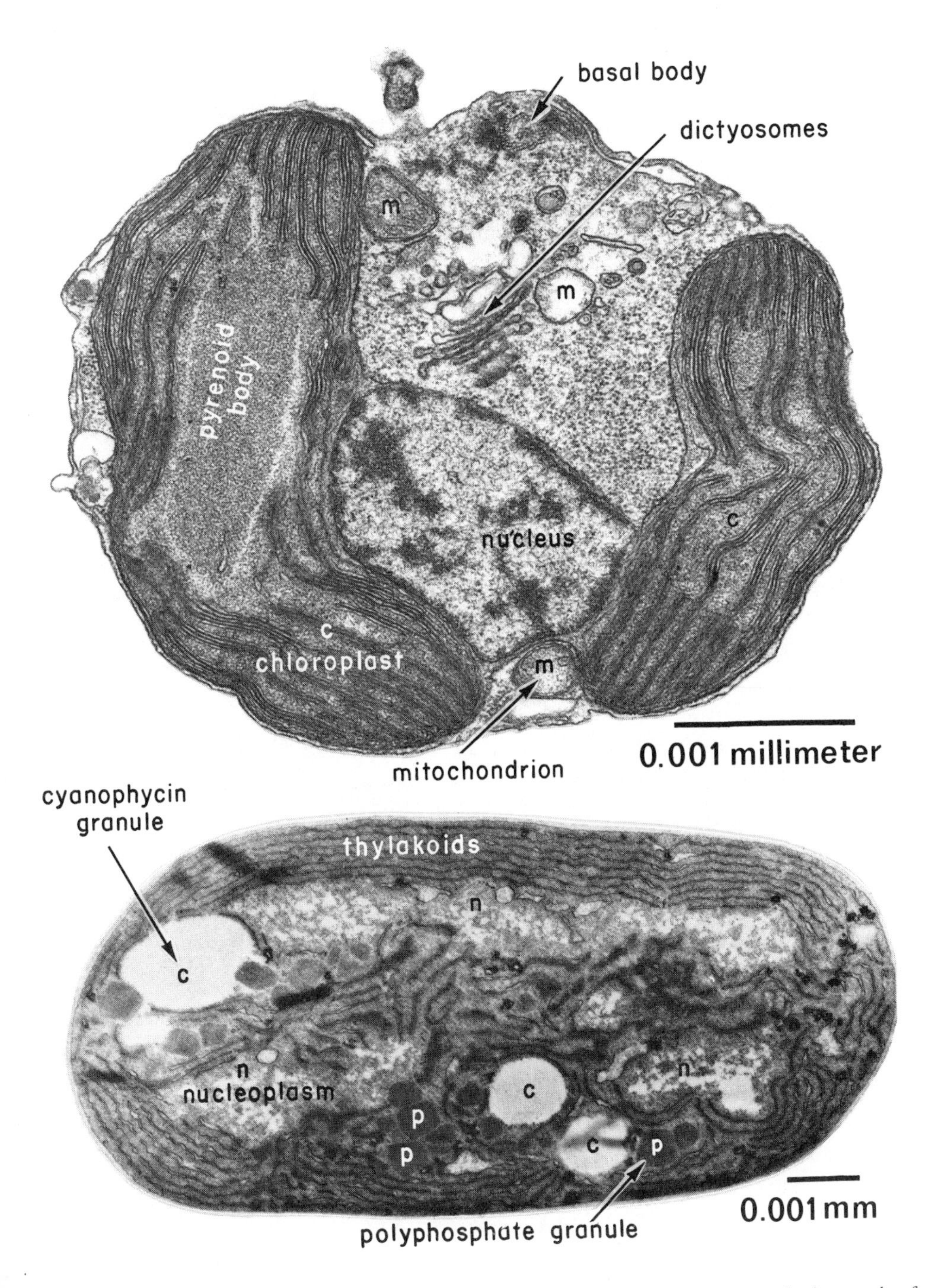

Figure 18. Structure of modern procaryotic and eucaryotic cells as shown in electron microscopic photographs of ultrathin slices. *Below:* normal blue-green algal cell, *Anabena,* a procaryote (after N. J. Lang and P. M. M. Rae, 1967, *Protoplasma,* vol. 64, no. 1; reproduced by permission of the authors and Springer-Verlag). *Above:* simple, small unicellular eucaryote, the golden brown alga *Chrysochromulina.* (Copyright © 1970 McGraw-Hill Book Co. (UK) Ltd. From "The Algae" by Dr. J. D. Dodge in Robards: *Electron Microscopy and Plant Ultrastructure.* Reproduced by permission of the publisher and the author.)

Thus it is of interest to examine briefly the evidence for the first appearance of a eucaryotic level of cellular organization. Unfortunately the ultramicroscopic detail that would bear conclusively on this point is lacking in the fossil record. Nevertheless there is supportive evidence that speaks with a degree of persuasion. One of the oldest occurrences of such evidence, if not the oldest, is preserved in the Beck Spring Dolomite of eastern California, from which branching, septate filaments and spheroidal unicells of relatively large diameter are illustrated in figure 19—taken from the work of G. R. Licari, done in my laboratory at the University of California. Procaryotic cells and filaments having diameters greater than about 0.02 millimeters are uncommon among living organisms, while diameters above 0.05 millimeters are rare indeed among procaryotes. Eucaryotic cells are commonly larger. An abundance in the Beck Spring microbiota of filaments and unicells above 0.03 mm, some of which exceed 0.05 mm, and a few of which attain sizes up to 0.062 mm, strongly implies a eucaryotic presence. The possibility suggested by critics that the filaments might be the tracks of boring algae rather than true filaments is eliminated on two grounds. First by the presence of cross partitions such as that shown at the upper right of the lower photograph in figure 19. Second because the branches grow *upward* from a basal mat, as also shown in the lower part of figure 19.

The age of the strata involved is not precisely established but is believed to be between 1,200 and 1,400 million years, based on indirect evidence. And, of course, the earlier-mentioned, older red beds tell us that there might have been enough oxygen around to support a eucaryotic level of oxidative metabolism as far back as 1,800 million to 2 billion years ago. But 1,300 to perhaps 1,400 million years ago is as far back as current evidence allows us to infer the presence of a eucaryotic level of organization with reasonable confidence.

The next clue we have as to oxygen levels is the appearance of multicellular animal life *(Metazoa)* about 680 million years ago, marking the beginning of what geologists call the Phanerozoic Eon of geologic history (see front endpapers). Oxygen by that time may have attained around 6 to 7 percent of the present atmospheric level. That number is derived by extrapolation from the minimal oxygen

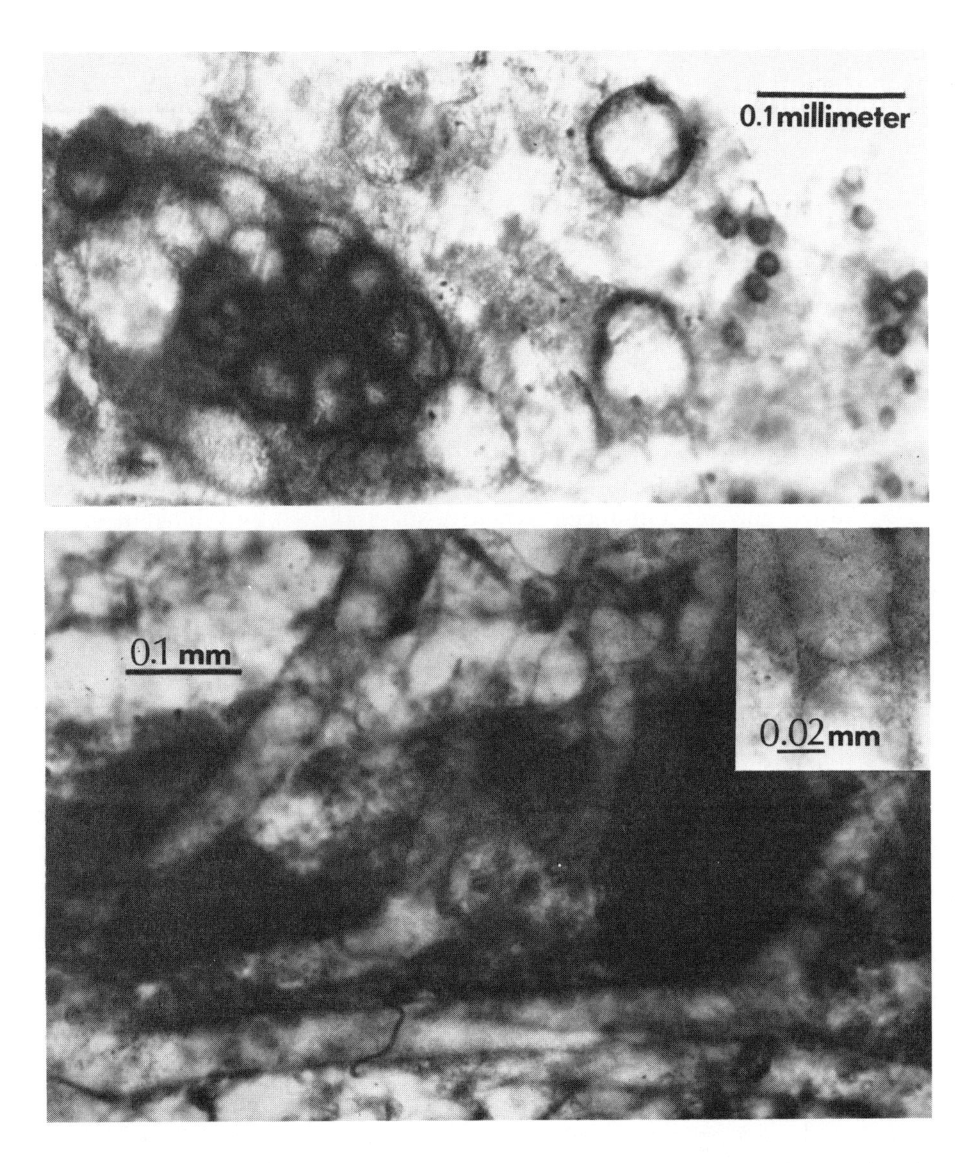

Figure 19. Some of the oldest known organisms of probable advanced (eucaryotic) cellular type, from the roughly 1.3 billion year old Beck Spring Dolomite of eastern California. Spheroidal unicells, above, and branching, sparsely septate filaments, below, attain diameters commonly greater than 0.050 (and occasionally up to 0.062) millimeters, which is much larger than nearly all procaryotic cells.

requirements calculated for thin-bodied Metazoa able to obtain their oxygen supply by simple diffusion, unaided by oxygen-transfer systems and unimpeded by external skeletons or other coverings that would retard or block simple gaseous transfer across their external surfaces. The subsequent evolution of internal oxygen-transfer systems would have permitted the evolution of external skeletons, which appeared in the geologic record around 600 million years ago, by which time atmospheric oxygen may have been around 10 percent of its present atmospheric concentration (see figure 20).

Coming up the scale of geologic history toward modern times, the late L. V. Berkner, a geophysicist, and his mathematical colleague, L. C. Marshall, have suggested that stages in biologic evolution might be related to increasing levels of oxygen and enhanced oxidative metabolism. Few biogeologists accept the exact relations proposed by Berkner and Marshall, but a somewhat less precise and more general relation seems likely. We can feel sure, in addition, that there were swings in the amounts of atmospheric oxygen through geologic time, related to episodes of major volcanism that introduced new sinks of carbon monoxide and other reduced gases into the atmosphere—first diminishing oxygen levels, then swinging back to enhanced levels with continued sedimentation of carbonaceous materials. A possible scale of oxygen increase in relation to biologic evolution, smoothing out swings in oxygen level, is suggested in figure 20. (This scale is arrived at by taking the appearance of land vertebrates and insects as indicative of modern levels and running the line backward to 1 percent present atmospheric level at 2 billion years ago, through 10 percent at the time of the oldest shelled animals and 7 percent at the first appearance of Metazoa.)

In such a manner might the air have become and remained breathable for oxidative metabolizers like us.

How long it will remain breathable is a question less easily answered. The burning of fossil fuels is returning carbon to circulation at a great rate, thus reducing oxygen and increasing carbon dioxide. Because most buried carbon is dispersed in shales and other sedimentary rocks, however, while only a very small fraction takes the form of fossil fuels, it is not likely that our atmosphere will

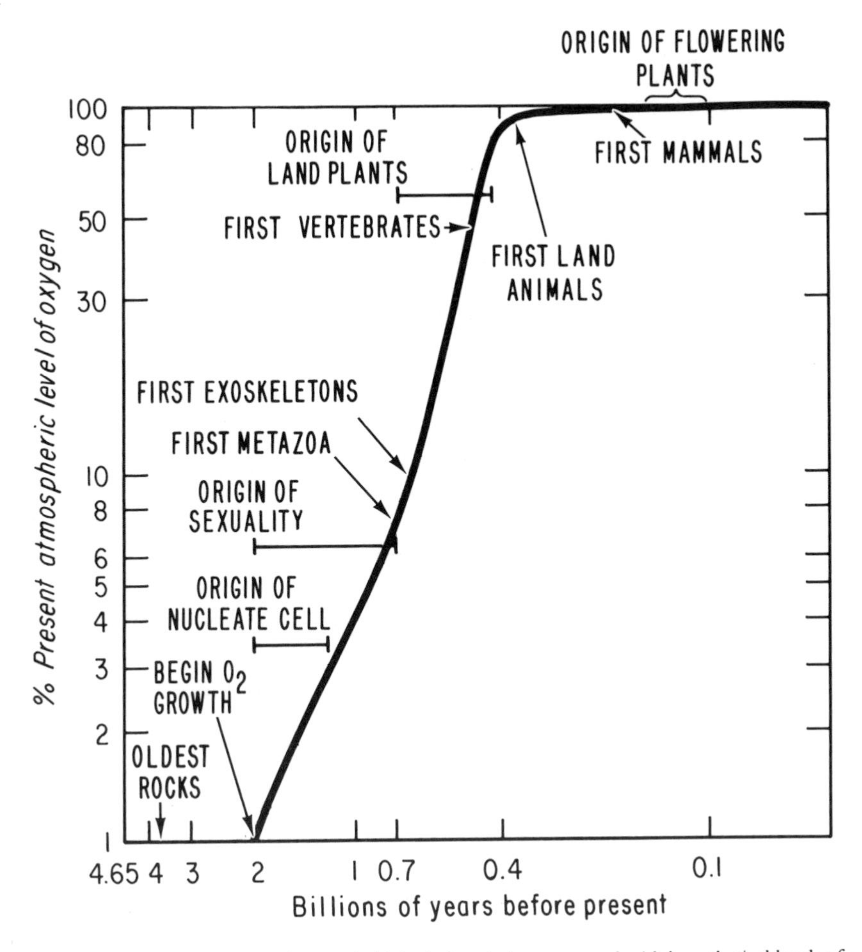

Figure 20. Apparent timing of events in biological evolution compared with hypothetical levels of atmospheric oxygen. Both scales are logarithmic in order to fit data into space available. (Adapted from P. Cloud, *Paleobiology* 2, no. 4, 1976.)

become significantly less breathable on a global scale from a reduction of oxygen in consequence of the burning of coal, oil, and gas. Even if we were to burn them at many times the present rates, with disastrous environmental consequences, other limitations on human activity would intervene long before oxygen-depletion became noticeable, except indirectly as a consequence of climatic changes brought on by increasing carbon dioxide. The balancing processes are all in place and functioning. The pivotal factor in maintaining something close to the present level of free oxygen in the atmosphere appears to be the balance between the rate of exposure of recycled carbon by erosion and the rate of burial of newly produced carbon by sedimentation. Unless rates of erosion or sedimentation are significantly increased or decreased, or unexpected new oxygen sinks are introduced, it seems likely that free oxygen will persist somewhere close to present levels far into the distant future.

For Further Reading

Cloud, Preston. 1974. *Encyclopaedia Brittanica,* 15th ed., s. v. "Atmosphere, development of."

Cloud, Preston, and Gibor, Aharon. 1970. The oxygen cycle. *Scientific American* Offprint No. 1192, 12 pp.

Granick, Sam. 1953. Inventions in iron metabolism. *American Naturalist* 87. 65–75.

Rubey, W. W. 1955. Development of the hydrosphere and atmosphere with special reference to the probable composition of the early atmosphere. Geological Society of America, Special Paper 62, pp. 631–50.

Urey, H. C. 1959. The atmospheres of planets. In *Handbuch der Physik*, vol. 52, Astrophysik III, Das Sonnensystem, Springer Verlag, pp. 363–418.

III

LIFE

11

THE BEGINNINGS OF LIFE

Elements and stars, planets and time, air and water—what would these things be without intelligent life to illuminate them with perception and insight?

We who possess the fleeting gift of life and consciousness nevertheless are properly more concerned with passing our days decently than we are with defining what it means to be alive or to be intelligent. Most of us, indeed, have little doubt that we can tell whether something is living or not, or that we can judge the relative intelligence of living things by their behavior. Yet the debate over whether fossils were once living lasted for hundreds of years, and normal men and women, at all times, have wished to understand what meaning their own lives might hold. Some have sought even more universal meanings. Those who have pursued that search scientifically have come to the very frontiers of life itself—and found no clear boundary between the living and the nonliving. Instead, cosmochemistry, laboratory experiments, and analyses of the early sedimentary record combine to favor the idea that there was an evolutionary continuum from precursor organic molecules of purely chemical origin to organized structures having the properties we attribute to living cells.

The relevant laboratory experiments go all the way back to 1828, when the German chemist Friedrich Wöhler made the first synthetic organic molecule. They began in earnest in 1951 with the production of several kinds of *amino acids* in an experiment by Stanley Miller and Harold Urey, both then at the University of Chicago. I will have more to say about amino acids. Here it suffices to recall that they are the building blocks of protein. In their experiment Miller and Urey, simulating lightning, sparked electric discharges as a source of energy in a synthetic atmosphere of chemically pure methane, ammonia, and steam in a sealed container. Numerous experiments since theirs (and at least one before) have produced amino acids as well as more complicated organic molecules, starting with sealed experimental atmospheres not only of ammonia and methane, but also of various mixtures of carbon monoxide, carbon dioxide, water vapor, and nitrogen, without as well as with methane or ammonia. The source of energy in these experiments has also varied, but with emphasis on ultraviolet radiation. In all instances where experiments have sucessfully produced important organic molecules under sealed conditions and starting with chemically pure reagents, *oxygen was excluded.*

It is now known that important organic building blocks are also produced by natural processes in interstellar space. The eyes by means of which these molecules are seen most clearly are radio telescopes. Soon after radio astronomers such as Arno Penzias and R. W. Wilson of the Bell Laboratories turned their instruments on the heavens, they were astonished to detect, by means of radio spectroscopy, not only the expected variety of light atomic particles, but also chemicals consisting of more than one element. More than thirty such molecules or molecular fragments were known by the end of 1977, with new ones turning up every few months. Among them are the thirteen important organic molecules listed in table 1, below.

Table 1
Thirteen Organic Molecules Found in Interstellar Space

hydrogen cyanide	HCN	isocyanic acid	$HNCO$
formaldehyde	$HCHO$	acetonitrile	CH_3CN
thioformaldehyde	H_2CS	methanol	CH_3OH
formic acid	$CHOOH$	methyl acetylene	CH_3C_2H
cyanimide	NH_2CN	methylamine	CH_3NH_2
formamide	NH_2HCO	acetaldehyde	CH_3CHO
cyanoacetylene	HC_3N		

This jaw-breaking baker's dozen of organic molecules is interesting for more than its tonal unfamiliarity. In all experiments starting with plausible primitive atmospheres, hydrogen cyanide, formaldehyde, and formic acid, all listed in table 1, have been found to be critical intermediate molecules between simple atmospheric components and amino acids. Methylamine, also on the list of observed interstellar molecules, reacts with formic acid to make the important protein-building amino acid glycine. And still another, cyanimide, reacts with water to form urea, the molecule synthesized by Wöhler. Ammonia, also detected in interstellar space, reacts with cyanimide to form an even more important compound, guanidine, one of several predecessor molecules for DNA, which has become a household abbreviation for the tongue-twisting genetic replicator molecule called deoxyribonucleic acid.

The building stuff of life is in and between the stars. How could it have been lacking on the primitive earth? Yet it is a far cry from amino acids to a living cell. Amino acids must be joined to make proteins. Mechanisms for reproduction and for energy capture and transfer must be provided. And all these things have to be linked together in the right way, preferably within a cell membrane.

When we stop to consider in an analytical way what life is, we quickly become aware that the most distinctive properties of the things we consider to be living are their beautiful adaptive organization, individual diversity, continuation from generation to generation in time, and transition to closely related forms. A more scientific way of saying it states that life is characterized by self-replication, genetic change, the replication of genetic change, and evolutionary continuity.

The crystals that make up rocks and add color to geological exhibits in museums are also beautifully organized and well suited to the environments in which they are formed. Yet they lack the other essential attributes of life. All crystals of a given kind are fundamentally identical except in size. Quartz does not evolve into geochemically related minerals or vice versa. Except in the case of natural radioactivity, there is no mutation among minerals, and even among radioactive isotopes the descendant isotope does not replicate itself.

The individuals among species of organisms, on the other hand, vary among themselves, so that, under the selective pressures of changing environments or geography, species tend to change through time. Because some genetically variant individuals in a given population are more successful than others in surviving and reproducing themselves, their descendants may evolve into a different species, or divergent populations may branch off through a variety of genetic isolating mechanisms into two or more species.

The basic chemistry of life is not terribly complex. Yet the genetic diversity of life, real and potential, is enormous. Possible variations in the combination of human chromosomes alone, for example, are more numerous than the total number of elementary particles in the physical universe (the latter being about 10^{80}, or 10 followed by 79 zeros). The interesting thing is that only a fraction of the possible combinations has been realized. It took the genius of Charles Darwin to see that natural variation among organisms was probably controlled and channeled by processes of natural selection (discussed in chapter 12).

It looks as if a similar kind of natural selection may have taken place in screening out elements and compounds that are well suited for the construction of organisms, thus channeling chemical evolution toward the emergence of life. There are excellent chemical reasons why hydrogen, oxygen, nitrogen, and carbon and not four other elements are the key elemental building blocks of life. These are the four lightest elements that regularly exchange or share one, two, three, and four electrons. Hydrogen is the ideal energy broker in that it serves to move electrons, the basic energy source, from one site to another. Oxygen, nitrogen, and carbon are the only elements that regularly form double and triple chemical bonds with one another and with other elements. Because of this they make flexible structures of high chemical-bonding energy, good for building cell walls, muscle fibers, and DNA—the self-replicating molecule that contains the genetic code. Other elements that are regularly involved in living organisms (such as phosphorus, calcium, and sulfur) also have special properties well suited to their functions—in energy exchange for example.

Modern biochemical experimentation, the molecular components of interstellar space, and studies of the primitive Earth thus reinforce one another in the supposition that a kind of chemical selection played a key role in the evolution of the large organic but nonliving molecules from which the first living cells were made. Sunlight, and perhaps lightning and other sources of energy, acted on the primitive oxygen-free atmosphere and hydrosphere to produce watery solutions of particular molecular building blocks, among which further reactions could take place. Yet the emergence of life was not without hazard. The primary energy source for the conversion from nonliving to living was probably the same high-energy ultraviolet wavelengths (around 2,400 to 2,600 angstroms) that even today cause disruption of DNA, malfunction, and death. It was necessary, as seen earlier, that the basic chemical building blocks (and later the primitive cells to which they gave rise) accumulate in places where they would be shielded from such radiation—as beneath about ten or more meters thickness of water, below a protecting film of sediment, or in permanently shaded places. Thus prebiologic chemical evolution and the origin and development of life had to await the origin of appropriate amounts of air and water to produce water bodies and habitats capable of meeting the delicate requirements of creation.

Although lively disagreement has swirled around the question of the nature and time of origin of the essential air and water, the chemical evolution of many of the organic molecules basic to life is surprisingly simple. A variety of primitive atmosphere models, exposed to a variety of energy sources, yield such molecules as long as no free oxygen is present. Formaldehyde and hydrogen cyanide appear in all such experiments, as they do in interstellar space, and these are exactly the substances from which it is easy to start building amino acids. Indeed a number of the twenty common protein-building amino acids have now been made in the same experiments that produce the basic organic molecules.

An *amino acid*—made of the four life-building elements hydrogen (H), oxygen (O), nitrogen (N), and carbon (C)—consists of an amino group (NH_2) and an acid group (COOH), stuck together in a particular way. The protein-building ones are all of the type called alpha amino acids, having the form

```
  H H O
  | | ||
H—N—C—C—OH
    |
    R
```

There is an interesting reason for showing this molecular structure. All protein-building amino acids have the same basic configuration and composition except for what is called a side chain. The side chain, designated by R in the above structure, decides the nature and the name of the amino acid. Where R is hydrogen, for instance, the amino acid is glycine. It is easy to see that this is no more than formaldehyde (HCHO), hydrogen cyanide (HCN), and water (H_2O) stuck together in the appropriate chemical arrangement. Were R to be represented by the methyl group (CH_3), which is simply methane (CH_4) with one hydrogen missing, we would have the amino acid alanine—and so on. If one lines up several of these amino acids end to end, as below:

```
    H  H  O         H  H  O         H  H  O         H  H  O
    |  |  ||        |  |  ||        |  |  ||        |  |  ||
H—N—C—C—(OH H)—N—C—C—(OH H)—N—C—C—(OH H)—N—C—C—OH etc.
       |               |               |               |
       R               R               R               R
```

and then simply dehydrates them by driving off water molecules as indicated by the dashed ovals, something called a polypeptide is produced, which is like a small protein molecule. String enough amino acids together in this way and you have, in effect, a protein. The strong but flexible bonds made between the carbons and nitrogens of the adjacent amino acids as a result of such dehydration are called peptide bonds. They are a very important feature of living systems.

S. W. Fox of the University of Miami and others have produced peptide bonds and made large protein-like molecules simply by heating up solutions containing a mixture of amino acids so as to drive off water. At first only elevated temperatures were used, but similar results have been found at room temperature and they also occur in dilute aqueous solution. Such dehydration processes are basic to other important organic chemical transformations on the way from simple to more complex molecules and eventually to living cells. One of the interesting results of Fox's experiments is that dehydration also causes his protein-like molecules to cluster into microscopic spheres that resemble simple unicellular organisms, and which, under certain circumstances, also divide and form long chains of microspheres.

Among people who study organisms, planets, or stars, structure and form are clues to function and origin. But objects with no more character than Fox's microspheres have been identified as fossils in ancient rocks and meteorites. At such a simple level of complexity, morphology can be a trap. The easy clue to the nonliving nature of the microspheres is that they commonly (but not invariably) display a relatively wide range of sizes, whereas known microorganisms which appear similar tend to cluster within a narrow size-range.

To be alive, or to have been alive, requires a mechanism for coding and transferring the genetic informationwhich forms the basis for self-replication and for the selective changes that are the basis of evolution. Catalysts must be around to assure that life reactions go on at some reasonable rate. And fats should be available to make tough, flexible cell walls.

Catalysts and fats are easy enough to come by. Some minerals, such as iron compounds, play a natural catalytic function and may have been the original biological catalysts until suitably specialized small proteins called enzymes arose. The fatty acids, building stuff of fats, are similar to hydrocarbons except for having oxygen instead of hydrogen as their terminal element, and both can be formed synthetically.

The coding and transfer of genetic information, however, are the most basic of life functions. Both are performed by the *nucleic acids* DNA and RNA (called nucleic because found in the cell nucleus). The now familiar DNA does the coding by means of specific arrangements in genes and chromosomes, while helicoidal molecules of RNA (ribonucleic acid) serve as messengers and handmaidens. DNA and RNA are chemically similar in that each consists of four special kinds of molecules called nucleotide bases, strung together in precise linear array on spiral chains of sugars and high-energy phosphate molecules. DNA consists of two such strands linked together to make the famous double helix of James Watson and Francis Crick, in which its particular four kinds of basic molecules are precisely matched in a definite joining order, like partners in a square dance. RNA is a single chain, also involving four basic molecules strung on sugar-phosphate chains, of which three are the same as in DNA. The DNA is capable of unzipping itself into two helical strands on each of which it then assembles appropriate matching molecules from its surroundings into coded sequences of new DNA or RNA. In this way it performs its basic reproductive and information-transmitting functions.

DNA never leaves the nucleus or other centers of nucleic acid concentration within the cell. Instead, as we know from the work of M. W. Nirenberg and J. H. Matthaei at the National Institutes of Health, it sends messenger RNA about the cell to put together amino acids and assemble them into polypeptides, enzymes, and larger proteins according to instructions furnished. The amino acids in a protein are put together in a sequence that is determined by the nature of the messenger RNA. Each protein-building amino acid, in turn, is separately coded for. In this system DNA is the architect, builder, and communication and replication center, RNA is the messenger and subcontractor, and the rest of the cell provides the goods and services to carry out instructions given by the messenger RNA in three-letter genetic words. Genes, located on chromosomes consisting mainly of DNA, working through RNA and enzymes, control all cell activity and, subject to environmental modification and constraint, dictate every step of an organism's development. This is the true essence of life.

How did such a system arise? No one knows. No one ever can know exactly. But although we do not know, we can at least visualize how the parts might initially have been put together. The building blocks of the nucleic acids are sugar, phosphate, and those special kinds of molecules that are strung along the sugar-phosphate chains. Sugars are simply multiples of formaldehyde. Five formaldehydes, for instance, arranged properly, make one molecule of pentose sugar—the kind used in making RNA and DNA. Phosphates would have been available from the solution of rocks. And the special molecules between the sugar-phosphate chains (the nucleotide bases) can be built from simpler chemical predecessors. One of them, for instance, called adenine, is simply five hydrogen cyanides organized in the right order. From adenine, linked with sugar and phosphate, arises the fundamental biological energy-transfer system. Added to other essential components, it makes up the nucleic acids. DNA has now been made in the laboratory from bottled components, with the help of enzymes, and it seems only a matter of time before we will see it made from scratch, using starting materials such as were probably present on the primitive Earth.

Thus one can visualize in broad terms how life on Earth might have originated as a consequence of reaching and eventually crossing some threshold in a sequence of chemical events. Such a sequence is seen as beginning with air, water, and a suitable energy source in the absence of free oxygen. Experimental results have repeatedly demonstrated that such a set of conditions leads readily through suitable precursor molecules to amino acids, polypeptides, and something approaching the structure of proteins. Formaldehydes might then combine to make sugars. And nucleic acids like DNA and RNA could arise from appropriate combinations of hydrogen cyanide, sugar, phosphate, and other molecules. At last it must all be put together and bundled up in the right order in hydrous solution within a defined space that is characteristically surrounded by a fatty cell wall. That is both the critical final step toward a living system and the least well understood of all.

Results from the last quarter century of research on the origin of life, nevertheless, strongly support the hypothesis that the life about us today did evolve from remote ancestors that originated on Earth as a result of natural processes of prebiotic chemical evolution. What is still in doubt, and what can never be known in detail, is by what steps the evolution from nonliving to living actually progressed, and at what point in the continuum a qualified observer would have pronounced the stuff alive. It is frustrating, although by no means an unusual situation in science, that even if objects unequivocally alive were to be produced in a test tube, that would not prove that the life we know originated in exactly the way observed. We would know only that this was one of the possible pathways that could have been taken by chemical evolution on its way from the nonliving to the living.

But chemical evolution evidently did cross the threshold to biological evolution one way or another, at which point there began a panorama of change that has continued unfolding until the present. Unlike chemical evolution, which had the effect of selecting from many elements those few with just the right chemical properties for making living things, the results of biologic evolution are seen as an ever-increasing diversity and complexity of form, physiology, and ecologic style. The contrast between these two types of evolution is suggested by figure 21, modified from an original sketch by the British biologist N. W. Pirie.

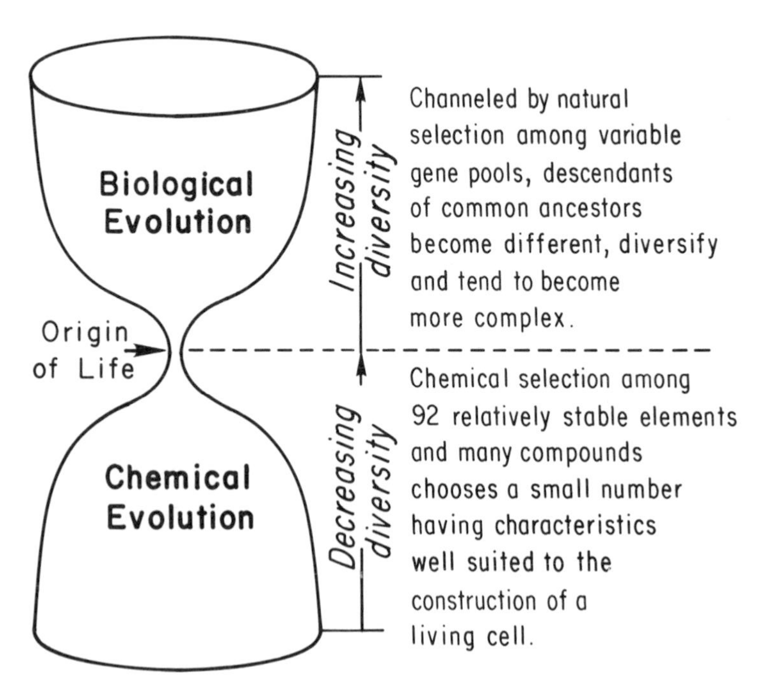

Figure 21. Comparison between chemical and biological evolution. (Inspired by a 1961 blackboard sketch by N. W. Pirie.)

Biological evolution began a very long time ago, exactly how long ago no one knows, but certainly a great deal more than 2 billion years ago. When examined microscopically, the oldest known well-preserved sedimentary rocks (the approximately 3.4-billion-year-old rocks of the lower Swaziland System in southern Africa) reveal tiny carbonaceous spheres that crudely resemble living bacteria and unicellular algae, as well as Fox's earlier mentioned nonliving microspheres. Discriminating morphology does not proclaim them to be undoubted traces of life, but intuition urges that they be accepted as such, and recent studies by A. H. Knoll of Oberlin College and E. S. Barghoorn of Harvard University have reinforced this judgment with evidence suggesting a biological type of cell division. Indirect evidence of very early life has also been noted in the form of banded iron formation in the 3.76-billion-year-old metamorphosed sediments of southwestern Greenland (figure 16) and in the presence of distinctive laminated rock structures believed to be made by successive colonies or "mats" of blue-green algae in 3-billion-year-old rocks of northern Zululand, in eastern South Africa (figure 15).

Such very ancient and uncertain (although strongly suggestive) records leave much to be desired, but they do imply very simple beginnings. They are followed by later records of unequivocal fossils beginning 2 billion years and more ago (figure 17), and they are consistent with the growing belief among students of the problem that chemical evolution leading to life is to be expected whenever and wherever conditions were favorable. Favorable conditions can be defined as the presence of an energy source, an atmosphere, and a fluid matrix (for example, water) suitable for the essential interactions; the absence of free oxygen; and appropriate means for shielding both molecular antecedents and living products from destructive radiation.

The first life on Earth, being not only intolerant of oxygen but also incapable of making its own food, would necessarily have subsisted on the organic but nonliving products of preceding and contemporary chemical evolution and the degraded components of contemporary organisms. But life could not evolve generally beyond that simple early state until an organism capable of manufacturing its own food arose. Indeed, life may have originated and died out more than once before a line appeared that gave rise to self-sustaining forms, and hence to the critical balance between the energy-consuming synthesis of carbohydrates by such forms and the energy-releasing assimilation and use of the same carbohydrates by other organisms.

Such a system works locally against the general tendency toward disorder. Life is the only system that regularly does this, and even living individuals become disordered (die) in the end. Once such a system emerged, however, its continuance and continuing change were assured for as long as its driving force, the sun, continues to shine and global catastrophe does not occur.

The necessity for the emergence of a self-sustaining form makes it certain that, although there could have been a number of independent and unsuccessful emergences of life on Earth up to that time, once such a self-maintaining system arose there could be no further successful ones. Any new living things would most likely have been promptly gobbled up by those already well established, unless they were biochemically unassimilable, in which case they would have to have given rise to their own self-maintaining system, which we do not see. And after free oxygen became generally prevalent, beginning about two billion years ago, the prospects for even temporary new emergences of life would be essentially zero. Thus it seems likely that biochemical similarities are correctly interpreted to mean that every individual organism has a genetic linkage, remote though it may be, with every other individual that has ever lived or will live.

Exactly what the first self-sustaining cell was like and how it operated we do not know. However, arguments based on chemical complexity and oxygen-demand suggest that it did not sustain itself by utilizing minerals as an energy source, as some unusual organisms are able to do now. Instead it almost certainly performed its obscure but far-reaching process of energy capture and storage by means of photosynthesis. In photosynthesis, light energy is employed to change carbon dioxide and water (or some other source of hydrogen) into carbohydrates, combining these with mineral and gaseous products from the surroundings to make carbohydrates, plant proteins, and so on. Only one of the three known types of photosynthesis produces free oxygen, so oxygen disposal need not have been a problem at the beginning.

Primitive single cells and chains of cells probably abounded in appropriate environments of ancient seas. Their DNA was not organized into sets of well-defined chromosomes, however, as it is in advanced cell types, nor was it enclosed in a well-defined nuclear membrane. Although they contained the nucleic acids DNA and RNA, they possessed neither true nuclei nor other membrane-bound intracellular bodies. And they did not experience the elegant choreography of *mitotic cell division*, whereby chromosomes first gather at the center of the nucleus, then reconstitute themselves into twice the characteristic number, and finally pair off into equal and identical sets, one for each daughter cell—the basis of information transfer among advanced organisms. Instead they simply divided in two, or the chains of cells broke up and added new cells terminally.

Such a mode of reproduction is called vegetative. This kind of simple cell division is not restricted to but is characteristic of the blue-green algae and bacteria. Evolutionary change in these primitive organisms depends almost solely on mutations—abrupt changes of the genetic potential of individual cells. Such mutant cells, if they survive, then replicate themselves almost exactly until they undergo a further mutation. Such a system probably worked reasonably well as long as enough high-energy ultraviolet radiation was getting through to stimulate an abundance of mutations. The mutation rate, however, would be reduced under an oxygen-rich atmosphere because of the ozone layer it generates, causing the higher energy ultraviolet component of solar radiation to be screened out. The rate of evolution would thus also be reduced among organisms where mutation was the main or sole cause of individual variation. The likelihood that this is what happened on the primitive Earth is suggested by the fact that the earliest types of organisms, the bacteria and blue-green algae, seem to have undergone very little change in form since the appearance of oxygen in the atmosphere. Many such organisms still alive are essentially identical morphologically (although probably not biochemically) to ones that lived two billion years and more ago.

With the reduction of the mutation rate as a result of the presence of atmospheric oxygen and ozone shielding, sex became the ideal supplement to mutation in assuring sufficient variability on which the creative processes of natural selection could work to bring about the channeled, progressive change we refer to as evolution. Organisms capable of sexual reproduction and thus of genetic recombination consist of, or are, cells that have double nuclear membranes, complex, membrane-bound intracellular bodies called organelles, and well-defined paired chromosomes of a regular number. Their regular cell division is mitotic and they undergo a special kind of reproductive cell division called *meiosis*. In meiosis the paired chromosomes are divided so that only half the usual number goes to each sperm and egg cell. These then join in the fertilized female egg cell or ovum, giving rise to a scrambled set of chromosomes of the conventional number in subsequent cell divisions of the growing offspring—one of each pair from each parent except as they may have varied by overlap or "crossing over" of the linear chromosomes. This process underlies the striking variability of organisms that reproduce sexually. Because of it there is much opportunity for new combinations of genetic characteristics and a wide variation among individuals.

Table 2
Main Events Enroute to Higher Organisms

Event or Precondition	*Likely Stimuli*	*Probable Consequences*	*Inferred Time of Event (billions of years ago)*
12. Tissues, organs	Sexuality, competition	Metazoa, higher plants, land biota, intensified weathering, subsequent evolution	about 0.7
11. Sexuality (meiosis)	Origin of advanced cell type, increased free 0_2	Complete eucaryotic hereditary mechanism	more than 0.7, less than 2
10. Advanced cell structure and division (mitosis)	Increasing 0_2, 0_2 shielding of intracellular functions	More 0_2, carbonate, and sulphate. Less $C0_2$. Higher algae	more than 1.3, less than 2
9. Fully oxidative metabolism	1% present free 0_2, enzymic neutralization of corrosive 0_2 derivatives	Increased 0_2, ozone shield, red beds, membrane-bound nuclei	about 2
8. Blue-green algae, (or close ancestral forms)	Light-energized breakdown of H_2O, limited 0_2 mediation	Limited 0_2 tolerance, BIF, finally 0_2 increase and fully oxidative metabolism	about 3.8
7. First self-sustaining cells	Competition for components and energy	Bacteria-like blue-green algae, subsequent evolution	more than 3.8
6. First life	Energy transfer, genetic code	Procaryotic diversity, biogeochemistry	more than 3.8
5. DNA, RNA, ATP	Chemical evolution, autocatalysis	Reproducibility, energy brokerage	more than 3.8
4. Sugar-phosphate bonds, nucleotide bases	Chemical evolution, autocatalysis	DNA, RNA, ATP	more than 3.8
3. Amino acids, polypeptides	Chemical evolution, autocatalysis	DNA, RNA, ATP	more than 3.8
2. Simple organic molecules (HCN, HCHO, etc.)	UV irradiation of 0_2-free atmosphere	Amino acids, polypeptides	more than 3.8
1. First air and water	Outgassing of Earth	Chemical evolution	more than 3.8

As will be discussed in the following chapter, natural selection, acting on that variation, is considered to be the guiding process of evolution. And many are the ways that different species have invented to play the hazardous game of natural selection successfully. Having many offspring, protecting the young in various ways, specializing for a particular mode or modes of feeding or taking shelter, mutual assistance arrangements of various sorts, concealment devices, lures for attracting prey, and a host of other elaborate devices have been employed over the last 680 million years or so, as the long and diversified process of multicellular animal and plant evolution unfolded.

The origin of life, the evolution of mechanisms for producing organic nutrients from inorganic materials, oxygen-producing photosynthesis, the perfection of the oxygen-mediating enzymes, the appearance of the mitosing cell and sexual reproduction, the diversification of the algae, the origin of sexuality, and perhaps the origin of unicellular animal life were the great events in biologic evolution on the primitive Earth. Nothing we can unequivocally call an animal, however, is known until the beginning of Phanerozoic and Paleozoic history (see the front endpapers), starting with the Ediacarian Period around 680 million years ago. These earliest definitive animals were multicellular forms, some of which resemble lower invertebrate organisms of the modern world.

A possible program of biological evolution from earliest times to the appearance of multicellular animal and plant life, with some irreducible jargon and chemical symbolism, is given in table 2.

For Further Reading

Glasstone, Samuel. 1968. *The book of Mars.* U.S. Government Printing Office, NASA SP-179, 315 pp.

Kenyon, D. H., and Steinman, G. 1969. *Biochemical predestination.* McGraw-Hill Book Co., 301 pp.

Ponnamperuma, C. (ed.). 1972. *Exobiology.* North-Holland and American Elsevier Cos. 485 pp.

Schneour, E. A. and Ottesen, E. A. (eds.). 1966. *Extraterrestrial life: an anthology and bibliography.* Natl. Acad. Sciences–Natl. Rsch. Council, Publ. 1296 A, 478 pp.

12

HOW THE DESCENDANTS OF COMMON ANCESTORS BECAME DIFFERENT

The Evolution of Evolution

One of the most characteristic aspects of life is its change with time. The spore contains the blueprint for the fern, the acorn for the oak, the fertilized human ovum for the prima donna or the peddler. Each adult is a product of all that has gone before. Each "seed" contains a genetic inheritance that defines what form the adult may take and with what talents it may shape its life under given circumstances.

But there is a change with time that goes beyond the individual and the generation. We see changes also in *populations* of individuals across the generations, and sometimes in the kinds of *community* associations those populations form with other species of organisms. Breeders of animals and plants have known since long before the time of Charles Darwin that, by selecting their stock, they could shift the central tendencies of their flocks and crops in the direction of any characteristic that occurred naturally in wild or domesticated members of the species. The beginnings of such knowledge reach far into antiquity. They underlay the agricultural (or "Neolithic") revolution, which began in the Near East about 9,000 to 10,000 years ago. They gave rise to reflections on the nature and causes of biological change long before there was any

conscious body of rules for the systematic investigation of phenomena and the testing of ideas—rules of the sort we call scientific.

As early as the fourth to sixth centuries B.C., philosophers such as Empedocles and Anaximander were postulating concepts of biological adaptation and progressive change from lower to higher forms of life. Theologians such as St. Augustine (353–430 A.D.) and Thomas Aquinas (1225–74 A.D.) looked on the biblical accounts of creation as allegorical. And the Chevalier de Lamarck suggested a complete, testable, but, as it turned out, unworkable mechanism for evolution around the time of Darwin's birth.

That the kinds of organisms inhabiting the earth have, in fact, changed dramatically over vast spans of geological time is no longer doubted by reasonable people who have bothered to familiarize themselves with the evidence. That evidence is too vast and too complex to present in full detail, but the essential parts are summarized in the following three chapters and in the references there given. In a nutshell, the verifiable records of biological change, preserved in well-dated successions of layered rocks spanning the most recent two billion years of geologic history, display a progression, albeit an uneven one, from simple unicellular forms of little diversity to the complex and highly diverse assemblages of organisms we see about us today. That of itself, leaving out the nearly two billion years of indirect records of life that precede it, establishes that systematic change in the basic characteristics of living things through time is a fact that must be explained. Another word for such systematic change through time is evolution—in this case biological evolution.

Such change, of course, differs from that observed over the life span of an individual, no matter how dramatic. Individuals do not evolve. Only populations evolve. Thus, when I speak of biological evolution (or simply evolution) I refer to the cumulative change in the genetic composition and appearance of populations from generation to generation that eventually produces new races, species, genera, and higher categories of life.

It is a mistake to equate the fact of biological evolution with the problem of the mechanism or mechanisms that may have brought it about. *The question is not whether evolution happened but how it hap-*

pened. How did the descendants of common ancestors become different?

That problem occupied much of Darwin's life as it has occupied the lives of many others since him. And, although Darwin's concept of the origin of species by means of natural selection stands as the cornerstone of modern evolution theory, the processes of biological evolution are still not completely understood and seemingly non-selective processes may operate at some stages. Later I will summarize modern outlooks on the problem. But let us begin with the young Darwin—setting sail, at the age of twenty-two, as expedition naturalist on H.M.S. *Beagle,* which departed England in 1831 on one of the round-the-world cruises of exploration that were all the rage in those days. Imagine his excitement at the prospect! He had a sweeping title and charge. His diary of the voyage glows with joy at all the wonders of nature he was seeing—the rocks, the fossils, the rain forest, the colorful tropical birds and fishes, the vertical sequences of biological zones, and the remarkable fitness of organisms to their environments, whether on land or in the sea.

Although his grandfather, Erasmus Darwin, had reflected deeply on evolution and the concept was in the biological winds of the times, Charles himself was not predisposed in favor of it at the time of his departure. It seems to have been only after his arrival at the Galápagos Islands, west of Ecuador, that young Charles began to think on evolutionary lines. There a local official showed him that one could tell which island any particular species of the giant Galápagos tortoises came from by small but distinctive differences in the shape of its shell. The eager Darwin could hardly wait to find out whether similar differences were to be found among other organisms. He soon found that, while every one of the separate small islands had much the same kinds of plants and animals, each island also supported some distinctive species and varieties that were found only on that island and not on others. That seemed odd. It seemed improbable that the species had been created differently on each island. More likely, thought Darwin, they were descended from common ancestral stocks, but had followed different evolutionary paths as a result of being isolated from one another by watery and other barriers—somewhat like the sheep in different paddocks of the animal breeder back home in England, only more so, as they were no longer cross-fertile.

That was just a hazy idea, but it showed the way for inquiry. Now Darwin did something that was much more difficult and important than simply having an idea. He undertook a scientific investigation of that idea. He set out over the next quarter century to marshal the evidence that gave his thoughts a factual foundation, and he formulated a testable mechanism for how evolution might have taken place. This was the hypothesis of natural selection—somewhat like stockbreeding but without conscious goals, and done by nature—a hypothesis that, in its essential elements, has withstood all opportunities for disproof until the present day, and thus qualifies, in its currently modified form, to be considered as a theory in the sense outlined in chapter 1.

The Anglican canon C. E. Raven summarized the matter from the viewpoint of a scholarly theologian in a BBC broadcast in 1950: "That creation was a process, not an act, continuous, not intermittent, operating through the orderly sequence of natural events, upset the whole idea of a God outside the world who set it going, and then on special occasions intervened by miracle to alter its course. We can now see that such an idea was never satisfactory; and in fact it was not the belief of St. Paul or the first great Christian theologians".

Yet all was not won in the first inning. Although, in the long run, the theory of evolution by means of natural selection prevailed, its early progress faced stiff opposition. Many devout biologists and paleontologists of and after Darwin's time, among them the great Louis Agassiz, founder of biology in North America, strongly resisted Darwin's views. They chose to interpret the then-observed succession through time of different and increasingly complex life-assemblages as the result of numerous, time-separated episodes of divine creation—God's way of evolution, preparing the earth for man. Others, impressed as Darwin was by the apparently gradational aspects of some of the changes observed, sought gradational explanations other than natural selection. Biological predestination has been a recurrent minority theme—claiming that ends were somehow set from the beginning so that the evolving organism was foreordained to change in the directions observed. As this is not really a testable explanation, however, and as there is a good deal of evidence supporting the view that many or most changes reflect adaptation to ecological or climatological shifts, such views have never met with much favor.

Perhaps the most popular and long-lasting alternative to natural selection has been the notion of the inheritance of acquired characteristics—a mechanism that Darwin himself entertained off and on. A favorite illustration of how this is supposed to work is seen in the long neck of the giraffe. Successive generations of ancestral forms, it was supposed, attempted to reach higher and higher branches of the trees on which they browsed; each stretched the neck a bit longer, with the result that each generation inherited a slightly longer neck until the present length was attained. This idea, championed in particular by Lamarck, is actually a common one in the folklore of primitive peoples, as well as among practical animal breeders today. It recommends itself to us by the same commonsense but, as it happens, invalid reasoning that would have the sun revolve around the earth.

The idea of the inheritance of acquired characteristics has not withstood experimental tests. The many supposed proofs, some celebrated, then notorious, have all been exposed as frauds or self-deceptions. Science is not all that inexorable, and scientists, like other humans, are fallible. It is not necessary to review this sometimes sad, sometimes amusing history, however, because the inheritance of acquired characteristics is completely invalidated by the central discovery of genetics that the *inheritable* characteristics of an individual organism are established at conception and cannot be changed by subsequent events other than mutation. It is also weakened by the observed mixing of dominant and recessive manifestations of particular characteristics in hybrid stocks. In the case of eye color or color and surface-texture of seeds, for example, all visible effects are the products of *pairs* of genes, located on *pairs* of chromosomes, one from each parent. The manifestation observed, as will be elaborated, is invariably the one called for by the dominant gene, not only in individuals whose gene pairs include only the genes for the dominant manifestation, but also in those whose gene pairs include both dominant and recessive genes. The recessive manifestation appears only in those later generations where hybridization reintroduces individuals whose gene pairs for the characteristic in question are both recessive. Such results could not occur if bodily characteristics produced a parallel effect on the hereditary mechanism.

The long neck of the giraffe, it seems, is better interpreted as a consequence of the fact that genetically longer necked giraffes in each generation were more likely to survive bad browsing years so as to reproduce and pass along the genes for long necks to the following generations.

Current Views of Natural Selection and Why They Are Held

For the reasons mentioned, and many others, the now prevailing view in all branches of biology and biogeology is that biological evolution is in some sense almost always the product of natural selection, somewhat as visualized by Darwin, acting on genetically variable populations of organisms (apparent exceptions involving genetic drift in small populations or unusual modes of reproduction). This is seen as a way of shifting the central tendencies or distribution patterns of a succession of populations of any given organism toward different modes, at rates and to degrees that vary with circumstances. The sites of change have now been identified, the mechanisms are beginning to be understood, and the once prevalent insistence on strictly gradualistic processes is disappearing.

The selective processes now envisaged are far from ideas of the "struggle for survival . . . red in tooth and claw" that formed the basis for "social Darwinism" and the justification of Victorian inequity and colonialism—ideas that were explicitly rejected by Darwin himself. Indeed they do not even exclude the prospect that the meek shall inherit the earth—provided the meek are well adapted, reproductively successful, and do not overpopulate. For it is the fittest in a reproductive sense whose genes survive in descendant populations. Struggle in the conventional physical meaning is rarely involved. Even competition is not involved where disease, recurrent adversity, or predation keeps the populations of potential competitors below the numbers normally supportable by their environment.

The idea of fitness in an evolutionary sense is a very subtle one. Obvious differences in morphology—that is, form and structure—between adult populations may be far less significant for selection than differences in genetic makeup expressed only in larval or juvenile stages, in physiology, in behavior, in courtship, or in reproduction. Studies of the fruitfly *Drosophila* show that two reproductively isolated species can differ by as much as 20 percent of their genes and yet show no morphological distinctions other than minute differences in the male genitalia. On the other hand, species as contrasting to the human eye as the chimpanzee and *Homo sapiens* vary only slightly in their genetic composition.

Perhaps the most widely accepted definition of a species is that it consists of morphologically or physiologically distinctive groups of interbreeding natural populations that are reproductively isolated from other such groups, but which have the capability of interbreeding with one another. Although such a definition obviously cannot apply to species whose primary or exclusive mode of reproduction is nonsexual, it does apply broadly to the great majority of known species, including all of those that will be considered further in this book. Either because of physical, physiological, or behavioral isolation, species so defined do not naturally interbreed with other species; often they cannot interbreed; and, if they do interbreed, hybrids produced are generally not themselves capable of reproduction. In practice, however, most described species are based on morphology, behavior, and habitat. In paleontology, form and habitat are the mainstays of species recognition. Yet where opportunity has arisen to test the results of such traditional practices against distinctions based on genetic, physiological, or molecular criteria, tradition has almost invariably been validated. (Exception prevails among the procaryotes, however, where biochemical characteristics take precedence and where identical morphologies may harbor great biochemical diversity or biochemical identity may unite a diversity of external form.)

The morphology observed is known to be a consequence of the developmental interaction between a particular set of genes and a particular set of environmental influences. The full array of genes found on all of the chromosomes characteristic of the individual is called the *genotype*. The genotype is the blueprint for the adult morphological product or *phenotype*, the characteristics of which also depend on how and with what materials the instructions of the blueprint were carried out. The sum of all of the different genes possessed by all of the individuals in a population is the *gene pool*. Change in the composition of this gene pool as a result of the favoring or elimination of particular physical, physiological, or behavioral types and their corresponding genotypes is defined as evolutionary change.

Shifts of this kind, (or extinctions, or both) occur whenever and wherever changes in climate, local ecology, or geography place stress on existing populations, destroy old or open new ecologic niches, break up or disconnect areas of land or water that were formerly connected, join areas that were formerly disconnected, or create new, previously unoccupied areas of land or sea that invite colonization. Such ecological or geographical changes can confer adaptive advantage on particular, genetically determined characteristics that differ from those previously favored. Thus long intervals of slow and seemingly aimless evolutionary change among generally large populations in established communities may be punctuated by episodes of relatively rapid evolution among small populations that are peripheral to or become isolated from the larger aggregates of similar organisms. Where rapid evolutionary change takes place simultaneously with geographical, ecological, or behavioral events that lead to reproductive isolation, new, genetically distinctive species, genera, and even larger categories of organisms may arise relatively rapidly—not uncommonly following extinctions among previously established forms. The process is illustrated by the organisms of volcanic island groups such as Hawaii, the Bismarck Archipelago, and the Galápagos—new lands that emerged from beneath the sea, as did Surtsey Island in the North Atlantic in November 1963, and which, like it, are colonized by seeds or creatures that drift or blow from already inhabited places.

Consider the Galápagos Islands, scene of Darwin's first intimations on the nature of biological change. This group of 16 small volcanic islands, 1,000 kilometers to the west of Ecuador, sits on one of those oceanic spreading ridges we talked about in chapter 7 and 8. The oldest lavas known on these islands are dated by the potassium-argon method as about 2 million years old, but local Pliocene marine fossils suggest an antiquity possibly double or triple that. The Galápagos have never been connected to the mainland, from which they are separated by a profound geological discontinuity—a plate boundary. However, they lie in the path of the southeasterly trade winds and of easterly oceanic currents that blow and flow toward them from South America. The animal and plant life of the Galápagos is, as might be expected from such circumstances, similar to that of South America but much less varied, and it contains many species peculiar to the islands (endemics). Here we see evolution in action under textbook conditions. Of some 700 plant species, about 40 percent are endemics. Only 700 species of insects are found on the Galápagos, many of them endemics (as compared with some 20,000 species of insects known from Great Britain). There are no amphibians, only 5 species of reptiles, and only 9 species of land mammals—7 rodents and 2 bats. The giant Galápagos tortoises, similar to ones that were once widespread on the mainland, survive only on the Galápagos, where they differ from island to island.

Most remarkable, however, are the finches, studied in detail by the British ornithologist David Lack in the 1940s. Of the roughly 80 species and subspecies of birds that breed on the Galápagos, only 24 are land birds and 13 of these are finches. Although their closest relatives are on the mainland, none of *these* finches are found anywhere else in the world except on the Galápagos. They are true endemics. As a result of one or more early migrations across the 1,000 kilometers of ocean that now separates them from the mainland, the Galápagos were evidently colonized by a pregnant female or a small breeding set of mainland finches long enough ago for her or their descendants to have evolved independently, adapting to local habitats and food sources, and thus diversifying into the forms observed. It is unlikely that there have been regular later migrations from the mainland, as gene exchange resulting from that would have swamped developing endemism. Over the few million years or less that have elapsed since the primary colonization, the ancestral Galápagos finches have evolved into an entirely endemic population of 13 noninterbreeding species, comprising 3 endemic genera and 1 endemic subfamily (a fourteenth species and fourth genus commonly bracketed with the Galápagos finches is found on isolated Cocos Island, which is not a part of the Galápagos).

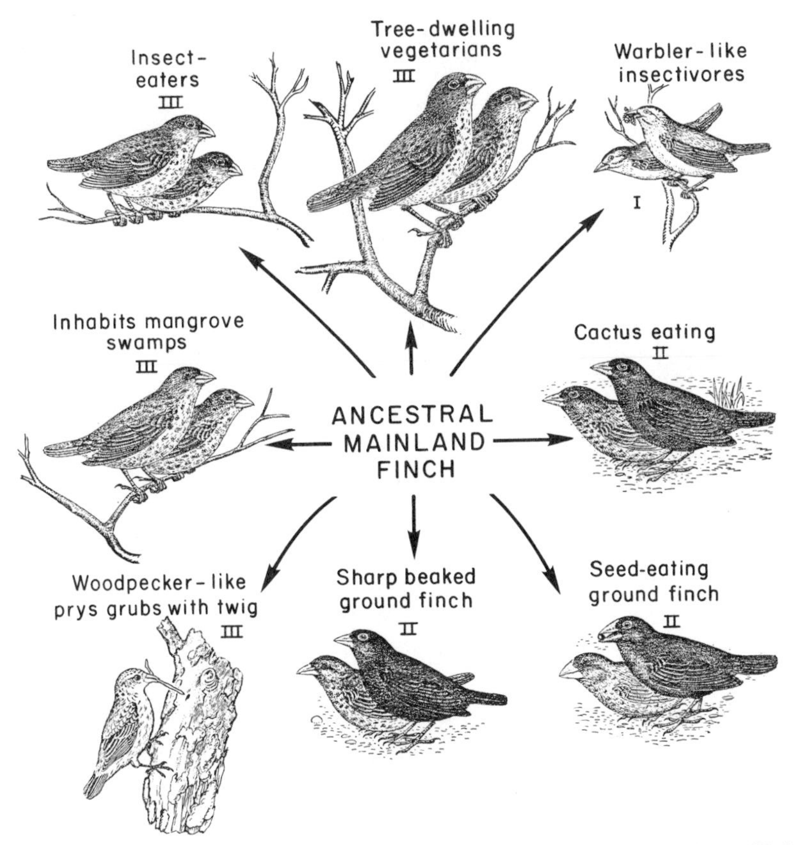

Figure 22. Evolution in action. Eight species of finches comprising three genera, I, II, and III, from the Galápagos Islands. (Adapted from published illustrations by David Lack, 1947, 1953, and Ledyard Stebbins, 1971.)

As all bird-lovers will recognize from figure 22, most of these birds are clearly finch-like, differing mainly in size and shape of the bill and in feeding habits. But one (upper right) resembles and behaves like a warbler, and another (lower left) has adopted the behavior and feeding habits of a woodpecker, using a twig or cactus spine in place of the woodpecker's long tongue for removing bugs and grubs from crevices in dead wood. They have occupied and adapted to the ecologic niches that were open to them on arrival. Not only do these little birds differ in appearance; they are also reproductively isolated as a consequence of habitat preference, feeding habits, and breeding behavior. On some islands several species exist side by side without interbreeding—up to 10 different species on one island. The differences in form between these birds are comparable at their level to those that separate *Homo sapiens* from his fossil relatives among the family Hominidae, while those that separate man from the chimpanzee are on about the same level as those between the Galápagos finches or the Hawaiian honeycreepers and their former mainland progenitors. And all of these differences have arisen within a comparable time frame—a range of time so brief that it would seem almost instantaneous in the older parts of the geological time scale.

Turning to marine invertebrates we find similarly rapid evolution and development of endemic species in similar settings. One of the geologically best documented instances is the high degree of endemism found among marine clams and snails on opposite sides of the Isthmus of Panama. The fossil record of these shell-bearing molluscs goes back about 18 million years, as worked out in a series of great monographs by W. P. Woodring of the United States Geological Survey and the Smithsonian Institution. At that time the Atlantic and Pacific waters mixed freely in this region, supporting a rich and generally cosmopolitan shelly fauna. The Panamanian isthmus began to rise about 5 million years ago and had emerged completely by about 2 million years ago. Thereafter it functioned both as a land bridge for mammalian migrations between the Americas and as a barrier to communication between the marine life on opposite sides. Where there was broad regional similarity among the marine clams and snails 2 to 5 million years ago, we now find provinciality. The approximately 4,500 species of the eastern Pacific Panamic province and the 1,000 species of the western Atlantic Caribbean province both show high levels of endemism. Many of the endemics occur as species-pairs, one on each side descended from a common ancestor. The time during which these differences arose is so brief geologically that it could easily pass unrecorded in known sedimentary rocks.

How does this process of natural selection work? How do variations in the gene pool arise? How can all of the seemingly random processes involved in individual survival and reproduction be brought to focus in a way that does, in fact, select out those characteristics that cause surviving populations to differ significantly from their ancestors and become reproductively isolated from related forms? Does this happen in different ways for different groups of organisms, or for the same groups at different times? And, granted that such processes are capable of generating new species, new genera, and even new families, how do orders, classes, and phyla of organisms arise? Those are the problems we confront when we ask not whether, but *how*, the descendants of common ancestors became different. They are by no means all solved to everyone's satisfaction, but we have come a long way since Darwin.

Darwin understood that *natural selection is a statistical process.* It is, however, not seen as random. Strictly random processes, if they exist in the biosphere, could not account for the observed succession of life forms. The production of orderly systems by processes that at first glance seem to be random, however, is not unusual. An example familiar to skiers and winter sports spectators is the generation on a heavily skied slope of the regularly arranged bumps known as moguls—illustrated in figure 23. Starting with either a previously unskied slope, a freshly snow-covered one, or one on which the previous set of moguls has been eliminated by slope-grooming equipment, a new set of regularly patterned moguls invariably appears, the size and spacing of which varies with angle of slope. To a casual observer the motions of the skiers appear to be completely random. All the individual skier wants to do is to get down the hill with as much style and joy as possible. Nor does he or she have much apparent influence on the path followed by other skiers. There is no plan or design for making the moguls. They just happen. But they happen as a consequence of a population of skiers moving downslope along individual tracks that are affected by gravity, the avoidance of obstacles, and the varied skill of the skiers. These are the selective forces that produce the moguled pattern. It is the product of a kind of natural selection.

The example of the moguled ski slope illustrates the fact that in order for natural selection to work it must *channel* variations in some way. If the motions of the skiers had really been entirely random, up hill as well as down, running off through the woods and without regard to the presence of other skiers or the tracks made by them, the slope would only have been chopped up. No orderly pattern would have formed. Similar examples can be multiplied in nature and reproduced in the computer, but the effects of antecedents and limited choices always counter the appearance of randomness.

Natural selection is, in effect, a *creative*, channeling, or generative process. In order for it to generate either new species or higher categories of life it is necessary that there be: (1) a sufficient genetic

Figure 23. Moguls on the ski slopes at Alta, Utah. (Photograph by Scott Goodfellow, reproduced with his permission.)

variability within the populations on which selective pressures are acting, (2) means of channeling variations in particular directions, and (3) isolating mechanisms. And in the past these mechanisms must have worked at such rates and at such a degree of effectiveness as to account for the roughly 2 to 5 million species now living, plus all of the larger categories in which we group them, as well as a vast but unknown number of other species that once lived, and the genera, families, and higher categories in which these now extinct forms arrayed themselves.

Thus the essential materials on which selection operates are genetically variable populations of organisms that interact with one another, with other species, and with their changing physical environments and geographical configurations to form the various natural associations of species in biological communities and food webs that make up the global ecosystem and its parts. Selection, of course, does not operate directly on the genes but on the morphological and physiological product of the reaction between genes and environment—the phenotype.

Genetic variability of populations is known to arise in two ways—by mutation and by exchange of genes as a result of sexual recombination. The oldest and most crucial, because no genetic variation can happen without it, is mutation—spontaneous change in the DNA of individuals as a result of external influence, commonly some potent chemical or physical factor such as cosmic or high-energy ultraviolet radiation.

Although mutation is the only known means by which actual changes of DNA structure can be achieved, a more common and less hazardous source of variation among higher organisms is genetic recombination. As previously noted, individuals of each sexually reproducing species have two complete sets of chromosomes, paired in such a way that every character in the genetic makeup of that individual is coded for by two sets of genes—one inherited from each of the two parents. The number of chromosome and gene pairs is the same in all individuals of a given species, but they code for a variety of responses. By inbreeding and selection this variation can be narrowed, as animal and plant breeders know. The normal breeding response, however, is sufficiently nonselective that genetic variation among the individuals of sexually reproducing populations tends to increase—more so in larger populations than in small ones.

Genetic shuffling as a consequence of sexual reproduction, however, has still another important aspect. The reader will recall that there are two types of cell division among sexually reproducing organisms. In the usual one, mitotic cell division (mitosis), first described back in 1882 by the biologist Walter Flemming, each new cell gets exactly the same complement of paired chromosomes as its parent cell. The second kind, meiotic cell division (meiosis), which occurs only at the time of reproduction, involves several stages. In the first division of meiosis the chromosome pairs have been observed to wind together so that they may exchange linked sequences of genes. Thus, during the second or reductive division of meiosis, when the chromosome number is halved in preparation for sexual recombination, the new chromosomes that are to become paired on sexual union may be mosaics of parts of the original chromosome pairs and their gene arrangements. In this way the opportunity arises not only for the recombination of whole chromosomes but also for the recombination of parts of the chromosomes in a great variety of patterns.

A large and diverse gene pool contains the seeds of response to such a wide range of ecological variables that it tends to maintain a kind of fluctuating balance with mildly changing environmental conditions, although the mean characteristics of successive populations may shift. The most impressive changes apparently take place where small isolated populations or subpopulations peripheral to or isolated from larger interbreeding groups evolve undiluted by recombination with elements of the main gene pool, eventually to diverge in new directions and become genetically isolated. This happens, for instance, where changing geography related to plate tectonics and changing sea level creates or eliminates migration routes, causes a continent or a piece of it to become isolated, breaks up a landmass into separate islands, or brings about climatic change, thereby both altering habitat diversity and leading to genetic isolation of surviving populations. Such events impose ecological pressures that channel limited gene pools in particular directions. The Galápagos finches and tortoises are good examples, as are the Panamanian marine molluscs and the different endemic species of Hawaiian honeycreepers (not to mention the Hawaiian land snails). The Hawaiian honeycreepers, to add an example, consisted, before extinctions caused by human intervention, of about two dozen endemic species, comprising an endemic family of birds related to the finches—an instance of local evolutionary differentiation comparable to that found among the Galápagos finches, but expressing a somewhat greater degree of diversity produced over a probably somewhat longer interval of time.

Genetic isolation is the final step. For if populations in the process of speciation are brought back together again before barriers to breeding have arisen between them, they may blend and revert to the original ancestral type. And many are the modes of sexual isolation of morphologically similar but genetically distinct species—courtship rituals, life at different levels in the forest or the sea, breeding at different seasons, variant habits and times of pollination among plants, chemical barriers to fertilization, and also nutritional preferences, different nutrient cycles, or foraging habits that lead to physical avoidance.

Linkage of variable genetic characteristics is similarly important in biological selection. George Gaylord Simpson, a major architect of modern evolutionary theory, has illustrated this point with the greatly oversimplified but conceptually helpful example of a pot full of all the letters of the alphabet from which the goal is to pick out, without looking into the pot, the three letters c, a, and t to spell the word *cat*. If each time three letters are picked the same three letters are thrown back into the pot before another three are pulled out, it might take a very long time to happen on the right three letters. If, on the other hand, each time one gets an *at* or a *ct* or a *ca* one is permitted to link those two letters together before returning them to the pot, it takes only a short time to get the three needed letters in one draw.

Thus, as a result of variations in the gene pools of populations, the linkage of genetic characteristics, and the guiding forces of natural selection which steer modal characteristics in directions responsive to changing physical conditions, new kinds or organisms evolve.

The Mechanism of Inheritance and the Origin of Major Groups

What Darwin did not know about are the principles of inheritance. Although originally worked out by the Silesian monk Gregor Mendel in 1866, his descriptions remained buried in an obscure publication until the principles were independently rediscovered by De Vries, Correns, and Tschermak at the turn of the century. Mendel's studies on sweet peas brought about, among other interesting genetic insights, the recognition of dominant and recessive genes. The work of the American biologists Thomas Hunt Morgan and Walter Sutton, early in this century, showed that the basic genetic information was somehow coded along the chromosomes. The unseen units of inheritance on the chromosomes were called genes. The genes, as we now know from studies and gene maps of the large chromosomes in the salivary glands of fruit flies, are in truth located along the chromosomes, and, like them, are made up primarily or wholly of DNA.

The idea of dominant and recessive genes is readily illustrated by the example of human eye color, where brown (including black) is the dominant and blue (including grey, green, hazel, and so forth) is the recessive manifestation. Each of us carries two genes for eye color, one on each of a corresponding pair of chromosomes. If any member of that pair of genes codes for the dominant color, the eyes are brown. Only where both members code for blue do we get blue. Thus brown-eyed couples may have blue-eyed babies but not the reverse. We speak of the gene bearer as being *homozygous* for a given character when both genes of the pair for that character code for the same manifestation, and *heterozygous* when they code for different manifestations. These terms allude to the genetic structure of the fertilized egg cell, or *zygote*.

Where an individual homozygous for blue eyes mates with one homozygous for brown eyes, *all* offspring in the second generation are heterozygous because all receive one of each pair of their chromosomes from each parent. And they *all* have brown eyes because brown is dominant in the heterozygous state. In the third generation, given a large enough recombinant sample, one fourth would be homozygous for blue and thus blue-eyed, but the rest would all be brown-eyed. Although only one-fourth of the whole are homozygous for brown, half are heterozygous and thus also display the dominant eye color. Thus (to consider only eye color and only three generations), where *B* signifies the gene for brown and *b* for blue, we see something like the following:

> BB × bb → Bb + Bb + Bb + Bb (all heterozygous, all brown), 2d generation
> Bb × Bb → BB + Bb + Bb (3/4 brown) + bb (1/4 blue), 3d generation

Even where additional characteristics are involved, the rules controlling the proportions of large samples that can be expected to display particular combinations of characteristics work out to be very regular as far as they have been taken. An idea of the potential complexities involved is given by the fact that about 1,000 genes are contained on the single circular chromosome of the simple bacterium that inhabits the human colon. In the case of *Homo sapiens* all of the information needed to manufacture the 16 trillion (16×10^{12}) or so cells of the adult and assign them to their proper places and functions is contained in the DNA of the single initial fertilized egg cell.

In seeking to gain some insight as to how all this works, the sensible thing is to start with something simple, but not too simple. The ideal organism for genetic studies relevant to advanced evolutionary states is one having short generation times, so that genetic mechanisms can be observed over many generations. It should have sizable, easily cultured populations. And it needs a sufficient degree of complexity so that results might apply to metazoan and vascular plant evolution. Thus most such work has been done on small, rapidly reproducing plants and animals such as the fruit fly *Drosophila*. The principles discovered have proved to be so nearly uniform in all organisms studied that they probably apply generally.

One must not become so entranced with the idea of sexual recombination and hybrid outbreeding, however, as to lose sight of the fact that mutation remains the only known way in which fundamental hereditable *changes* in the gene structure can be made, as distinguished from the acquisition of genotypic diversity by sexual recombination. Summing up, we can see that the basic raw materials for evolution are a product of gene mutation, while genetic recombination rearranges this material in the much more diversified patterns seen in the gene pools of cross-breeding populations. Working synergistically, and supplemented by genetic linkage resulting from the chromosomal organization of the genes and its variation, such processes, under the influence of natural selection, can produce orderly arrangements of variation within the gene pools. Thus natural selection, responding to ecological and climatic change, channels appropriately variant populations in new directions, to which limits are set by reproductive isolation.

Rates of change seem to be affected by the sizes of populations. Both geological and biological observations suggest that changes in large populations tend to be slower than in smaller ones. In the long perspective it seems that the evolutionary record has featured two alternating states involving two kinds of populations. One of these states was characterized by long intervals of relatively stable ecologic, climatic, or geographical diversity, during which community structures were dominated by large populations of more or less cosmopolitan, slowly evolving or genetically drifting organisms. The other consisted of shorter intervals of changing ecologic, climatic, or geographical diversity, during which sexually isolated smaller populations or small peripheral populations of more populous species evolved rapidly while radiating into new, previously unoccupied, or recently vacated ecologic settings. The multitude sets and defends the style, but the pioneer makes the breakthroughs.

This bears on one of the major remaining puzzles in the current version of natural selection theory. Why is it that forms intermediate between major biological categories are so often missing? For although unquestionable intermediate forms are known, as in the case of *Archaeopteryx* between dinosaurs and birds and the mammal-like reptiles between orthodox reptiles and mammals (see chapter 15), they are generally missing between the major groups of invertebrates. This is apparently in part a consequence of rapid evolution of the major invertebrate phyla between about 680 and 550 million years ago, an interval of geologic history for which the preserved record is very incomplete as compared with that of later times. But it is probably also a function of similarly rapid evolution in small populations at other times of unusual ecologic opportunity, as after intervals of extinction of previously dominant forms of life. The mammals, for instance, flowered only after the disappearance of the dinosaurs, while modern types of reef-building hexacorals appeared and flourished only after the extinction of their Paleozoic tetracoral predecessor.

Intermediate forms, moreover, are almost by definition of short duration and limited number. Thus they are statistically much less likely to be preserved as fossils (and to be found) than are successful forms of long duration, the more so as the geologic age and the opportunity for destruction of the sediments that enclose them increases. And destruction of fossiliferous deposits of the past may result either from erosion or from descent beneath the continents on the downgoing plate of a convergent pair. It seems, in fact, that important intervals of Earth history are not represented or are very inadequately represented by known sedimentary rocks. Taking all of this into consideration lends force to the view, often expressed, that only a small fraction of all the species that have ever lived is known (of the roughly 1.5 million species so far described, only about 200,000 are fossils)—either because they lacked preservable hard parts or because they were not preserved, were lost by erosion or in plate convergence, or have not as yet been found. Considering the odds against their preservation and discovery, the fact that we know as many intermediates as we do and that new ones continue to turn up from time to time seems more remarkable than their common absence.

One interesting aspect of evolution that has not been mentioned earlier involves the role of multiple chromosome numbers, a state referred to as *polyploidy*, almost limited to the higher plants but common among them. This is one way of making new intermediate species in a single step by stabilizing hybrids that would otherwise be infertile. In this process, the chromosomes do not undergo reductive division at the time of meiosis. Thus, where a hybrid would be sterile only because its chromosomes are too dissimilar to pair properly, both sets of chromosomes may be retained to pair among themselves in subsequent reductive meiotic divisions. Once formed, the polyploid breeds true for its intermediate characteristics. Polyploid species are known that have various multiples up to eight times the original number of chromosomes. It is estimated that between one-fourth and one-third of all species of flowering plants are polyploid. Common examples include a cross between radish and cabbage, a variety of crop plants, and familiar weeds, wild flowers, and grasses. Polyploidy is rare among animals, however, and is almost confined to forms that reproduce asexually. Although it might seem a likely candidate for that role, it is not at this time known to have any bearing on the origin of higher categories of life.

To sum up; a very large body of evidence supports and none opposes the conclusion that species, genera, and even families do originate as a consequence of observable selective processes and stabilization by reproductive isolation. The steps by which higher levels of life arise are less clear, although the Galápagos finches and Hawaiian honeycreepers seem to point the way. Factors to be considered include the great length of geologic time and the prospect that major changes could take place in small peripheral populations through extension of the DNA chain to accommodate additional genetic material resulting from mutation and recombination at times of rapid ecological (including climatic and geographic) change. Also important is the incompleteness of the geologic record, the frequent parallelism and convergence of separate lines of evolution among lineages preserved, and the fact that intermediates *are* found and continue to be found between important major categories, such as the classes of the vertebrates. From such considerations, among others mentioned, arises a constructive focus for the still hotly pursued goal of understanding in detail the mechanisms of evolution. This is the very high probability that processes of variation and selection similar to those actually known to exist and demonstrated to be capable of producing separate, genetically variant populations are sufficient (given appropriate ancestry, ecological circumstances, and the lengths of time involved) to explain the fact of evolution at *all* levels without postulating unverifiable processes.

This conclusion has consequences that can be tested in an interesting way because of the work of a number of modern molecular geneticists. If gene mutations are the basic stuff of evolution, then differences between organisms should be reflected by differences in the structure of their DNA. The farther apart they are morphologically and physiologically, the greater these differences should be. Experiments in molecular genetics show that, where two strands of DNA have been separated along their zipper-like line of linkage, they will, given proximity, zip back together again. Where separated strands of DNA from different species are juxtaposed they will zip together only in the regions where they are complementary, not where they differ. Thus it is possible to quantify the similarities and differences of the DNA of different species by separating strands of it from both and allowing them to zip together as they will. The extent of pairing thus observed conforms to the degree of relationship previously worked out on other grounds. Such experiments, for instance, show no matching between the DNA of bacteria and man and only slight similarities between man and the lower vertebrates. A 99 percent similarity, however, is found between the DNA of man and the chimpanzee. Similar relationships are found in the sequences of amino acids observed in the comparable proteins of different organisms, confirming judgments of relationship based on conventional morphological criteria.

In this way genetics and molecular biology support predictions from evolutionary theory. The observed fact of evolutionary change acquires a firm theoretical foundation in the modern synthetic theory of evolution based on gene mutation, sexual recombination, natural selection, and genetic isolation. Gaps are explained by the failure of short-lived intermediate populations to leave a geologic record. *Biologic evolution primarily by means of natural selection acting on variable gene pools thereby becomes the unifying underlying principle that binds the different aspects of the biological sciences into a single coherent whole.*

A Brief Preview of Biologic Evolution

Every being that has ever lived has been a part of a population which itself was only one piece of a community of associated species. That community, moreover, was or is in some way connected to every other community in its global ecosystem. The world of life, the biosphere, can be viewed as a single, interconnected but almost endlessly diversified system. Whether in desert, prairie, forest, or sea, each living thing reacts with other members of its population and with other populations in the community as prey, predator, host, parasite, scavenger, or primary producer. The primary producers are the plants that transform radiant energy from the sun into the chemical energy of carbohydrates and plant proteins through the process of photosynthesis, based on chlorophyll and utilizing carbon dioxide, water, and mineral nutrients as raw materials. The sun is the pump that pushes life uphill against the predictions of thermodynamics. Plants are the beginning of the *food chain*—the base of the pyramid of life. The animals that live directly on the plants are the first-order consumers, which in turn are eaten by second-order consumers, and so on up to the dominant carnivores or omnivores at the top of the pyramid. Actually this is a vastly oversimplified sketch of the reality of who eats whom in a biological community, which forms a much more complicated set of interconnecting lines better referred to as a web than a chain. A simplified insect *food web*, in which common broccoli is the primary producer, is shown in figure 24.

The sum of all these interconnecting food webs for all communities of the biosphere is a system of such exquisite complexity that it defies schematization. Yet there are unifying features—the limited number and commonality of protein-building amino acids, the dependence of all energy-transfer mechanisms on the same basic chemistry involving ATP (adenosine triphosphate), the general infrastructure of the eucaryotic cell, and the genetic code. Working with these raw materials and their variations through billions of years of geologic time, natural selection, responding to ecologic and climatic pressures on evolving communities of organisms, has shaped the biosphere we see about us.

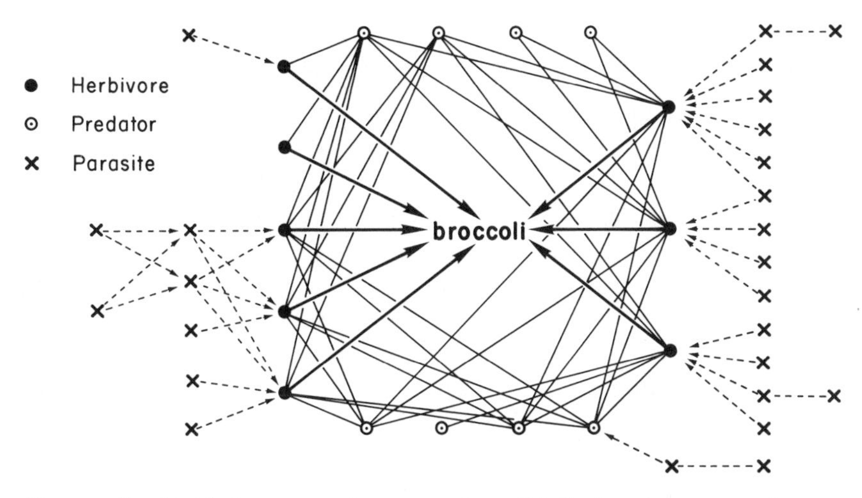

Figure 24. Simplified food web centered on common broccoli, indicating relations among the more abundant herbivorous insects that feed on it, their common insect predators, and parasites on all. (Adapted from D. Pimentel, 1961, *Annals Entomological Soc. America* 54.)

We have seen that major shifts in biologic evolution seem to happen rapidly at times of marked ecologic or climatic change and that they tend to be separated by intervals of gradual diversification and genetic drift when slower advances are made along a broad front and gains are consolidated. What we conventionally think of as animal life represents a substantial advance in body form and organization over the morphologically (but not biochemically) simple algal and bacterial cells and thread-like chains or clusters of cells that precede it. Even the unicellular or noncellular Protozoa were more complex than these. But the truly multicelled animals or *Metazoa* are differentiated still further into tissues and organs that carry out specific functions of metabolism, respiration, excretion, and so on. The story of evolution, as usually told, primarily involves the diversification through time of these multicellular animals and their distant relatives, the terrestrial vascular plants.

That elaboration began with the seemingly abrupt appearance of the first Metazoa, only about 680 million years ago, and this was followed by a geologically rapid diversification into major groups during the first hundred million years or so of the Paleozoic Era—the oldest of the three historical divisions of the Phanerozoic Eon, comprising the last 15 percent of Earth history (see the front endpapers). Most of the major nonbackboned, or *invertebrate*, groups of Metazoa with good fossil records were present by the end of the Early Cambrian, perhaps 550 million years ago. But the *vertebrate*, or backboned, animals did not arrive for another 60 million years or more, early in Ordovician history. The record of the roughly 440 million years of Paleozoic history is thus primarily one of a great elaboration of marine invertebrates and fishes, culminating, after amphibian origins, with the earliest reptiles, ancestors of the dinosaurs. The term Paleozoic points to the relatively archaic appearance of the organisms then extant.

The most recent 230 million years or so of the record of biologic evolution leading up to the present is one of progressive diversification and complexity, especially among the terrestrial animals and plants. Historically this includes the Mesozoic and Cenozoic Eras, or the times of "medieval" and "modern" life. Dinosaurs were prevalent during the Mesozoic Era, which represents the oldest two-thirds of post-Paleozoic time. Mammals and flowering plants appeared during their reign, but did not flourish. Mesozoic marine and land invertebrates gradually acquired a more modern appearance. The end of the Mesozoic Era was heralded by great extinctions of dinosaurs and marine invertebrates. The exact causes of these extinctions must remain forever unknown. They may, however, relate in some way to climatic cooling, to increasing continentality of climate and other effects resulting from the withdrawal of the shallow seas that flooded the late Mesozoic continental masses as a result of then expanding spreading ridges, and to a related collapse of the then prevalent food webs.

With the unfolding of the Cenozoic Era, the most recent 62 million years, flowering plants came to the fore, grasses appeared, mammals thrived and diversified, and the marine invertebrates underwent a steady modernization—a modernization so progressive, in fact, that one way of identifying the divisions of Cenozoic history is by observing the proportion of presently living species in rocks. Almost lost in all this welter of change is the scant record of small animals with short muzzles, some of whose descendants could see in three dimensions and had an opposable thumb for grasping and manipulating things. From such animals there eventually emerged an erect, flat-faced, big-headed species that calls itself "man the wise"—*Homo sapiens.*

During all this time the level of biological complexity increased. There are limits to the morphological variation possible for a single cell, or for strands or colonies of cells that reproduce by simple cell division and depend on mutation alone for the basic variability upon which selection acts. There are limits to what can be done with low-efficiency fermentation as a source of energy, as practiced by most bacteria and primitive blue-green algae or their ancestors. Sex was the great evolutionary leap forward, with the use of oxygen in energy conversions playing a significant role by increasing the energy release approximately tenfold over what could be obtained from fermentation of the same quantities of nutrients. The driving system of the present biosphere is photosynthesis, linked with oxidative metabolism, marshaling the sun's energy to sustain life and to impel it toward higher things by way of genetic variation channeled by natural selection.

The more salient features of the succession of biospheres ancestral to ours over the previous 680 million years are the subject of the next three chapters.

For Further Reading

Bonner, John Tyler. 1962. *The ideas of biology.* Harper and Row. 180 pp.

Darwin, Charles. 1859. *On the origin of species.* Facsimile edition. Harvard University Press, 1964. 502 pp.

Dobzhansky, Theodosius; Ayala, F. J.; Stebbins, G. L.; and Valentine, J. W. 1977. *Evolution.* W. H. Freeman & Co. 572 pp.

Mayr, Ernst. 1963. *Animal species and evolution.* Harvard University Press. 797 pp.

Raup, D. M., and Stanley, S. M. 1971. *Principles of Paleontology.* W. H. Freeman & Co., 388 pp.

13

THE GREENING OF THE LANDS

Passengers in a spaceship soaring among the planets could easily tell them apart by their distinctive colors and patterns in the sun's reflected light. Mercury is dull, cloudless, pockmarked with impact craters like Earth's moon, and laced with lobate fault scarps. Venus is completely shrouded in white clouds of carbon dioxide (or perhaps ammonium carbonate), beneath which radar shows a complicated blocky surface. Mars is pale red, with white polar caps of water ice veneered with frozen carbon dioxide and with cratered areas, sandy plains, volcanoes up to 25 kilometers high, and evidences of deformation and extinct river erosion. Jupiter, shining by its own radiant energy, is like a brightly striped beach ball, with its turbulent surface including a great red spot. Saturn and blue-green Uranus are circled by rings of small satellites. Earth's bold, irregular patches of blue, green, red, and white are like a canvas by Van Gogh. Blue is dominant because more than 71 percent of Earth's surface is covered by sky-reflecting water. White represents cloud banks, pinwheel-like storm centers, snow, and ice. Many deserts and cultivated low-latitude lands have a reddish hue. But green is the symbol of life on the solid surface of our planet—the forests, fields, prairies, and tundra-covered regions that make all animal life above sea level possible.

During most of Earth's history these areas of green did not exist. Before oxygen accumulated to concentrations sufficient to bring about the generation of a shielding ozone-screen in the lower stratosphere, the disruptive effects of high-energy ultraviolet radiation on DNA would have made most of Earth's surface above sea level uninhabitable. For life, as noted earlier, almost certainly originated in the sea and remained confined to the sea until such time as atmospheric conditions became favorable to its existence apart from the radiation-absorbing shield of overlying water.

A world without land plants would be vastly different from the one we know. Its lands would be devoid not only of plants, but also of all animal life—for even carnivores, scavengers, and parasites depend ultimately on plants or other organisms that feed on plants. Such a world would be dry, monotonous, and subject to extremes of heat and cold. Without land vegetation there would be nothing to bind a soil in place and little to retard the gravitational downslope flow of such rain as might fall. There would be no plants to assist the sun in pumping water from the soil to the atmosphere between rains, and little to moderate the merciless solar radiation. Indeed the loose material at Earth's surface would be soil only in the sense in which that term is used (or misused) to refer to particulate matter at the surface of the moon or Mars. In the absence of plant acids, weathering would be limited to purely physical and physical-chemical processes—heating and cooling, ice plucking and grinding, the beating of rains, downslope movement due to gravity, the physical and chemical effects of running and standing water, and the effects of atmospheric gases and carbonic acid in rainfall. There would be no alteration by plant acids and perhaps not even bacterial decay, while any organic matter in Earth's surface debris would most likely be of meteoritic, cometary, or otherwise physical-chemical origin. Rainfall would flow immediately and pretty much unimpeded toward the sea, flushing everything portable before it in flash floods that roared and finally trickled down braided, barren, intricately channeled stream floors. No overbank vegetation would slow the currents so as to trap and hold fine sediments in place, creating the fertile floodplain deposits from whose alluvium so much of the world's civilization has arisen. No water would be transpired from plants to the atmosphere to supplement the land-based cloud cover nourished by evaporation alone, or to moderate the harshness of the continental climate. The scenery would be unrelieved by plant life.

By what means was all this severity transformed into the generally clement and verdant earth we know today? How was the way prepared for the greening of our planet, for the moderation of its atmosphere?

Thinking back to that primitive Earth before there was enough oxygen to make an ozone screen, one can't help wondering whether, even then, there might not have been some forms of land life in some special sheltered places. It is not hard to imagine that, even before conditions were conducive to land life generally, there may have been places that were both moist enough to sustain life and permanently shielded from damaging radiation. Could not some simple early forms of life have established themselves, if only temporarily, in deep crevasses, below poleward-facing or overhanging cliffs, within soils or sediments, or even in standing fresh water bodies of sufficient depth? Even today we know soil-dwelling bacteria and sediment-binding blue-green algae that enjoy a degree of shielding from radiation thereby, and algae of this type have a very long geologic record (see figure 15).

Although it seems possible, therefore, that some types of soil- and sediment-binding bacteria or blue-green algae could have occupied suitable but presumably limited sites of life above sea level almost from the time of the origin of life onward, we have no good evidence for or against this possibility. And, if such forms existed, they could have contributed only on a small scale to weathering and a crude kind of soil formation by the production of biologic acids. They would also have had little effect on the transpiration of water from soil to atmosphere or on climatic feedback or segregation of coarse and fine components of stream-borne sediments. Indeed, if such things did exist they could hardly have been land plants in any significant modern sense, although they could have pioneered the migration into the fresh waters and moist, low-lying places from which the first true land plants spread out.

Consider the barriers to the establishment of life on dry land. Nearly all advanced multicellular forms of life reproduce by mechanisms that call for the union of sperm and egg in a fluid medium. Marine organisms normally shed their sperm directly into the water near the female and enough egg cells become fertilized by this haphazard process to assure the continuity of the species. Land organisms must evolve special systems to protect the delicate sperm and unshielded eggs from death by desiccation. Water, as Coleridge's Ancient Mariner remarked, is everywhere at sea, so that marine plants and animals have an ample and persistent supply of it. Terrestrial organisms, on the other hand, must get theirs from rain, flowing or standing fresh water, the food they ingest, or the soil on which they grow. In addition their surfaces must be waterproofed by special adaptations such as scales, thick skins, bony coverings, waxy coatings (on leaves), or bark, to prevent undue loss of the water they are able to get.

Because the body fluids of animals are salty, like the sea, animals that live in fresh water must develop special mechanisms to prevent the fatal dilution of their body fluids and bursting of cells that would result from osmotic intrusion by surrounding fresh waters. Getting enough oxygen for oxidative metabolism is another problem. Aquatic organisms, except for those mammals that evolved on land and returned to aquatic life, absorb their oxygen from the surrounding water as it circulates through gills or other oxygen-collecting systems. Terrestrial organisms must develop special mechanisms to obtain oxygen from the air—extended surfaces and leafy structures in plants, and lungs or other special respiratory systems in animals.

Despite such formidable problems and the many adaptations required to cope with them, the potential advantages for organisms able to colonize the lands were also great—expanses of open space with plenty of sunlight for photosynthesis, ample oxygen for high-energy metabolism, and, once land plants were established, a steady supply of food for animals and a variety of previously empty ecologic niches. It seems likely that the transition from life in mother ocean to life on dry land involved first an adaptation to life in streams and lakes, then to moist places at stream and lake margins, and finally to the permanently emerged land itself.

It is easy enough to formulate ideas about probable evolutionary pathways from the comparative morphology, physiology, and ecology of modern organisms, but we must turn to the geologic record for clues to when the land actually began its greening. And, as always, such clues encompass not only the direct evidence of specific kinds of preserved organisms, but also the data provided by biogeochemistry and sedimentology. We are especially interested in stream deposits because before a soil-binding land vegetation became prevalent these deposits would be mainly or exclusively those of the complexly multichanneled types of streams described as "braided." Once sediment-binding land vegetation had become extensive, however, stream-deposited sediments of non arid regimes should show a less intensively channeled structure and a consistent gradation of types between end members consisting on the one hand of floodplain silts and clays, and on the other hand of cleaner channel sands and gravels.

A major difficulty is that the sedimentary and biogeological records of terrestrial processes and events are likely to be much poorer for any given interval of time than are records of the same interval for the sea. The lands are places of erosion and removal of sedimentary evidence, the seas of deposition and preservation of data. The record of events that took place in or adjacent to the sea or to large, long-lived lakes is, therefore, apt to be far more nearly complete than that of rivers or transient upland basins. It is important to keep this in mind as a limiting factor in the attempt to reconstruct terrestrial events. By the same token it is all the more impressive when good reconstructions of such events can be made.

Typical land plants have root systems to collect and transmit water and mineral nutrients upward from the soil, leaves rich in chlorophyll pigments to carry out photosynthesis in the open air, a fluid-transporting or *vascular* system to pump water and mineral nutrients upward from the roots, and a surface coating or structure that retards the loss of water from leaves and stems. These are the vascular plants, comprising all of the familiar trees and shrubs, the grasses, the cacti, and the herbaceous plants, as well as ferns, cycads, and the like. Of lesser complexity, and presumably earlier in the evolutionary time scale, are the amphibious plants or *bryophytes*, the mosses and liverworts, which lack the special adaptations of the vascular plants. Therefore, they are either limited to perpetually wet environments or they undergo intervals of dormancy or metabolic depression between wet episodes.

The bryophytes, nevertheless, display two features that are characteristic of all higher land plants and imply an ancestral relationship to them: they show an alternation of sexual and asexual generations, and they possess a structure called an *archegonium*, which protects their reproductive bodies from drying out in the sexual stage. Bryophytes and vascular plants, together comprising the higher plants, are believed to have evolved from green algal ancestors. Excluding the blue-green algae (which are actually more closely related to bacteria than to typical algae), the green algae are the only major algal group to have made the transition from salt to fresh water. They share uniquely with the higher plants a distinctive combination of chlorophyll pigments.

The time being long ago even by geological standards, and the chances of destruction by erosional and weathering processes in terrestrial environments and sediments being great, the prospect of finding direct evidence of the oldest terrestrial plants and of early terrestrial plant evolution is slight. We can infer something about these important biological events, however, by looking carefully at the kinds of sediments deposited.

Looking backward through time from the present at the scant older records of terrestrial sedimentation, it is possible at places to observe the kinds of sedimentary segregation that a rudimentary land vegetation might cause in sediments deposited by flowing streams. The evidence dates as far back as perhaps 460 million years ago (latest Ordovician). Such sedimentary characteristics are also approached by deposits laid down as much as 700 to 1,100 million years ago and perhaps foreshadowed in sediments as much as 1,200 to 1,400 million years old. Still older sediments, however, believed on other grounds to be those of streams, seem to find their modern counterparts primarily among the deposits of braided streams such as today characterize desert and subpolar regions.

Since at least a few fossils of land plants with crude rootlike systems and erect above-ground foliage are known from latest Silurian deposits, it is possible to understand why we see the sorting observed in stream sediments this old (about 410 million years). It is harder to explain the older ones. Late in pre-Phanerozoic history, however, before the oldest authentic records of multicellular animal life, there existed a variety of spore-like forms that have been interpreted as single-celled floating marine plants and also as the spores of land plants. (A *spore*, properly speaking, is a small, specialized reproductive structure that does not require fertilization). Except for some known contaminants, these spore-like bodies from pre-Phanerozoic rocks are undoubtedly not the spores of normal woody land plants, and all indigenous ones *could be* simple single-celled algae. Some of them, however, are not unlike the spores of mosses and liverworts, although no demonstrable remains of bryophyte tissues have yet been found in rocks older than about 375 million years. It is possible, then, that bryophytes—with rudimentary rootlike and leaflike structures and with specialized organs to protect their fruiting structures from desiccation, but without vascular systems—had already emerged in suitable wet places even as far back as the close of pre-Phanerozoic history. If so, and the proposition is very iffy, they would have been the first relatively conspicuous land dwellers. Such primitive and restricted land plants, mostly under 30 centimeters tall, could have caused the local segregation of the fine fractions of some of the later pre-Phanerozoic stream-borne sediments. As no direct evidence of such forms of life is known, however, I must stress that the idea of a pre-Phanerozoic wetland vegetation rests on very shaky ground.

Nevertheless, there was enough free oxygen in the atmosphere by 1,200 to 1,400 million years ago to generate the ozone needed to screen out the DNA-disrupting ultraviolet radiation that had impinged on the land surfaces of Earth up until about 2 billion years ago (figure 20). That would have fulfilled the most essential condition for occupancy of the land by plant life. And it seems to be characteristic of biologic evolution that new forms of life tend to appear reasonably soon after conditions favorable to them are established—the *principle of evolutionary opportunism.* Had there been a rudimentary wetland vegetation in later pre-Phanerozoic times, it would probably not have been enough to affect climate substantially, but it could have been enough to make a start at generating soil conditions favorable to the later emergence of authentic vascular plants. Those plants, of course, are the familiar everyday plants, with real root systems, erect trunks, branches and leaves, and the well-defined fluid-transfer system that is so distinctive and so important for nutrient-transfer and the pumping of water from ground to air during plant transpiration.

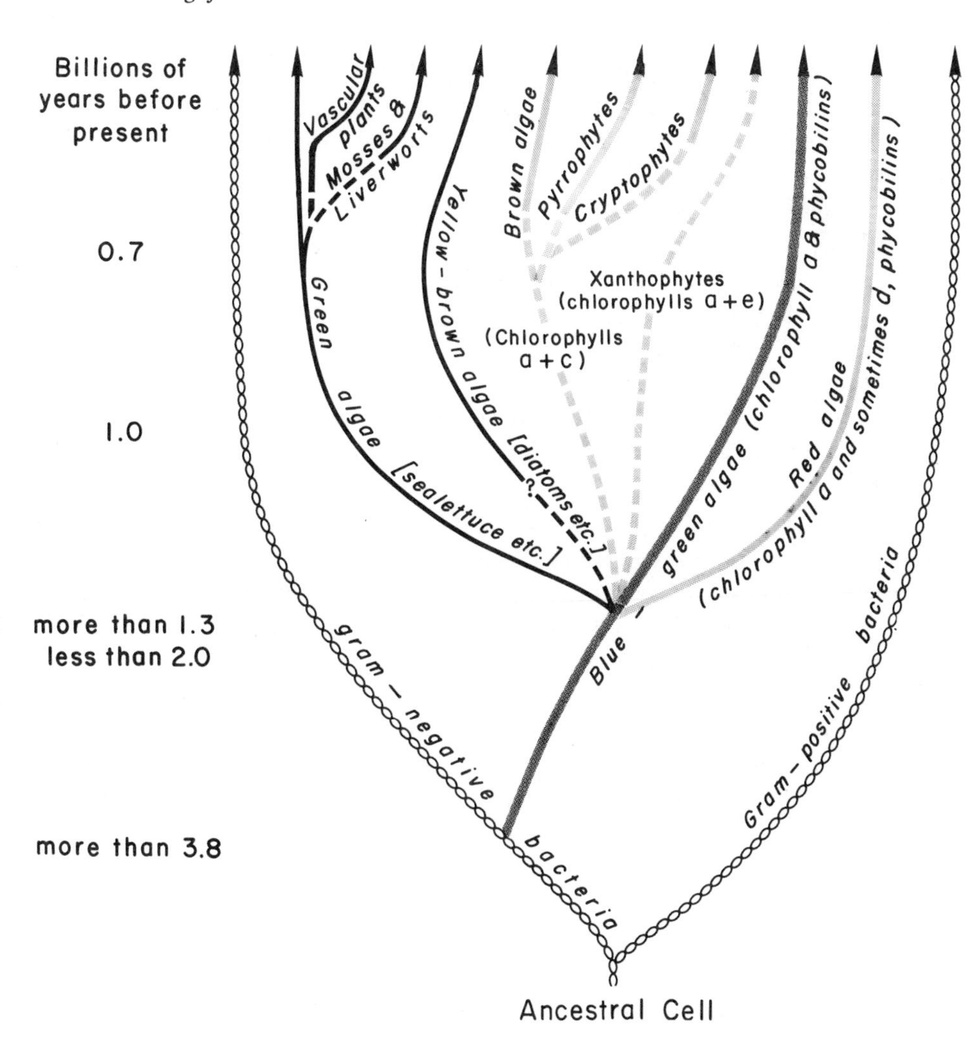

Figure 25. Possible pattern and timing of the evolution of the higher algae, bryophytes, and vascular plants. (Adapted from P. Cloud, M. Moorman, and D. Pierce, *Quarterly Review of Biology* 50, 1974.)

A possible scale of evolution from procaryotic ancestors through algal predecessors to the mosses, liverworts, and modern vascular plants is suggested in figure 25. The bracketed technical terms for plant pigments in this figure are not needed for general comprehension but only to indicate the biochemical grounds for the relations suggested.

In the remainder of this chapter and the next three chapters we also face the problem of discussing plants and animals that have no common names and of doing so in terms of their sequence in geologic time. I will keep the jargon to a minimum, but it would deprive the reader of a proper sense of the language of science to eliminate it altogether. At the same time it would be a bore here to enter into a detailed discussion of the implications of either biological or geological nomenclature. Instead, table 3 is provided to illustrate the nomenclatural structure of descriptive biology, while table 4 briefly summarizes the main principles used to order the sequences of events in Earth history (including the succession of rock strata, or *stratigraphy*) where radiometric ages are not available. The front endpapers include the basic nomenclature of historical geology.

Returning to the question of a land vegetation, and granted the validity of the thoughts articulated, one might expect to find fossil vascular plants at any time from perhaps 700 million years ago onward. Yet vascular plants that old are not found. At various times zealous searchers have reported their discovery in rocks as old as 450 or even 500 or 700 million years old, only to be discredited by more critical reexamination. For some time plant remains from eastern Australia were considered to be the oldest authentic vascular flora and to be of Late Silurian age (a little over 400 million years before the present). It is now accepted, however, that the sediments in which they were buried are less than 400 million years old, and that the plants, in fact, are earliest Devonian. The only universally accepted vascular plant so far known to be of unequivocal pre-Devonian age is the genus *Cooksonia* (italicized because it is in Latin form), a very simple, precariously rooted, forked branching, leafless form with photosynthetic pigments in the branches, with specialized (tracheid) cells for water conduction, and with spheroidal spore-cases at the ends of the branches containing spores with distinctive triradiate markings. The oldest known records of *Cooksonia* consist of poorly preserved remains from rocks of Late Silurian age in central Bohemia and definitely vascular material of the same type recently reported from the uppermost Silurian deposits of South Wales. Better preserved material of the same genus in the oldest Devonian beds of Wales, displaying all of the features noted above, assures us that it is beyond doubt the remains of an early vascular plant. A photograph of that *Cooksonia* is reproduced as figure 26.

Table 3
Nomenclatural Structure of Descriptive Biology

Domain Procaryota

(Consists of morphologically simplest living cells: lacking membrane-bound nucleus and organelles, lacking 9+2-strand flagella)

Kingdom Monera
- Phyla
  - Subphyla
    - Classes
      - Orders
        - Superfamilies
          - Families
            - Genera
              - Species

The domain Procaryota and kingdom Monera have the same content. Only two phyla of procaryotes are recognized: the bacteria and the blue-green algae, more closely related to bacteria than algae and sometimes called cyanobacteria. Fewer lower nomenclatural subdivisions are used than among eucaryotes because visible (morphological) variation is much less. Biochemical criteria are considered much more important than variation in form and structure in determining the proper classification of living forms. Among fossils, however, only morphological criteria are available for classification.

Domain Eucaryota

(Comprised of morphologically complex cells: having membrane-bound nuclei and organelles, having 9+2-strand flagella)

Kingdoms Animalia, Plantae, Fungi, Protista
- Phyla
  - Subphyla
    - Classes
      - Subclasses
        - Orders
          - Suborders
            - Infraorders
              - Superfamilies
                - Families
                  - Subfamilies
                    - Genera
                      - Subgenera
                        - Species
                          - Subspecies

The eucaryotes include four kingdoms: animals, plants, fungi, and the protists (single-celled eucaryotes, with or without chlorophyll). Phyla are variable in number, depending on preference of the classifier. As few as 9 and as many as 35 have been suggested for the animals, for instance (see figure 29). Divisions (often called taxa) below the rank of phylum are also variable in number for the same reason. Because organisms tend to become more diverse and more complex further up the evolutionary scale, larger numbers of hierarchical levels are generally recognized among the higher eucaryotes than among the lower eucaryotes or procaryotes.

Table 4

Summary of Ordering Principles in Earth History and How They Create a Sequence Geochronology

Earth history is a product of interactions between evolving air masses, water bodies, living systems, and internal motions of the earth, the crustal deposits they produce, and the resulting surface features of the earth.

The results of these interactions only become history, however, when placed in appropriate sequences of time.

Before nuclear clocks became available (figure 8) the only useful geochronology was a relative or *sequence geochronology*, wherein events recognized were placed in a sequence of before, after, and contemporaneous with, using ordering principles which are still the operational backbone of much historical geology.

These ordering principles are:

1. *The law of superposition.* In a sequence of stratified rocks the ones at the bottom are the oldest and those at the top the youngest, except where they have been overturned by mountain-building forces or where volcanic rocks have been injected parallel to the stratal surfaces (both of which events leave clear evidence of their workings).
2. *The principle of original lateral extension.* Where a sedimentary deposit neither thins out nor abuts older rocks in depositional contact it must once have continued beyond observed limits, as a distinctive rock in one wall of the Grand Canyon (figure 7) can be matched with counterparts in the opposite canyon wall and across the buttes and pinnacles between.
3. *The principle of crosscutting relationships.* If one rock cuts abruptly across another, the one that does the crosscutting is the younger. Scratch a line in a piece of modeling clay with your finger and then another cutting across it and there is no doubt which came first and which second. Intrude a granite into a pile of sedimentary rocks and crosscutting relationships will show that the sediments are older than the crosscutting granite.
4. *The principle of bracketing relationships.* A rock or event that falls between two events or rocks of known age is of intermediate age. If a sediment lies on the erosional surface of one granite and is crosscut by a granite of a different type, all three rocks can be placed in their correct relative sequence.
5. *The principle of biological succession.* Fossils in continuous successions of rocks show progressive changes upward through the time represented by the rock succession. Similar fossils in similar rock successions whose equivalence can be established by lateral tracing or closely spaced boreholes invariably show similar sequences. By piecing many such occurrences together, a succession of fossils has been worked out that is the same worldwide for the last 680 million years of geologic time. A beginning is now being made at extending a similarly unique sequence of life forms downward into much older rocks.

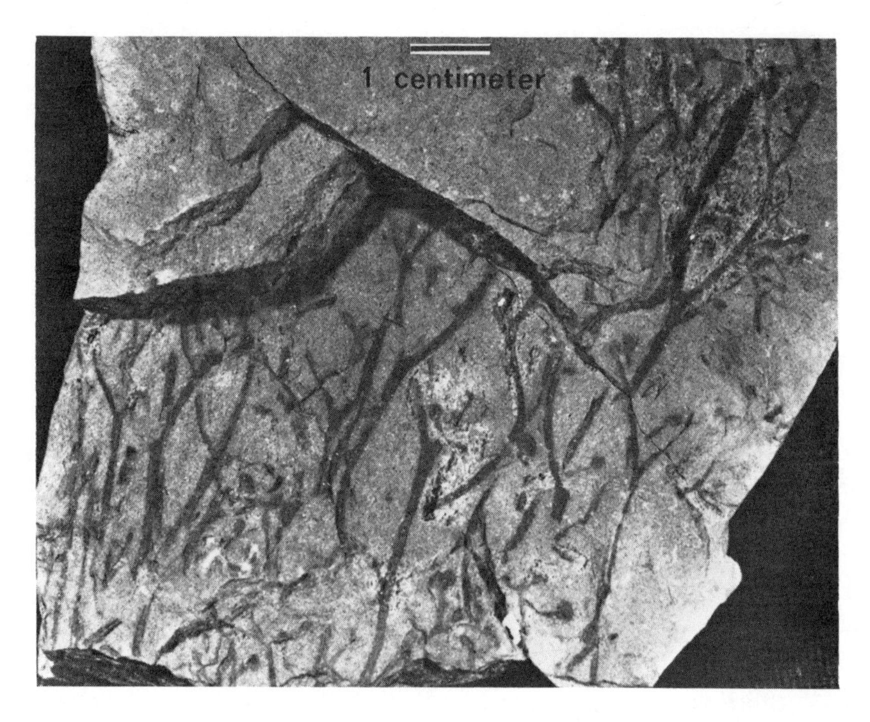

Figure 26. Oldest known well-preserved vascular plant, *Cooksonia caledonica,* showing forked branching and terminal spheroidal fruiting bodies. Photograph shows type specimen on bedding surface of 400-million-year-old sedimentary rock from Wales (by courtesy of C. D. Waterston and Dianne Edwards).

It's quite a trick, however, to be sure that you have the earliest or the latest of anything. Paleontologists are always finding fossils in rocks deposited before they were supposed to have appeared and after they were thought to have become extinct. In the case of land plants, Jane Gray and Arthur Boucot of the University of Oregon and Oregon State University have found spores of Early Silurian and Late Ordovician age, roughly 450 million years old, that they interpret to represent those of vascular land (or semiaquatic) plants, although "a bryophyte origin cannot be excluded." Yet the evolution of Late Silurian and Early Devonian vascular plants shows a pattern that is characteristic of first appearances of new types of both plants and animals throughout geologic time—rampant early diversification, followed by a streamlining of types. The 30 million years following the oldest unequivocal record of vascular plants (*Cooksonia*) slightly more than 400 million years ago saw a great diversification of form. This diversification was characterized by the evolution of well-defined leaves, both with and without veins, by the evolution of solid vascular strands, by the development of a clear distinction between a main axis of growth and lateral brances, and by the introduction of two-thirds of the kinds of true spores that are to be found over the entire 60 million years or so of Devonian history.

Suddenly, as it seems on the clock of geologic time, plant life emerged from the cocoon of history with all the basic attributes of a true land flora. Such a pattern is consistent with the prevalent judgment that true vascular plants cannot have appeared much earlier than perhaps 410 million years ago. Yet, as remarked in the next chapter, ancient representatives of the tiny spider-relatives called mites are known from rocks even older than the oldest vascular plants and their presence is strong presumptive evidence of a preceding land flora of some kind. What kind of a flora could that have been?

And if, as prevalent opinion holds, the first vascular plants really are only slightly more than 400 million years old, why the big delay—perhaps as much as 800 million years—between the first hint that simple wetland plants could have been about and the apparent debut of the vascular plants? That is a subject calling for further study, with interest focusing on fine-grained carbonaceous sediments associated with nonmarine deposits of Early Silurian and older age. The discovery by Lisa Pratt and associates at the University of North Carolina of primitive land plant-like remains in just such continental deposits of Early Silurian age at the north end of Massanutten Mountain in Virginia reinforces the work of Gray and Boucot, who had inferred a land-plant origin for similar material in marine deposits of the same age. Pratt and associates conclude that their material is pre-vascular but clearly land plant. Thus there may have been a substantial interval of Early Silurian and older pre-vascular land plant evolution leading to fully vascular plants in Late Silurian time—a prelude of fuzzy green fringes along the waterways readying the drab lands for the original wearing of the green and preparing the way for the first simple land animals. The earlier mentioned suggestion by Berkner and Marshall that oxygen levels were simply not high enough to support the metabolism of land plants before about 400 million years ago seems unlikely to be correct, for the presence of a terrestrial fauna and advanced types of fishes shortly thereafter implies that atmospheric oxygen was then already approaching present levels (figure 20), and was probably substantial for some time previously.

But, even if it were true, as it may well be, that vascular plants first appeared late in the course of Silurian history; and whether they did so because oxygen in the atmosphere then first attained sufficient concentrations or merely because the land was there and it was inevitable that it would be colonized sooner or later, that is not the end of the story. When we speak of how the land got its plants we have something more impressive in mind than the barely rooted, weak, knee-high to prostrate Late Silurian and Early Devonian plants alluded to above. Such little fellows might serve to hold some soggy river flood-plain and swampy soils in place, but they are hardly the stuff with which hillsides are clothed, cloud-cover created, and climates significantly altered.

A more ample vegetation was not long in coming. Important changes that were going on between about 370 to 340 million years ago included the evolution of leafy fronds and woody structures; plants now attained heights up to 8 or 10 meters and constituted the first true forests (see front endpapers). The earliest vascular plants thus gave rise to three other main groups of plants, all of which have living relatives. These three groups are the club mosses, the horsetails or scouring rushes, and the true ferns—the dominant forest plants of Devonian history.

But, because these primitive plants all lacked seeds, they were still restricted to moist environments. Their familiar leafy stalk or trunk is a nonsexual, spore-producing phase whose spores give rise to tiny, separate, leafless, *sexual* plants. An individual plant of the sexual type produces eggs or sperm or both, but the motile sperm must travel to the egg in at least a thin film of water. Despite its unprepossessing appearance, the sexual phase of the plant's life cycle controls the fate of the species, for the larger, spore-bearing plant can grow only from the fertilized egg of the sexual generation. Although spores may be widely scattered by the wind, the obscure sexual plant can reproduce the more conspicuous nonsexual phase only in a moist environment. Thus the Devonian uplands were probably not forest clothed. Nevertheless, club mosses, giant horsetails, and tree ferns are found in sufficient abundance at widely enough scattered localities to indicate that dense semiwetland forests of Middle and Late Devonian age were sufficiently widespread to be regarded as a proper land vegetation, with all of the effects to be expected from this.

The familiar modern forest trees, grasses, and shrubs had to await the clever seed, with its advanced protections against desiccation. In seed plants the unfertilized female egg cell or embryonic seed is usually borne on a cone or in the ovary of a flower that is to become a fruit, seedpod, or berry. It becomes fertile on contact with the sperm, which is carried to it in a *pollen grain* produced by the male plant or by male flowers on the same plant. The pollen is designed for ready transportation by wind or animals and is protected against drying up by a tough outer coating. Proximity to a female egg cell of the same plant or another of the same species causes a pollen grain to generate a moist tube that penetrates to and punctures the wall of the egg cell. How it knows to do this no one can say, although the stimuli are surely chemical. The sperm then passes through this tube into the embryonic seed to fertilize it, but only under suitably moist conditions will it germinate to grow into a mature plant. Among the flowering plants (as distinct from conifers and other gymnosperms), the sexual parts do not occur as independent plants. Instead they are reduced to small features within the moist tissue of the plant that gives rise to the seed-producing ovaries and pollen grains.

The appearance of the seed is considered so important and so distinctive that the five main kinds of seed-producing plants are generally classified together as a single major group of the vascular plants. The oldest known seed appears late in Devonian history, perhaps 340 to 350 million years ago. It was found associated with an extinct fern-like plant of the type called a seed fern. As might be expected, such seeds were not of fully modern type, imbedded within or covered by a fruit, berry, or pod. Instead they were attached unprotected to seed-bearing surfaces such as those of a pinecone. The

Greek word *gymnosperm* designates all of the different kinds of *naked-seeded* plants now recognized (seed ferns, cycads, gingkos, and conifers). The word is useful in contrasting such plants with the *angiosperms*, or *flowering plants*, whose seeds are enclosed in coverings of various sorts.

The evolutionary advantages of flowering plants—seed protection, flowers and nectar to attract insects (and later some birds), pollen to transport sperm by wind or other carrier, and a variety of adaptations for transportation and delayed germination of the fertilized seed—account for their dominance in the present land vegetation. Of about 260,000 living species of vascular plants—the overwhelming majority of all plants—about 96 percent are angiosperms. The remaining 10,000 known plant species are mainly ferns. Gymnosperms include only about 700 living species, despite the importance some of them have as forest trees in dry, high, infertile, and cold regions.

Yet the flowering plants are relative newcomers. The forests of the great coal-forming epochs of late Paleozoic history (figure 27), 340 to 260 million years ago, consisted entirely of gymnosperms, including conifers, which may have had an independent origin from seedless ancestors. Where were the flowering hardwoods, shrubs, and grasses? The record of plant life between about 130 and 250 million years ago is regrettably sparse. A few fragmentary and uncertain records from this interval of Earth history suggest to some paleobotanists that flowering plants may have arisen gradually beginning in earliest Mesozoic or latest Paleozoic times. Yet the oldest probable remains of flowering plants known to the present time are from rocks that are scarcely 130 million years old, and the oldest widespread and diversified remains of flowering plants are barely 100 million years old. There are, in addition, no demonstrably transitional forms from other groups to angiosperms. But there is good reason to suppose that they arose from gymnosperms over a relatively short interval during the middle part of Cretaceous history in partial response to a parallel evolution among insects. P. J. Regal of the University of Minnesota, for instance, takes the recent discovery of insect-associated bisexuality among certain fossil gymnosperms to imply a role for insects in transforming the soft, cone-like, spore-bearing masses

Figure 27. Reconstruction of a Pennsylvanian coal-swamp forest made to scale from actual fossils. (Photograph reproduced with permission of the Field Museum of Natural History, Chicago, Illinois; © F.M.N.H.)

of some Cretaceous gymnosperms into flowers, while other aspects of the angiosperm style can be explained by calling on a well-known process of early sexual maturation and juvenile reproduction among their gymnosperm ancestors.

During later Cretaceous history (100 to 62 million years ago) came the expansive evolutionary radiation and diversification of flowering plants into all major groups now known. Grasses evolved more recently than 62 million years ago (during early Cenozoic time), completing the roster of major vegetative types and setting the stage for the diversification of horses, other grazing mammals, and their predators. Modernization of Earth's plant cover was essentially complete by about 35 million years ago. Earth was then radiant with color, varied in its raiment, pristine, and virgin, as it paused before the far-reaching evolutionary events of later Cenozoic and modern history.

All of the changes that a plant cover could ever bring about on Earth were then possible and presumably in force. The last 35 million years has seen only relatively minor variations on the modern plant theme—at least until the Neolithic revolution and the succession of great unplanned experiments set in motion by man's ever increasing efforts to subdue and populate the earth.

For Further Reading

Andrews, H.N. 1961. *Studies in paleobotany*. John Wiley & Sons. 487 pp.

Banks, H. P. 1970. *Evolution and plants of the past* (2d ed.). Wadsworth Publishing Co. 170 pp.

Bold, H. C. 1977. *The plant kingdom* (4th ed.). Prentice-Hall. 310 pp.

Delevoryas, Theodore. 1966. *Plant diversification*. Holt, Rinehart & Winston. 145 pp.

Gray, Jane, and Boucot, A. J. 1977. Early vascular land plants: proof and conjecture. *Lethaia* 10:145–74.

14

THE PROGRESSIVE UNFOLDING OF EARLY ANIMAL LIFE

Like many of nature's gifts, the power of the life-giving sun can be a curse or a blessing, depending on circumstances. It becomes life giving for animals only where there are primary producers such as plants to focus its energy on the production of the nutrients that ultimately power the biological machines of the animal world. The greening of the lands was, therefore, an essential prelude to the development of terrestrial animal life. The unfolding of such life is a subject of special interest to one of its most unusual products—the only one known to be capable of taking such an interest. The beginning of that unfolding is the subject of this chapter, with special attention paid to the early land vertebrates. But first let us review the important prior events that set the stage for the beginnings of animal evolution, the invasion of the lands by animals, and the development of the terrestrial vertebrates.

The first obscure living thing could scarcely have been more than a minute blob of naked living stuff. The popular shorthand for such stuff is the term amoeba. But, rather than being lobate like a true member of the genus *Amoeba*, its shape was more likely that of a simple spheroid. And, unlike *Amoeba*, it was intolerant of oxygen. It was also very probably incapable of making its own food and was thus dependent on assimilation of the accumulated products of chemical evolution for sustenance. As discussed earlier, its survival and the continuation of evolutionary progress to the present depended on the emergence of organisms capable of manufacturing their own food and thus of serving as a nutrient source for associated organisms that could not do so. Most important among such organisms are those that utilize sunlight, carbon dioxide, and water to make carbohydrates and release free oxygen. Records available (see, for example, figure 15) imply, as biological evolution would have predicted, that the earliest of such photosynthesizers were blue-green algae or forms ancestral to them. Details of the ultrastructure of the cell wall of living bacteria and blue-green algae, determined by electron microscopy, indicate that the ancestors of the blue-green algae were most likely bacteria having the distinctive thin wall-structure called "gram-negative" (figure 25). All of these earliest organisms were necessarily aquatic and presumably marine because they had no protection against desiccation out of water and they needed a shielding thickness of water above them as protection against the then intense ultraviolet radiation.

The first blue-green algae (or similar predecessor forms) found themselves in possession of a poisonous by-product, oxygen, against which special defenses were needed before its metabolic and evolutionary potential could be realized. These early oxygen-producing photosynthesizers are believed to have depended on reduced iron compounds and other oxygen-hungry chemicals in their surrounding environments to maintain oxygen at tolerably low levels until similar organisms with efficient oxygen-mediating enzymes arose—much in the way that some living oxygen-sensitive blue-green algae today thrive only in the presence of oxygen-consuming hydrogen sulfide. A complete complement of advanced oxygen-mediating enzymes is believed finally to have arisen around 2 billion years ago, to judge from the abundant appearance in younger rocks of detrital continental and marginal marine sediments that are rich in the oxidized iron mineral hematite, attributed to the effects of free oxygen in the contemporary atmosphere.

The first modern types of cells were probably simple green algae, which are related to common living filamentous seaweeds. The fossil record suggests that these arose from blue-green algal ancestors between 1.3 and 2 billion years ago (see table 2 and figure 25). And they, in turn, gave rise to the higher plants, and, indirectly, to simple invertebrate animals hundreds of millions of years later—the first tentative evolutionary step toward the still far-distant human condition.

Given an advanced level of cellular organization, sexuality, and a sufficient level of free oxygen, it is not hard to imagine an evolutionary transition from many-celled algae to higher plants. After all, the basic photosynthetic, metabolic, and oxygen-diffusing processes are similar among all green plants, large and small, in or out of water. Nor is it difficult to visualize the transition from a single-celled alga to a protozoan as a consequence of simple loss of pigmentation.

To go from a single-celled protozoan or "plant" (together the Protista of table 3) to a many-celled animal, a metazoan, is quite another matter. Simple diffusion cannot provide the oxygen needed for metazoan metabolism except in the flimsiest soft-bodied forms. To get oxygen to the inner parts of a thick-bodied animal, or one enclosed within an outer shelly covering or impervious integument, requires circulating body fluids of the right composition and structure to transport oxygen to the sites of need and remove waste products. There must also be an opening for taking in food, a place for the discharge of wastes, a digestive system for extracting vital energy-carrying and tissue-building molecules from the food and moving the wastes along, and much more. The organisms we call Metazoa, or simply animals, are far from being just a cluster of cells. They have a particular detailed structure that includes circulatory and nervous systems as well as other tissues and organs of varying levels of complexity that have to do with perception of the animal's place in the environment, food gathering, reproduction, interaction with other organisms, and detection and avoidance of hazard.

The Emergence of Animal Life

Science's passion for simplicity is such that conventional hypotheses for explaining where the Metazoa came from tend to postulate a single premetazoan ancestral type, although they differ on the preferred form. Family trees that start from single ancestral types are said to be monophyletic. People who make up monophyletic family trees are, of course, usually knowledgeable about the groups of organisms they discuss. Suppose, however, that *all* or several of the

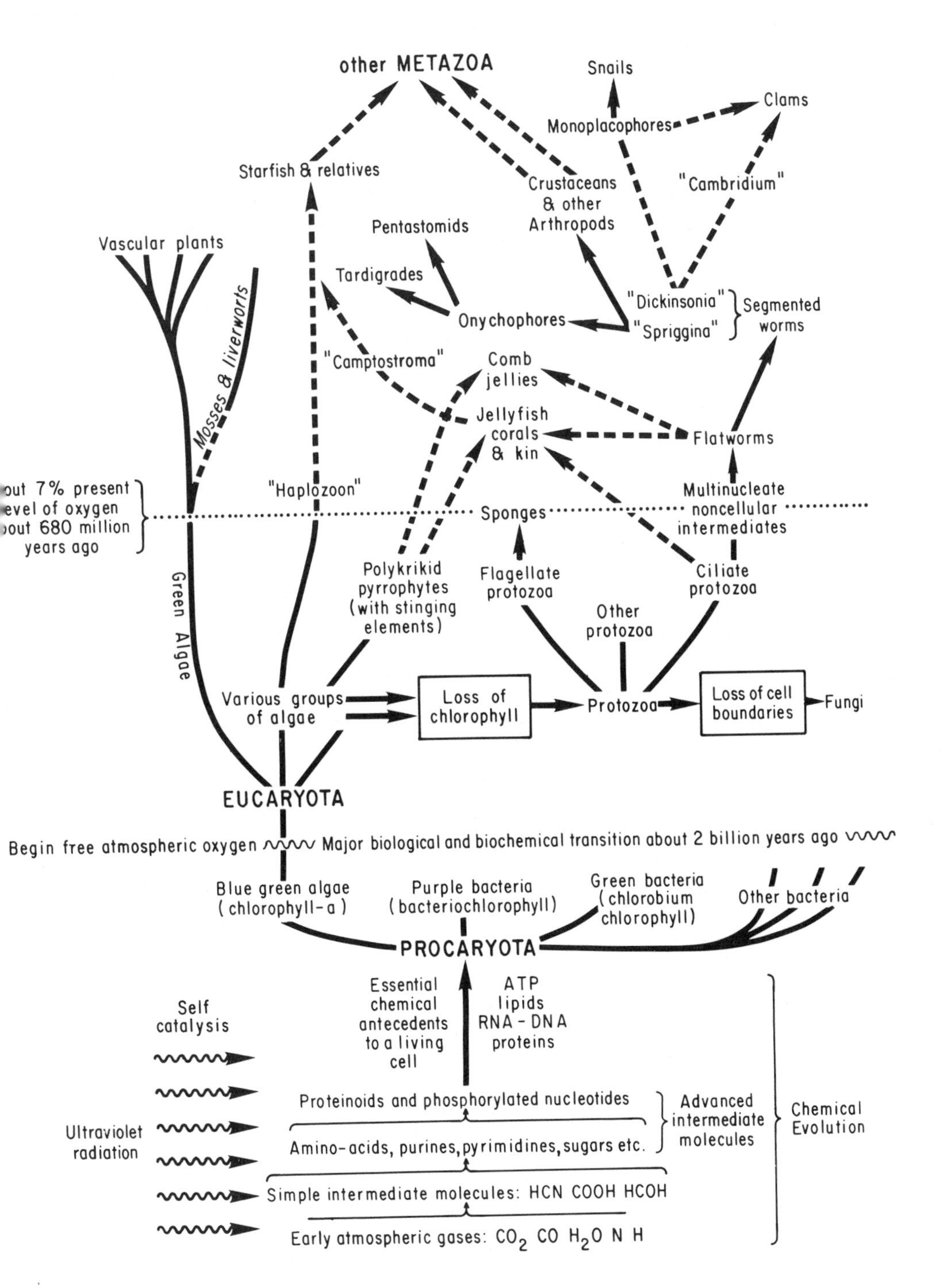

Figure 28. Possible lines of evolution leading to Metazoa, hinting at complexities involved. Unfamiliar names lack useful common equivalents; those in quotation marks are fossil or recent genera of types similar to intermediate forms visualized. See also figures 25 and 29. (Adapted from P. Cloud, in E. T. Drake, ed., *Evolution and Environment,* New Haven and London: Yale Univ. Press, 1968.)

different proposed metazoan ancestors were indeed at the base of one or another branch of the many-branched animal family tree. If that were the case, differentiated multicellular animal life of the sort bracketed under the term Metazoa, the Kingdom Animalia of table 3, would have had several different and independent premetazoan ancestors. Such an ancestry is said to be polyphyletic.

I have examined the possibility of polyphyletic origins for the animal kingdom and found it likely, as figure 28 suggests. A strong point in favor of two or more independent origins is that a broad and essentially simultaneous radiation toward animal multicellularity in different ancestral protistan groups is more consistent with the initial, geologically rapid evolution of invertebrate animal phyla observed than is simple linear evolution from a single distant ancestor. In addition, the several kinds of oxygen-carrying pigments in the blood of different metazoans would be consistent with independent origins, just as the main different types of chlorophyll imply at least three separate origins of photosynthesis.

One group of many-celled "animals" about whose origin virtually everyone agrees is the sponges—mostly sea animals, but with a few aberrant freshwater forms. Sponges, in fact, are not truly Metazoa in a strict sense because they do not show a differentiation of their cellular components into organs or even true permanent tissues. Sponges, in fact, consist entirely of one type of chameleon-like cell that has the remarkable ability to change its shape and function, depending on where it is in the body wall of the sponge—whether within it, facing outward, or facing inward toward the central cavity which opens barrel-like to the surrounding water. Cells that face outward or inward are flask-shaped cells having a single whip-like structure called a flagellum at the open end. They are identical to some of the individual cells of the group of motile protozoan microbes called Flagellata. Cells within the sponge wall have multilobate, or amoeboid, shapes. And cells change shape on removal from an amoeboid site to a flagellate site or the reverse. Sponges are thus barely advanced beyond colonial Protozoa (unpigmented microbes of the Kingdom Protista of table 3). There is little doubt that they were derived directly from flagellate protozoans, and there is no evidence that they ever gave rise to anything more complicated than themselves. Thus the sponges are a dead end—a kind of living fossil from one of the oldest known experiments in animal multicellularity.

Leaving the sponges aside, what are the most likely candidates for truly metazoan ancestry? One of the more probable and commonly suggested metazoan root-stocks is another group of motile protozoans, called "ciliates," because they move about by the synchronized beating of thousands of tiny hair-like features or "cilia" instead of by the sculling action of a flagellum. (*Paramecium* is a well-known example of a living ciliate.) Certain colonial ciliates, by first losing their cell boundaries and then reorganizing into outer and inner body walls and open interior, might be converted into multicellular flatworms as visualized by the Yugoslav zoologist Jovan Hadži. Other metazoan groups may have a direct or ultimate flatworm ancestry, as suggested in figure 28. Metazoa with stinging cells (the corals and other coelenterates), may also have arisen from flatworms, but more likely came directly from an independent ciliate ancestor, or perhaps from a group of chlorophyll-bearing protists related to the infamous red-tide microbes whose periodic population explosions make stinking, noxious, rust-colored waters where seas had been blue before. Such organisms also have stinging cells something like those of the corals and array themselves in colonial structures. Still a different origin may account for starfishes, sea urchins, their relatives the sea lilies, and related marine invertebrates with spiny skins or covers (*Echinodermata*). These may come from a jellyfish-like ancestor via something like an extinct Cambrian echinoderm of jellyfish-like construction known as *Camptostroma*, or they may have emerged directly from a joint ancestor to corals, jellyfish, and echinoderms.

These possible origins may all be likened to Churchillian riddles, inside of enigmas; wrapped in mystery. Like the origin of life, the origins of the Metazoa are effectively unknowable in detail. Yet it is not idle to think about them. As long as paleontologists were burdened with a single route of gradually unfolding evolution for metazoan origins they were also burdened with a time problem. How, they asked, could the different major types of marine invertebrates known from the oldest Phanerozoic rocks have arisen in linear sequence without a long but unrecorded ancestry reaching far back in time before the first known metazoans? The hypothesis of several independent origins is more consistent with the growing recognition that the diversity of invertebrates observed in the oldest Phanerozoic rocks did, in fact, arise over a relatively brief interval of little more than 100 million years.

This is increasingly seen as the logical outcome of the biological and physical events that prepared the way for a wave of multicellularization among various potential metazoan ancestors, beginning about 680 million years ago. When the first Metazoa arose, all of the types of habitats that could possibly be occupied by Metazoa were unoccupied. The elaboration of metazoan diversity seen in the oldest Phanerozoic rocks (the Ediacarian and Early Cambrian of the front endpapers) may thus be interpreted as a function of the many-channeled *adaptive diversification or radiation* of relatively small, early metazoan populations into a variety of previously unoccupied ecologic niches under low competitive but strong selective pressures. These included ample prospects for reproductive isolation—altogether a made-to-order situation for rapid, large-scale evolution and adaptive radiation.

Biological evolution among sexually reproducing organisms was earlier described as the product of mutation-induced genetic novelty and sexually generated variation, under the influence of environmental and other selective constraints that permit certain groups of individuals to survive and replicate themselves while others do not or do not do as well. Historical geology tells us that *biological evolution is likely to be opportunistic*: it is closely keyed to the evolution of the physical environment and to other components of the associated biological community. Biological innovations seem commonly to first flower when physical conditions are ripe for them, and where essential prior advances in related biological evolution have taken place. Thus advanced cellular microstructure and sexuality could not and did not arise until after free oxygen became a perceptible and persistent part of the atmosphere. Metazoa had as prerequisites for their origin the prior appearance of the nucleated cell, sexual reproduction, and a sufficient level of free oxygen. The ancestors of land-dwelling animals could not come ashore until land plants had prepared the way. The evolution of land vertebrates is linked with that of insects. And insect and plant evolution are interrelated in important ways. Life itself may have appeared soon after oxygen-free air and liquid water were both present in sufficient quantity and the temperature was right.

The oldest known Metazoa were thin, fragile, and soft bodied, as one would expect of organisms that obtained their oxygen by simple diffusion from an aquatic environment where oxygen content was still relatively low (figure 20). As oxygen levels continued to increase and circulatory systems evolved, body surfaces no longer needed to be in contact with an external source of oxygen. Thus mutations involving body-shielding and stiffening by the growth of tough external coverings or shells, previously lethal because such coverings inhibit the diffusion of oxygen, would then become advantageous. Solid outer shells (exoskeletons) provide seats of attachment for muscles and ligaments and thus pave the way for ascent to new levels of organization that involve articulation and movement of shells and body segments, locomotion, and aggressive feeding patterns (they may also serve an excretory function). This step in evolution marks the boundary between the first two divisions of Phanerozoic and Paleozoic history— the Ediacarian, characterized by a soft-bodied invertebrate fauna, and the succeeding Cambrian, in which shelly invertebrates first appear on a large scale (see front endpapers).

Backboned Animals Appear

Metazoa that lack rigid *internal* structures of bone or cartilage are called *invertebrates*, referring to their lack of backbones. Some invertebrates eventually grew quite large and fierce looking, with external armor of hard or leathery shell (like the Silurian arthropods called "eurypterids"). Others—the agile octopus and squid, and perhaps their shelly Paleozoic ancestors—developed eyes as good as ours and well-developed nervous systems, thereby becoming able to see and react to their surroundings almost or quite as efficiently as many vertebrate competitors of large size and aggressive disposition. Yet the internal bony or cartilaginous skeleton of the *vertebrates* allows for a degree of flexibility, of strength with lightness, and of levering and sheltering for the muscles and nervous system generally which surpasses that of the invertebrates and is approached among them only by the arthropods. Such attributes favored the evolution of more highly organized, larger, stronger, more agile, more versatile, and more cunning life forms.

As the emergence of such forms initiated the line that leads to man, interest focuses on them. It would be tiresome to most readers, given the orientation of this book, to pursue the details of invertebrate evolution, briefly suggested (without reference to a time framework) in figure 29. Suffice it to say that all of the major phyla and many of the classes of invertebrates likely to be preserved as fossils were already extant by the end of the great diversification of invertebrate types that occupied the first couple of hundreds of millions of years of Phanerozoic time—many of them probably at least in part the products of soft-bodied evolution not preserved in the fossil record. Thereafter, and up to the present, the record of mainly shelly, bottom-dwelling forms preserved, with all its fascinating and important variation, suggests in general a relative stability of community structure and broad biological design among the marine invertebrate faunas. That generalization is not invalidated by a number of episodes of extinction and a great diversity of newly evolving species, genera, and families, replacing types that for one reason or another became extinct.

Shifts in the community structure of marine invertebrates during most of the past 450 million years are more in composition than in grand design. They tend to be characterized mainly by episodes of ecological substitution, with some episodes of adaptive radiation among novel but subordinate groups of animals. This appears to be related, among other things, to the reconstitution of communities as a result of spillovers and retreats of the sea across the continental framework brought about by tectonic, and to some extent by volcanic and climatologic, events. Even after the great extinctions at the end of the Paleozoic (preceding about 230 million years ago), and again following the extinctions that terminated the Mesozoic (preceding 62 million years ago), the very substantial changes observed in the composition of the marine invertebrate faunas, including shifts in the dominant community components, introduce few really major new groups of organisms and no fundamentally different

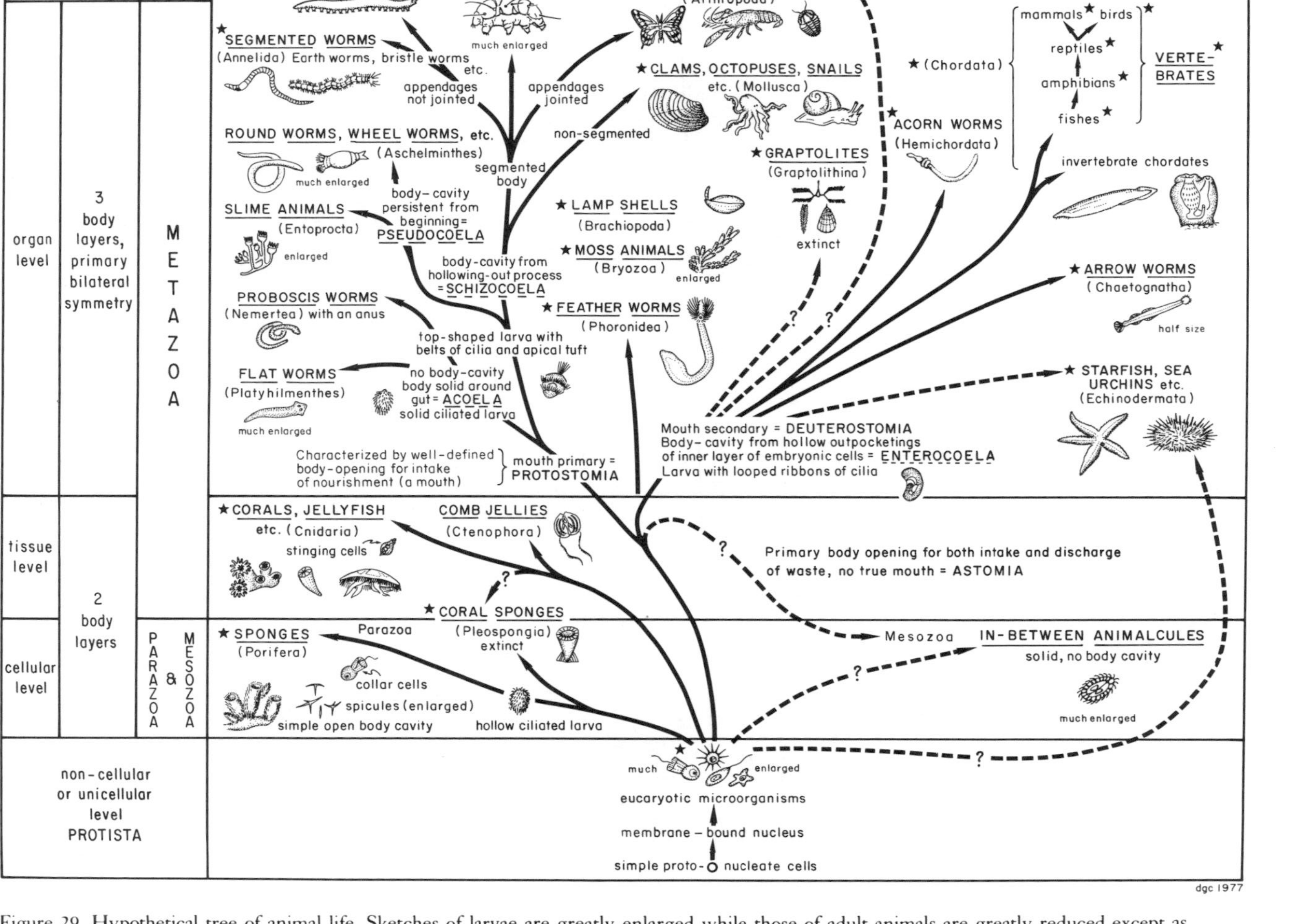

Figure 29. Hypothetical tree of animal life. Sketches of larvae are greatly enlarged while those of adult animals are greatly reduced except as indicated. Capital lettering with solid underscoring indicates common names for phyla noted in parentheses beneath; without underscoring it indicates major body types; with broken underscoring it indicates subordinate body types. Starred forms known as fossils.

community structures. Thus the heyday of the evolution of marine invertebrate community structure seems to have terminated with the establishment of the first coral reefs in Ordovician times.

Few later major changes of marine invertebrate community structure occurred. One of these involved the massive adoption of a mode of life *within* submarine sediments (and later those of lakes and streams) by several groups of post-Paleozoic clams that, rather than displacing or replacing other organisms, simply moved into previously unoccupied habitats. This involved new ways of eating and taking shelter. These clams (some 14 of 18 post-Paleozoic superfamilies according to Steven Stanley of the Johns Hopkins University) evolved a specialized mode of feeding by means of siphons that circulate nutrient-bearing waters through the gill chambers, allowing them to sit safely within the enveloping sediments while moving their food supply past their food-collecting surfaces. A different and later major change in the marine invertebrate community structure followed the relatively recent return of land plants to the littoral zone of the sea, giving rise to the first mangrove and sea-grass communities and their numerous associated invertebrates.

As for the vertebrates, that is quite another story. They, the land plants, and the insects have provided most of the notable evolutionary action over the past 400 million years or more (although the insects are poorly recorded). Indeed the evolution of these three groups seems to have been pretty closely integrated in the sequence of simultaneously evolving terrestrial (as opposed to marine) community structures.

Following the origin and diversification of the early invertebrate metazoans, then, the next important step up the evolutionary ladder toward man was the evolution of the interlocking vertebrate skeleton and the associated nervous system. And it happened within 200 million years after the first primitive Metazoa known. The oldest vertebrate animal remains we know of, roughly 490 million years old, are small denticles of primitive fishes from Lower Ordovician rocks near Leningrad, USSR. The next oldest fish remains are fragments of armored types from Middle Ordovician rocks of the Rocky Mountain region, perhaps 470 to 480 million years old.

No experienced paleontologist, to be sure, thinks that these bony fishes are necessarily the very oldest of their kind or can say what their ancestors were or where they lived. But we can piece together a story from the evidence of predecessor organisms, enclosing sediments, and the comparative anatomy and embryology of related modern organisms. We know modern organisms that share features with simple vertebrates but themselves lack backbones or, like the sharks, have cartilaginous instead of bony vertebrae. Most remarkable are a number of little animals that have a solid gelatinous rod called a notochord instead of a jointed backbone of either bone or cartilage. This notochord resembles a structure found at early embryonic stages in true vertebrates but replaced in them by separate cartilaginous or bony vertebrae at a later stage of growth.

Because they are similar in their most fundamental attributes, all notochord-bearing animals, whether or not they also have bony or cartilaginous vertebrae, are lumped together into a single major group of organisms. That is the Phylum Chordata, comparable to the grouping of clams, snails, and octopuses as molluscs or of insects, crabs, spiders, and the like as arthropods. In addition to the vertebrates, living chordates include three groups of inconspicuous notochord-bearing or invertebrate organisms of equivalent biological rank. Such animals presumably included the ancestors of the vertebrates, but where are the fossils of such organisms and what were their ancestors? Annelid worms and arthropods are favorite candidates for chordate ancestors because they are elongate, display a mirror-image likeness across a central plane of symmetry (like the two sides of a face or opposite hands), and possess segmented bodies that are reminiscent of vertebrate segmentation. Difficulties arise with these and related arguments, however, and embryological studies of early stages of development, together with biochemical similarities, have focused interest on a different possible prechordate ancestor which at first glance seems unlikely—a hypothetical ancestor they are thought to have in common with the radially symmetrical echinoderms (starfish, sea urchins, and the like). This idea arises from the fact that the larval stages of certain primitive chordates are very similar to those of echinoderms, which in turn are unlike those of all other invertebrates.

Now, although starfish themselves are rare in rocks of Early Ordovician age and absent in older ones, echinoderms of a variety of shapes and forms are known amongst the oldest shelly invertebrates, and the group may have included naked forms not found as fossils or not recognized as echinoderms (for instance, the Ediacarian fossil called *Tribrachidium*). One of these, embryology suggests, could have been the ancestor not only of the more advanced echinoderms, but of the invertebrate chordates as well, and thus indirectly of the true vertebrates.

The characteristics of the sedimentary deposits in which the oldest vertebrate remains are found, and their associations within these rocks, bear on the question of the environment in which vertebrates originated. The most extensive set of early occurrences known is found in fine-grained Middle Ordovician sandstones, over an area of more than 5,000 square kilometers (roughly half the size of Yellowstone National Park) in central Colorado. These finds consist of individual plates and fragments of the bony external armor of fishes, associated with trilobites and other marine fossils. They are angular, relatively unworn, and very much larger than the rounded grains of the enclosing sandstone. At least one good cluster of scales is reported by E. C. Olson of UCLA. These fossils, therefore, more closely resemble disaggregated after-death remains than fragments that could have been transported for any great distance. As the American geologist Robert Denison has shown, they must be the remains either of saltwater fishes that lived nearby in the great marine embayment whose deposits surround them or of gas-lightened freshwater fishes that floated dead to their marine burial places, where they sank and became disarticulated. Similar fish remains of about the same age are found over a large area centered on the Black Hills in Wyoming, eastern Montana, and South Dakota.

Vertebrate paleontologist A. S. Romer of Harvard University unwaveringly supported a nonmarine origin for these and other early fish remains of similar associations. But, Denison asks, what food could have nourished these fishes in a continental environment some 70 million years before the oldest known land or freshwater plants? And, granted a food supply and the possibility that a contemporaneous land area may have existed between the central Colorado site and the Black Hills, stream mouths around such a hypothetical land area would have been 100 to 500 kilometers from most of the fish localities—a long float trip for a lot of dead fish. A marine origin, therefore, seems more probable than a nonmarine one, particularly if fish ancestry is to be found among or shared with the echinoderms, all of which are extremely sensitive to reduced salinities.

So much for the first bony but jawless fishes. In actual fact their vertebrae were probably cartilaginous, as in sharks and lungfishes today, because all we find are bits of the bony external armor. But the armor plates were real bone, they were real fishes, and there is little doubt that the other known classes of fish arose in some way from them during post-Ordovician history.

Animal Invasion of the Lands

At this point it seems advisable to interrupt the story of vertebrate evolution in order to return to the question of the colonization of the lands. I have already described how the land plants gained a foothold, diversified, and eventually prospered. But what about land animals? Although animals had flourished under aquatic conditions from the time of their first appearance, their invasion of the continental surface had to await preparation by other events.

Like the first fishes in inland waters—or like troops securing a beach—the first truly land-dwelling animals depended on an advance wave of life to prepare the way for them. But in addition to their dependence on land plants as a primary (even if indirect) source of nutrients, they also had to solve problems of water retention, respiration, and reproduction. In order to survive as a species, a new form of life must not only establish itself somewhere; it must endure there. In order to endure as a species it must reproduce and its progeny must survive and reproduce. Although reproduction in water is easily achieved by the simple discharge of eggs and sperm in close proximity, reproduction on land involves formidable problems. As in the earlier discussed case of the land plants, adaptations must evolve for bringing sperm and eggs together in a moist environment and to prevent desiccation of fertilized eggs and developing larvae. Among the invertebrates, only three important groups include forms that have become fully adapted to life on dry land: the jointed-legged arthropods (insects, spiders, mites, and a few crustaceans); snails; and the ubiquitous round worms. Earthworms (annelids) are common enough in moist soil habitats, and obscure organisms like the so-called bear animalcules or tardigrades have adapted to even more specialized terrestrial environments, but none is as widespread as roundworms, arthropods, and snails.

Despite the poor chances insects, spiders, and mites have of being preserved in the fossil record, and their consequent rarity, mites are reported from Silurian rocks that are even older than the oldest records of vascular plants, which tends to confirm the suspicion that a land flora and fauna of some sort was already present. Moreover, both mites and insects, as well as crustaceans, have long been known from the Rhynie Chert, a bog deposit of late Early Devonian times in central Scotland, while a rich variety of closely associated and magnificently preserved marine, brackish, and terrestrial species—including plants, crustaceans, arachnids, insects, and clams—has been described by Norwegian paleontologist Leif Størmer from black shales of similar age near Alken, in Germany's Mosel Valley. Land snails were already common by a bit less than 340 million years ago, in Early Carboniferous sedimentary rocks.

Both arthropods and snails were, in a sense, preadapted to life on land and needed only a food supply to succeed there. Both had nearly impermeable outer carapaces and shells for protection against drying out, and both were mobile enough to move about in search of food. Together they illustrate the marvelous flexibility of life in their adaptations for breathing air and oxygenating their blood. Land snails developed lung-like organs, which is why such snails are called pulmonates. Land-dwelling arthropods (the insects and arachnids), on the other hand, developed perforations (tracheae) that conduct oxygen through the outer covering directly to the internal tissues. They are also capable of storing oxygen for brief emergency periods. Although some aquatic fossil arthropods attained relatively large sizes (up to a meter or more), tracheal breathing in the open air limits the maximal size to which an organism can grow because the efficiency of the tracheal diffusion of oxygen and the exclusion of carbon dioxide decreases as size increases. That is probably why even the giant insects of Upper Pennsylvanian age attained their zenith in dragonflies with sixty-centimeter wingspans and cockroaches thirty centimeters long.

Two other important facts about the air-breathing insects and snails can be mentioned. Both have survived and diversified right down to the present time, and insects have been so successful that, although only about 12,000 fossil species are known, they now account for a very large proportion of all living species. Of these the beetles and the so-called social insects (ants, termites, and the more highly organized bees and wasps) represent by far the greatest number, variety, and complexity of adaptations. The fossil record implies that the social insects appeared at about the same time as or soon after the oldest widespread flowering plants, near the middle of Cretaceous time, and that there was some connection between the evolution of the flowering plants and the evolution of the insects. There were evidently also relationships between the latter and the evolution of the vertebrates, including the emergence of birds and mammals, which brings us back to our chordate cousins.

Some paragraphs back we left the early bony-plated but jawless fishes looking for descendants. By 400 million years ago, at the end of Silurian time, they had evolved into forms with one movable jaw and were poised at the verge of a burst of evolutionary adaptive radiation so dramatic that the Devonian Period (340 to 400 million years ago) is widely and aptly known as the "age of fishes". By Middle Devonian time all major groups of fishes had evolved, including true bony fishes and a formidable variety of sharks with cartilaginous vertebrae but true bony dermal plates and teeth.

The big evolutionary and ecological leap to the land surface, however, was yet to come. Late Devonian continental sediments of Greenland and eastern Canada bear witness to the beginnings of that event. Here a succession of skeletal remains has been found that includes all stages of bony organization from very amphibian-like fish to very fish-like amphibians. Far-reaching modifications are involved in the transition from fish to a four-footed animal that could clamber about on land. The cauldron of ecologically linked evolutionary change had to boil hard to produce them. In addition to solving the difficult problems of respiration, reproduction, and (although less of a problem for early bony-scaled forms) water retention, these early amphibians also had to solve the problem of locomotion. This is reflected in changes in the structure of the backbone and limbs, as well as, to a much lesser extent, in the skull. Of the major kinds of bony fishes, only the lobe-finned or crossopterigian fishes had a bone and muscle arrangement that was suitably preadapted to the development of four-footed locomotion. The lobe-finned fishes had another characteristic that favored the adoption of a semiterrestrial habitat. They lived in pools along streambeds and in wet-season ponds that became deoxygenated or vanished during dry seasons, like the billabong in which Waltzing Matilda's swagman drowns himself in the rollicking Australian song. Such factors may have led to air-gulping at the surface and the evolution of enlarged, oxygen-absorbing surfaces as auxiliary lungs with which these early fishes could extract oxygen from air instead of water during times of stress.

As the lobe-finned fishes may also have been able to drag themselves about on their lobate stumpy fins, it is not difficult to imagine them making their way from pool to pool along stream courses during periods of prolonged drought—some dying en route, some burrowing into moist sediment in the manner of modern lungfishes (figure 30) to survive or become museum specimens of the future. From the kinds of sediments preserved we can infer with confidence that the Late Devonian terrestrial environments that saw the emergence of the amphibians were similar to modern tropical and subtropical environments of marked seasonal rainfall punctuated by intervals of drought. It is in just such areas that we find the surviving lungfish of modern times (figure 30). Thus did the lobe-finned ancestors of life on the dry lands, spurred by the need for oxygen and moisture and lured (as UCLA's E. C. Olson suggests) by a growing taste for insects, crawl across the boundary from fish to amphibian.

Emergence of the Reptiles

Still, amphibians were bound to the water. They could not get far from it without drying out, and they had to return to it to lay their eggs—tiny lumps in a slimy, tapioca-like, gelatinous mass which could survive and hatch only in an aquatic environment.

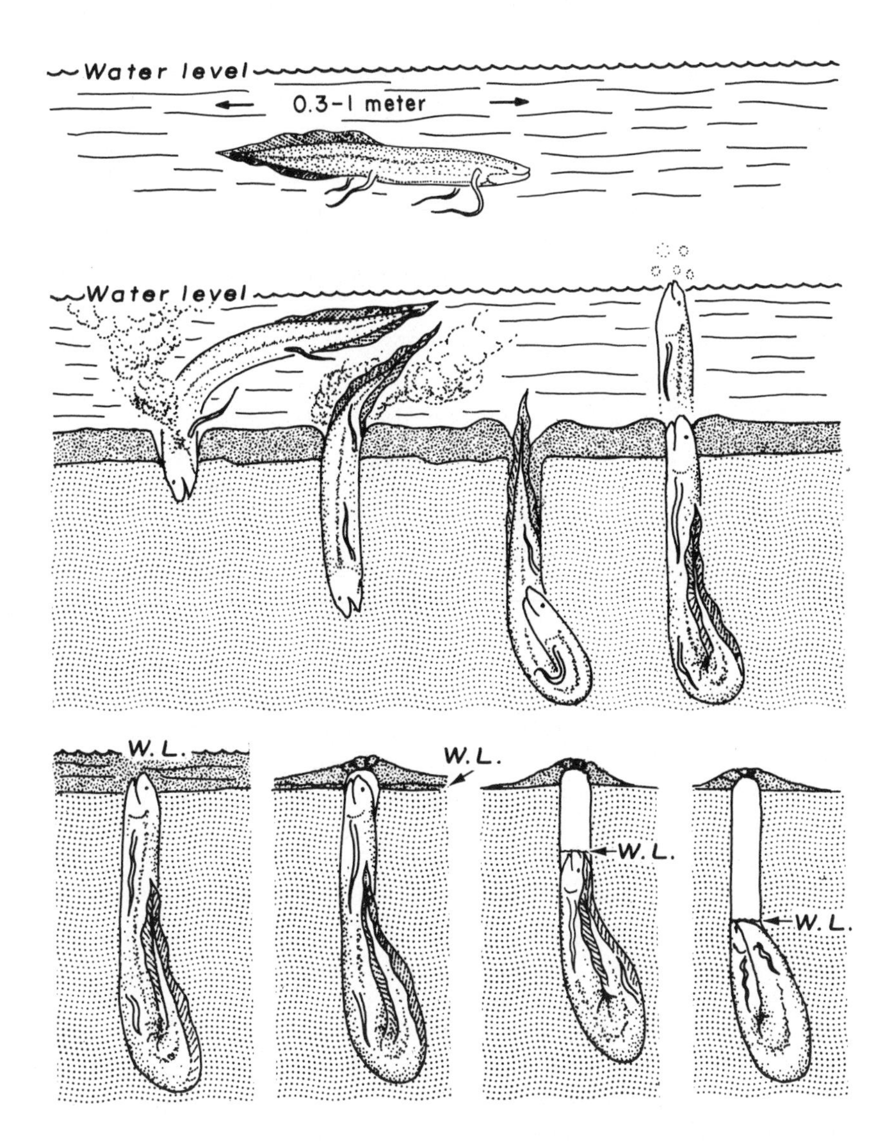

Figure 30. Modern west African lungfish in Gambia River (above), preparing a burrow as water level drops (middle and below), and sealed in a cocoon (lower right). (Adapted from A. G. Johnels and A. O. Svensson, 1955, *Archiv Zoologii,* vol. 7, no. 7, with additions by author.)

The big advance from amphibian to reptile was the evolution of a shell-covered and nutrient-filled egg, freed from dependence on an external watery milieu and protected against desiccation, within which the embryo could develop to an advanced state. Indeed the skeletal features of reptiles are not very different from those of amphibians, especially the early ones. Rather it is the shelled or *amniote egg* and the scaly, water-retentive armor that makes a reptile out of an amphibian. The oldest such egg known as a fossil is from semidesert or desert deposits about 275 million years old (Early Permian) in west Texas, but it is not certain when the first amphibian laid an amniote egg and so became the ancestral reptile. There are however, minor differences in skull and vertebral bony structure between the latest amphibians and the earliest reptiles from which it is possible to conclude that the root reptilian stock (the captorhinomorphs) developed from amphibian progenitors about 300 million years ago, during the latter part of Carboniferous history—about 25 million years before the oldest known amniote egg. The dependence of the evolving vertebrates on permanent water bodies was thereby much reduced and their access to the then rapidly evolving insect fauna as a source of food was greatly increased.

Reptiles rapidly became important land animals. As has happened to various groups of organisms throughout geologic time, following virtually every major evolutionary advance that opened up new habitats or ecologic niches to colonization, reptiles underwent a geologically rapid evolutionary radiation during the last 70 million years or so of Paleozoic history. They thus set the stage for still further diversification and territorial expansion, establishing a reptilian and near-reptilian dynasty that was to endure for another 170 million years, to the end of Mesozoic time about 62 million years ago.

When the first reptile slithered on to the dry warm lands some 300 million years ago, it literally had the place to itself, along with all those juicy bugs. Diagrams published by Olson show a very close parallelism between insect and early reptilian evolution during later Paleozoic history. The two also lived in the same places. The insects at all stages of life could have been prey for the earliest reptiles, whose skulls, jaws, dentitions, and locomotor equipment, according to Olson, all imply the ability to catch and eat insects. The land plants at this time offered little direct dietary support to reptiles but provided a good food source for insects. Along with the evolution of these early insectivorous reptiles, however, other reptiles were evolving as tropical semiaquatic herbivores, feeding on the plants in and around freshwater bodies. In time the nonaquatic terrestrial vegetation also began to support herbivorous reptiles, but the main trends of reptilian evolution—carnivorous as well as herbivorous, strictly terrestrial as well as semiaquatic—eventually came to focus on a food source based primarily on the plants of aquatic environments. A beautifully balanced community structure had evolved, and it lasted without basic modification until about 260 million years ago. Meanwhile a trend toward increasing size shows up. By the end of the Carboniferous, about 280 million years ago, some adult pelycosaurs were more than a meter long, and the trend continued.

Increasing size brought its own problems. Large volumes meant heavy weights to be supported and moved about. The lizard-like slithering of the early small reptiles gave place to a slower, more plodding gait on legs that came farther under the heavy bodies for greater bearing strength.

The habitats in which the earliest reptiles evolved were all tropical ones, more or less continually wet and humid. But with time, increasing continentality, and spreading glaciation this gave way to a seasonal variation of wet and dry intervals in which availability of plant and insect food changed with the seasons. As a result, about 260 to 250 million years ago, there was a change in the composition of terrestrial vertebrate faunas that brought into prominence groups that supposedly had originally evolved at higher elevations under conditions of seasonal change. Reptiles predominated in these new communities although amphibians remained moderately abundant.

Increased later Permian seasonality of climate, related to the joining of land areas into one or two supercontinents, may have provided the ecologic stress needed to generate selective pressure toward the emergence of adaptations for coping with climatic variation—things like internal temperature regulation ("warm-bloodedness"), fur, and feathers. Such adaptations are ordinarily associated with mammals and birds, but there is evidence in the records of historical geology to suggest that internal temperature regulation may actually have originated among animals that would have to be called reptiles on the basis of their gross skeletal morphology. That subject will be the focus of the next chapter.

For Further Reading

Barnes, R. D. 1974. *Invertebrate zoology* (3d ed.). W. B. Saunders. 870 pp.
Beerbower, J. R. 1968. *Search for the past*. Prentice-Hall. 512 pp.
McAlester, A. L. 1977. *The history of life* (2d ed.). Prentice-Hall. 167 pp.
Olson, E. C. 1971. *Vertebrate paleozoology*. Wiley-Interscience. 839 pp.
Stahl, B. J. 1974. *Vertebrate history: Problems in evolution*. McGraw-Hill Book Co. 594 pp.

15

DINOSAURS, BIRDS, AND MAMMALS

The most important feature that advanced types of animals like birds and mammals have in common is their ability to maintain their internal temperature at relatively constant levels. Because of this they are said to be warm-blooded, in contrast to reptiles, amphibians, and fishes, which are not capable of such internal temperature regulation and are accordingly called cold-blooded. Internal temperature regulation is also aided by body coverings of feathers or fur. Warm-bloodedness is the secret of mammalian and avian success in remaining active under extremes of heat and cold where reptiles or amphibians would freeze or die of heat prostration. But the terms warm- and cold-blooded are misleading. The blood of "cold-blooded" animals may heat up to lethal temperatures in the open sun. More precise terms are needed in order to discuss the evolution of internal temperature regulation unambiguously. The ability to maintain a *constant* internal temperature is called *homeothermy*. Another term, *endothermy*, refers to internal temperature regulation that takes place without regard to whether that temperature is constant or not. Correspondingly, *ectothermy* signifies the prevalence of external factors in causing variable internal temperatures, as in the lower vertebrates and invertebrates.

The Hot-Blooded Dinosaurs?

Internal temperature control probably began to evolve in animals that were still reptilian in appearance and bony structure.

Indeed it was suggested as far back as the beginning of this century, in a passing reference by the Yale paleobotanist G. R. Wieland, that dinosaurs may have been endothermal if not homeothermal. And the idea has reappeared in the writings of several vertebrate paleontologists over the past couple of decades, with a view to the thought that the development of endothermy in premammalian stocks may have been important both for the evolution of the reptiles during times of climatic stress and for the origin of homeothermal mammals and birds.

R. T. Bakker of Johns Hopkins University has gone farther than anyone else in assembling the relevant evidence, organizing it clearly, and stating it forcefully in readily available publications—too forcefully some would say. Bakker advances three arguments in favor of the hypothesis that endothermal trends began among reptiles or reptile-like organisms with the strongly seasonal climates of later Permian history that prevailed after about 250 million years ago. These arguments involve (1) differences in bony structure; (2) latitudinal zonation of the suspected endothermal groups; and (3) estimated ratios of predator to prey, which tend to be higher where the predator is an endotherm because of its greater metabolic needs. The bone of mammals and birds, endotherms all, is actively involved in the formation of blood cells and the maintenance of balance between calcium and phosphate, which is important to the tone of muscles and nerves. This bone characteristically has an open, porous structure, rich in blood vessels and canals. It contrasts with the bone of living reptiles, which is much less active in body chemistry, and characteristically has a compact structure. Such ectothermic bone also shows growth rings in the outer layers in strongly seasonal climates where reptiles become dormant during cold or dry seasons—somewhat like the familiar annual rings of trees under similar conditions.

By studying the bone structure of fossil reptiles and their offshoots, Bakker claims, one can judge which were probably ectotherms and which endotherms. He has done this for most major groups of early land vertebrates. And he finds that bone structure, estimated ratios of predator to prey, and latitudinal range are all consistent with an endothermal nature for important groups such as the dinosaurs, the so-called mammal-like reptiles, and some others. Indeed this inquiry leads him to limit the Class Reptilia to the ectothermal descendants of the early reptiles (turtles, snakes, lizards, crocodiles, and alligators) and to propose two new class names: one for the mammals and mammal-like reptiles taken together; a second for birds, dinosaurs, and several other groups taken together. In Bakker's overall picture of the terrestrial vertebrate world, the dinosaurs are not extinct. They live on as birds!

These ideas have provoked animated discussion and some vigorous objections. Marianne Bouvier of Duke University, for instance, points out that many mammals and some birds lack porous-structured (Haversian) bone whereas many clearly ectothermal reptiles have it. Similarly Bakker's argument based on predator-prey ratios has drawn fire. C. R. Tracey of Colorado State University counters that if meat-eating dinosaurs were secondary carnivores (or cannibals) they could easily have been ectotherms and still have the observed ratios to prey. And so on. Still the view that "warm-blooded" animals were about on the land before anything that would have been called a mammal or a bird is a durable one of much appeal. And it is consistent with the now rather clear lines of ancestry from basically reptilian stocks to birds and mammals.

It is remarkable that although evidence suggestive of a trend to internal temperature regulation began back in Middle Permian history, mammals did not emerge for another 60 million years or birds for another 110 million years. The delay apparently was simply a matter of accommodating all the antecedent physiological and structural changes as well as the intermediate evolutionary probings that finally accumulated the basic foundation of genetic information needed to specify mammal on the one hand and bird on the other. It took that much time for the perhaps endothermal, mammal-like reptiles and dinosaurs to leave their "coldblooded" reptilian ancestry behind and play out their part on the stage of life.

During the Mesozoic Era dinosaurs, mammal-like reptiles, and all manner of other reptilian types together deployed into almost all conceivable habitats and life-styles on and above the continents and in the surface waters of the seas. One group of dinosaurs, with bowl-like (reptilian) pelvises, included the largest herbivorous forms as well as the largest, best equipped, and presumably fiercest carnivores. A second group, with forked (or bird-like) pelvises, included mostly herbivorous species, among which some were protected against their carnivorous cousins by extravagant tank-like armor and protective horns or spines. Such were the horned ceratopsians and the bizarre stegosaurs, with rows of bony plates projecting upward from either side of their backs and mean-looking spikes on their tails. As the dinosaurs claimed the lands, their flying and swimming cousins—the giant, bat-like pterosaurs, the fish-like ichthyosaurs, and the crocodiles and crocodile-like phytosaurs—were establishing sovereignty over their own watery and atmospheric domains.

From about 230 to 62 million years ago, the span of Mesozoic time, the planet was ruled by dinosaurs and related reptiles—on land, in the sea, and in the air. Mesozoic history is the history of their rise and fall, paralleled in the sea by the perhaps indirectly related ascendance and decline of floating microscopic plants and bizarre shelly relatives of the octopus, the squid, and the clam.

Suddenly, as it were, over an interval of only a few million years, all this power and glory came to an end, lowering the curtain on Mesozoic history and foreshadowing the emergence of previously obscure descendants. Mammalian evolution was to dominate the last 62 million years of geologic history—that of the Cenozoic Era.

The only survivors of the perhaps endothermal, or even homeothermal, dinosaurian hegemony are the birds. Why did all the others disappear? What could possibly have wiped out the mighty *Tyrannosaurus rex*, king of the ruling reptiles, or the seemingly impregnable horned ceratopsians and bristling stegosaurs, while leaving feeble, feathered creatures; crocodiles; lesser reptiles like lizards, snakes, and turtles; and scurrying, fearful, pygmy mammals to inherit the earth? No one knows. But informed guesses can be made. The most prevalent suggestion is that this great dying of the "terrible reptiles" had something to do with climatic change, degree of internal temperature regulation or lack of it, collapse of food pyramids, or some combination of such factors. As is less well known, a nearly contemporaneous wave of extinction took place in the sea as well, and there is much appeal to the idea that all the dyings may be connected somehow.

One thing we do know is that the great inland seas that characterized much of the later Mesozoic (Cretaceous) scenery and climate were being drained, mountains leveled, and new mountains raised over the same historical interval during which the dinosaurs suffered their most dramatic decline. The moderating climatic effects of extensive inland seas were thus lessened and eventually removed, giving way to more continental types of climate, with greater extremes of heat and cold. There is also some paleoclimatic evidence to suggest a progressive decline of average temperatures, which was to continue for another 58 million years until the onset of the last great ice age, beginning perhaps 4 million years ago.

Large animals that were not fully homeothermal would have been at a disadvantage under prolonged cooling or seasonal extremes of heat and cold. Large reptiles—unlike their smaller cousins the snakes, turtles, and lizards—cannot burrow or crawl into holes to escape extremes of heat or cold. Cold would make them sluggish. Herbivores might have a hard time getting enough to eat and carnivores might lack the energy to hunt effectively. Once a decline started it would be hard to reverse. Wherever in the food chain it struck, and for whatever reasons, collapse of community structure above that level would follow and extinctions would take place. In the case of the dinosaurs, their generally high degree of specialization probably worked against them.

The few reptiles that pulled through the great terminal Mesozoic dying seem to have been either more generalized, more adaptive types or else types that occupied relatively unvarying habitats. Crocodiles survived by continuing to exploit the relatively stable environment of major tropical river systems, while their fellow survivors, the lizards, snakes, and turtles, made it through by virtue of being small, fertile, collectively adaptive, capable of warming up and cooling down quickly, and able to burrow or slither away from temperature extremes. The meek had inherited the earth.

As for the roughly simultaneous late Mesozoic marine invertebrate extinctions, that's a story too far from the theme of the vertebrate progression for pursuit here, except insofar as the two sets of extinctions may be related. On that point I will note only that the marine extinctions too have been related to drainage of the epicontinental seas, with related decreases in habitat variation and availability, as well as in rates and volumes of delivery of nutrients to the seas—leading to the decline of primary producers and the collapse of food webs. It seems very likely that plate tectonics had something to do not only with the widespread Late Mesozoic extinctions, but also with the preceding Mesozoic diversifications of both marine and terrestrial faunas. The initial breakup during Triassic time of the ancestral megacontinent called Pangaea (figure 13) would have ameliorated climates in two ways: first by reducing the contiguous continental areas; second, by the growth of new spreading ridges. The latter would bring about decreases in the volume of the ocean basins, the effect of which would be to flood low-lying areas of the continents with heat-storing and heat-transferring ocean waters. That would have created many new shallow epicontinental marine environments for marine invertebrates at the very time that continental environments were being moderated from their Late Permian and Early Triassic harshness by the same marine waters, thus facilitating the adaptive radiation of reptilian and related stocks. On the other hand, terminal Cretaceous adjustments within the ocean basins, such as the growth of deep-sea trenches related to plate collisions and leading to drainage of the formerly extensive epicontinental seas, could have set in motion the chain of events that ended with collapse of both terrestrial and marine food webs and the consequent extinctions. It seems almost too logical to be true, but such is the illuminating power of plate tectonics.

The Transition from Dinosaurs to Birds

Let us return now to the origin of mammals and birds, both of which have long but mostly obscure Mesozoic histories, neither becoming prominent until Cenozoic time.

As in the case of the transitions from fish to amphibian and from amphibian to reptile, a combination of luck, persistence, and incisive scholarship has established convincing evidence of gradations from reptiles through intermediate forms to birds and to mammals. A major prerequisite for both was the acquisition of the physiology of internal temperature regulation, with its high cost in metabolic energy. The nearly perfect stabilization of internal temperature that we see in birds and mammals allows them to remain active over a much wider range of temperatures than is possible for their "cold-blooded" predecessors and distant cousins, the reptiles, amphibians, and fishes. Together with this homeothermy, the fur, milk glands, relatively large brain, and other special adaptations of mammals gave them the means to become preeminent on Earth, even though initially they were much outnumbered and outweighed by other organisms. Likewise the acquisition of feathers and wings, and the extinction of the flying reptiles, endowed birds with a new dominion above Earth's solid surface.

Flight is of course not unique to birds, for it offers a number of advantages (escape from predators, unimpeded mobility) that were bound to be exploited by other groups with the potentiality to develop winged structures. Birds and insects, to be sure, have been the most successful flyers in terms of numbers and survival as species. But reptiles (pterosaurs) and mammals (bats) have also developed wings and the capability of sustained flight.

On the basis of various anatomical features of living species ornithologists have long suspected that birds were derived from reptilian ancestors. Confirmation of that insight, however, rather than being the result of a deliberate, focused, research project, came about, as scientific discoveries often do, in a most indirect way. It started more than a century ago with the growth of the publishing industry and the demand for lithographic stone to illustrate magazines, journals, books, and atlases. This stimulated a unique industry in the lovely rolling farmlands of Bavaria, near the ancient villages of Solenhofen, Eichstätt, and Pappenheim. In this area there occurs a remarkably pure, smooth, fine textured, cream-colored limestone that has ideal ink-absorbing properties. When paper is rolled across its inscribed and inked surface, it yields copy of excellent pictorial and cartographic quality. Quarries for this lithographic limestone and associated roofing slabs sprang up in the area to supply a growing world market. The rock, as it turned out, itself contained an exquisitely illustrated chapter of middle Mesozoic (Late Jurassic) history. As the quarrymen turned these pages of stone they found imprints, tracks, trails, and remains of a remarkable assemblage of superlatively well-preserved shallow-marine animals that had been stranded here in sticky calcareous muds on the tidal flats of a 150-million-year-old tropical lagoon.

Although a paleontologist named von Schlotheim reported feathered fossils among these treasures as early as 1820, their significance went unnoticed and their whereabouts is now unknown. In 1861, however, what was obviously a feather turned up, was preserved, and was described in a scientific publication. Some months later quarrymen working the same quarry near Pappenheim found a dis-

Figure 31. *Archaeopteryx.* (Photograph of compressed specimen in Berlin Museum, by courtesy of K. Fischer, Paleontological Museum, Humboldt University, East Berlin.)

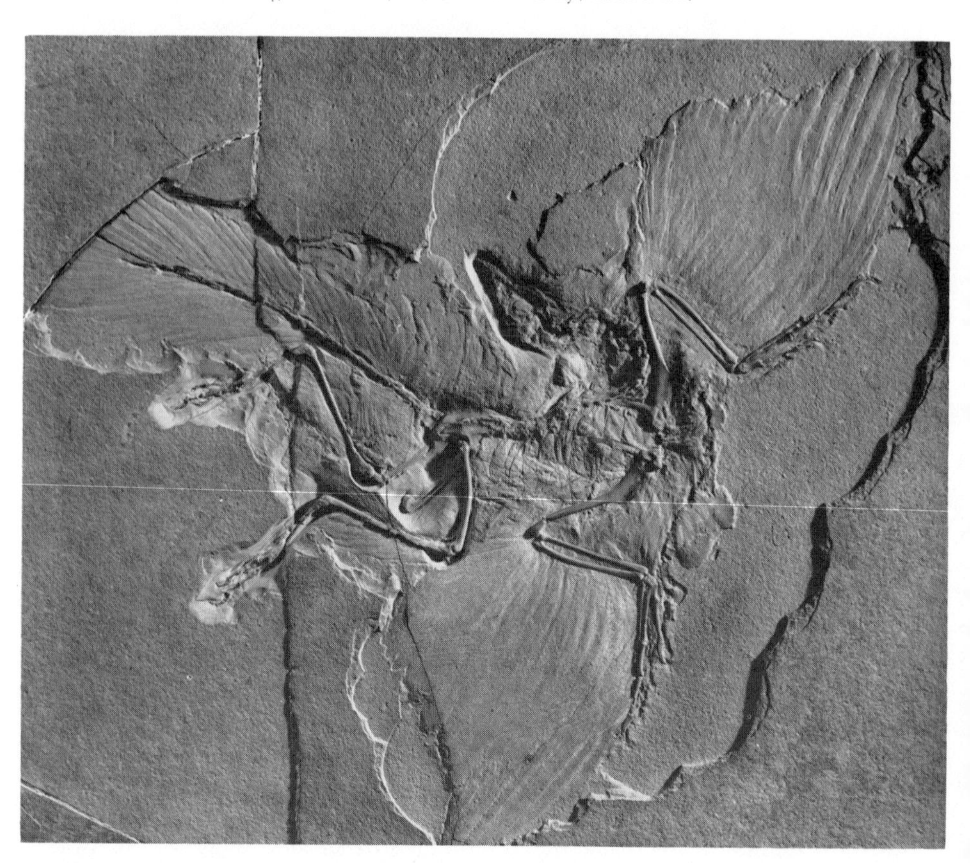

torted but almost complete specimen of what appeared to be a small reptile, except that it was associated with clear imprints of unmistakable feathers. That was the animal later to be called *Archaeopteryx*, interpreted by some to be reptile because of its skeletal structure and teeth and by others to be a bird because of its feathers. It was taken to a local physician and collector of fossils, F. R. Häberlein, who sold it to the British Museum. A second, even better specimen, from another quarry sixteen kilometers from Pappenheim, was obtained in 1877 by Häberlein's son, who deposited it in the Natural History Museum (now the Humboldt Museum for Natural Science) in what is now East Berlin (figure 31). Three subsequently discovered but fragmentary specimens of *Archaeopteryx* are preserved in museums in Solenhofen itself, in Maxberg, and in Haarlem in the Netherlands.

These five skeletal specimens and the one isolated feather comprise the entire extant collection so far of what is generally considered to be the world's oldest bird. The history of its interpretation is worth recounting, particularly in view of the fact that its unqualified classification as a true bird is one of the central arguments advanced by leading dialecticians of fundamentalist Old Testament creationism against the occurrence of intermediate forms, and thus against evolution.

Debate among paleontologists of the late nineteenth century centered on the question of the proper systematic assignment of these fossils. Were they feathered reptiles or true birds? Coming as it did only a few years after the publication of Darwin's *Origin of Species*, the discussion took on strong religious overtones. One prominent fundamentalist paleontologist of the time, finding it impossible to imagine a "reptilian bird," declared the fossils to be those of a "feathered reptile," adding that such a conclusion was necessary "to ward off Darwinian misinterpretations of our new Saurian." Indeed, recent restudies of the problem by paleontologist John Ostrom of Yale University have turned up the interesting fact that the Haarlem and Eichstätt specimens had long lain unrecognized in their museum cases because they were believed to be the remains of small reptiles.

Ostrom studied the details of the vertebrae, pelvises, skulls, shoulder arches, and limbs of all five skeletal remains of *Archaeopteryx* in the light of a half-century of advances in the study of small Mesozoic dinosaurs with bird-like pelvises. His studies, published in several papers since 1971 and comprehensively in 1976 (see recommended readings at the end of this chapter), show that "virtually every skeletal feature of *Archaeopteryx* is known in several contemporaneous or near-contemporary coelurosaurian (theropod) dinosaurs." Stripped of its feathers and posed in the same orientation as similarly preserved small contemporary dinosaurs, the beautifully preserved skeleton of *Archaeopteryx* is essentially identical bone for bone with those dinosaurs, right down to the possession of teeth and detailed osteological peculiarities (such as a notched trochanter) by both. The only obvious difference is the somewhat greater length of the forelimbs of *Archaeopteryx*. Indeed it is altogether possible that some of these related small dinosaurs themselves possessed feathers. That such body shielding may have been in the making is suggested by the presence of long, keeled, overlapping scales, which may represent a structural stage in the evolution of feathers in some older small dinosaurs—just as contemporary birds preserve reptilian-looking scaly structures on their legs. An artistic reconstruction of this likelihood is compared with reconstructed *Archaeopteryx* in figure 32.

Thus, although *Archaeopteryx* had teeth and a skeletal structure almost identical to that of certain small dinosaurs, it also had undeniable feathers, a rudimentary wishbone, and may have been capable of limited soaring flight. What was it then? Was it dinosaur, bird, or part dinosaur and part bird? In fact *Archaeopteryx is a clear intermediate between dinosaurs and birds*. But man classifies animals by putting them in pigeonholes (see table 3). His pigeonholes do not allow for intermediates. In such a system this creature had to be either a reptile, a bird, or a new class of vertebrate animal neither reptile nor bird. In the judgment of most students of the problem, feathers win, and so *Archaeopteryx* is considered to be a bird—but it was a *very dinosaur-like bird*.

Its wing and tail structure suggest that it was a feeble flier, like the modern roadrunner (or paisano) of the southwestern states. It was probably not able to do much more than launch itself a short way into the air after insects or perhaps to glide out of danger or across obstacles, and it was quite incapable of sustained flight, let alone of

Figure 32. Reconstruction of *Archaeopteryx* compared with a hypothetical and even more reptilian contemporary or ancestral birdlike dinosaur. (Adapted from drawings made by Z. Burian under the direction of Joseph Augusta as seen in a 1961 Russian reproduction.)

struggling free from the lagoonal mud where the preserved specimens landed or fell.

Although *Archaeopteryx* is clearly an intermediate form between reptiles (dinosaurs) and birds, it has long been suspected by vertebrate paleontologists that it may not have been in the direct line of evolution to modern birds. It is of interest, therefore, that recent studies by P. M. Galton of the University of Bridgeport (Connecticut) have revealed a single fully avian bone (a tibiotarsus) from British Museum collections of Late Jurassic age made in Tanzania during the late twenties. Even though only one bone is known, comparative osteology shows it to be more avian than the feathered *Archaeopteryx*. The evidence for this is the smoothly keeled rather than notched end (or trochanter) of this particular limb bone. At about the time this bone came to light another seemingly fully avian bone (a fragmentary femur with a keeled trochanter) was reported from supposedly equivalent strata in western Colorado by J. A. Jensen of Brigham Young University. As both of these bones came from strata that are probably two to three million years younger than those enclosing *Archaeopteryx* they *could* be in a direct line of evolution from *Archaeopteryx* to modern birds. More likely, however, *Archaeopteryx* (the sole known member of the subclass of birds called Archaeornithes) represents an early offshoot in bird evolution, whereas the slightly younger birds from Tanzania and Colorado are already true members of the subclass (Neornithes) to which all other known birds, both fossil and recent, belong. They are the forerunners of the host of soaring, flying, and diving birds that brighten our landscapes, cheer us with song, awaken us in the morning, and inspire us to contemplation in nearly all corners of the earth today—the sole known survivors of the dinosaurian heritage.

Evolution of the Mammals

Because we are mammals ourselves, however, it is to those hairy, homeothermal, milk-glanded cousins that our attention inevitably returns. It is a remarkable manifestation of that interest, and a tribute to the vertebrate paleontologists who have worked it out, that we know as much as we do about the origins of mammals, in view of the fact that the relevant fossils are so scarce and evidence so hard to come by. Our distant ancestors, the probably endothermal but most likely hairless or thinly furred and milkless mammal-like reptiles (center of figure 33), are peculiarly specialized forms (therapsids). Their lineage is believed to have evolved from an earlier group of reptiles (the pelycosaurs), which emerged late in Carboniferous history and diverged into a number of stocks before giving rise to the mammal-like reptiles at the end of the Paleozoic. By earliest Mesozoic time the bony structure of these still reptilian-looking vertebrates had made enough progress toward a mammalian configuration that mammalian affinities with reptiles were originally recognized strictly on the basis of comparison between the homologous bones of the mammal-like reptiles and modern reptiles. When that conjecture was made the oldest known animals of unequivocal mammalian affinity were early Cenozoic. Few Mesozoic mammals were known and their classification was not agreed upon. Evolutionary theory, however, would have predicted the older mammals and intermediate forms that have now been found at various levels in Mesozoic rocks back to 190-million-year-old (Late Triassic) redbeds in the kingdom of Lesotho in southern Africa.

Being "warm-blooded" like their feathered distant cousins the birds, mammals must, like birds, eat amply and regularly to maintain their metabolic efficiency. Therefore adaptive modifications of jaws and teeth for the piercing, cutting, crushing, and grinding of different kinds of food are leading indicators of evolutionary diversification. In fact the metabolic demands of homeothermy, and hence related feeding habits and adaptations, are so important that the teeth and jaws of mammals and the bills of birds are both leading criteria for their classification into lesser taxonomic groups. Studies of tooth and jaw structures, and comparison between true mammals and mammal-like reptiles, while deepening the conviction that all mammals had a therapsid origin, also precipitated a great debate as to whether the ancestral mammals had only one or many roots. A recently prevailing view, of which E. C. Olson was the principal architect, had it that mammals were the product of a general trend toward mammalian characteristics among different sectors of the mammal-like reptiles. Regulation of blood temperature was seen as the central physiological trend, along with reduction of the number of bony elements in the jaw, enlargement of the brain case, specialization of the teeth for different diets, and the evolution of milk glands, hair, and other mammalian characteristics.

Still more recently, however, it has come to seem likely that the mammals can be connected directly to a single line of mammal-like reptiles—a conclusion due to A. W. Crompton of Harvard University, who earlier traced the mammalian jaw step by step and bone by bone from that of the mammal-like reptiles. The key evidence comes

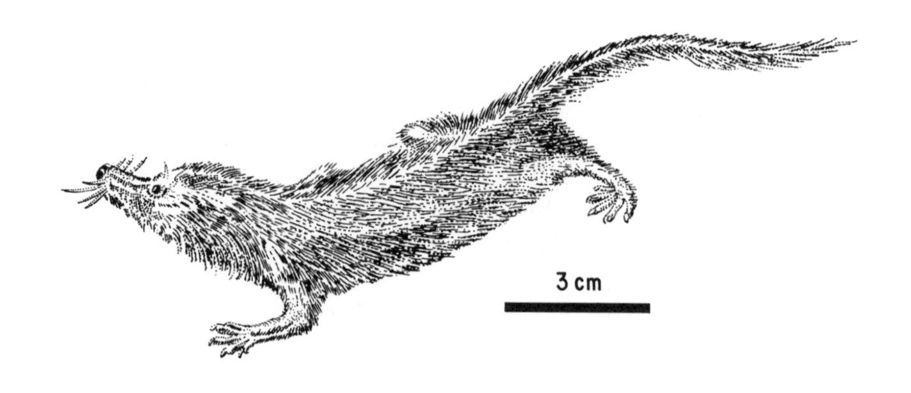

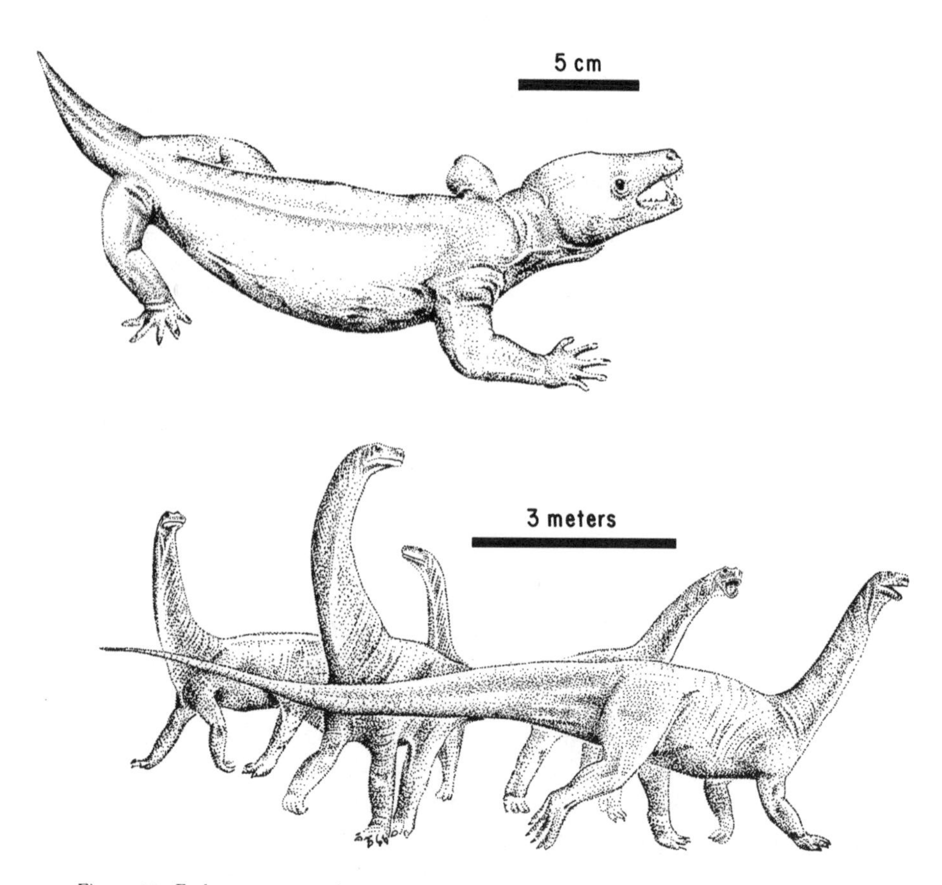

Figure 33. Early true mammal (above), compared with a mammal-like reptile (center) and dinosaurs (below) from the Triassic of southern Africa. *Erythrotherium* from Late Triassic Cave Sandstone of Lesotho, above; *Thrinaxodon* from oldest Triassic, center; and *Melanorosaurus* from older Triassic red beds of Lesotho and Hershel, below. (Adapted from drawings by R. T. Bakker in A. W. Crompton, 1968, *Optima,* vol. 18, no. 3, with permission of A. W. Crompton.)

from widely separated areas. Some, found by Crompton and his associates, involves fossils of the previously mentioned Late Triassic continental sediments of Lesotho. Supporting data, largely attributable to K. Kermack, comes from fossiliferous Late Triassic and Early Jurassic fissure-fillings in England, as well as from similar material now known to be widespread in rocks of similar age in Europe and the Yunnan province of China. The Lesotho fossils include teeth, skulls, and post-cranial bones of a small, rat-like animal that in many respects resembled the dominant reptilian remains at the same site, but which, in its most distinctive skeletal characteristics, was an undoubted mammal (top of figure 33). Analysis of these fossils and their comparison with contemporary and older mammal-like reptiles assures us that *a therapsid ancestry for mammals is not open to rational doubt.* The only substantial question is whether one or more than one line of therapsids was involved. Although the transition was so gradual that some intermediate forms might, with equal logic, be assigned either to the mammal-like reptiles or to mammals, it was in progress throughout Triassic history and was completed before the dawn of the Jurassic.

The mammals, nevertheless, remained obscure for 130 million years following their oldest known occurrences. In a sense, they appeared before their time. Reptiles were there first, and contemporary reptiles dominated the scene, occupying most of the ecologic niches available to four-footed animals. The mammals had gained a toehold but could not displace the reptilian establishment. They hung on to that toehold with little diversification until the end of the Cretaceous (of the 25 or so main groups or orders of mammals known, only 4 are known before the Late Cretaceous and only 7 then). How they survived remains mysterious, considering their small size, high metabolic demands, and the general saturation of available ecologic niches by the dinosaurs. Once the dinosaurs began to decline, however, the ecological niches they had occupied became available for occupancy by appropriately adaptive mammals (although some were never refilled). Thus there was what might be called an evolutionary scramble to fill up these niches with new mammalian adaptations, already beginning toward the end of the Cretaceous, about 70 million years ago. It is in sedimentary rocks of this age that we find the first primates or near-primates—the group from which man himself eventually arose.

That was the beginning of the great mammalian adaptive radiation. It occupied much of the last 70 and especially the last 60 million years, often taking bizarre forms. And, as if heeding the biblical edict to go forth and multiply, it succeeded in pretty thoroughly populating Earth with a spectacular variety of mammals. Among these were swimming, flying, burrowing, and climbing forms; fleet-footed herbivores and plodding ones; carnivores to prey on the herbivores; insect-eaters and scavengers; eaters of fruits and nuts; and finally omnivorous man himself—nearly all dependent in one way or another on the prior evolution of the flowering plants, particularly the grasses and leguminous plants which sustain so many mammalian food pyramids.

It is not the goal of this discussion to treat all these ramifications in detail, fascinating though they are. The engrossing stories of horse evolution from small, four-toed browsers to large, swift, one-toed grazers; of elephant and rhinoceros; of pig and camel; of mammals that fly and those that go down to the sea must all go untold here. Instead the remainder of this chapter will focus on the emergence of the primates from insect-eating mammalian ancestors and their progression to an anthropoid stage of evolution, thus laying the ground for a more detailed discussion of the origin of man in the following chapter.

Primate Evolution

So what are primates and anthropoids? Recalling table 3 the reader will be aware that biological description entails a systematic arrangement of animal, plant, and other hierarchies. In fact, this part of biology is called *systematics*. In the systematic arrangement of the animal kingdom are thirty or so major divisions called *phyla* (singular *phylum*). Vertebrate animals comprise a subphylum of the Phylum Chordata. As indicated in table 3, phyla are divided into classes, orders, families, genera, and species, in descending order of systematic significance, with a sprinkling of prefixes (for example, *sub*, *super*, *infra*). The primates are an order of the Class Mammalia, which includes pre-monkeys, monkeys, apes, and men. There is no other name for this group or for several others of interest in the following discussion. As the story moves from primates toward man, therefore, it is well to keep in mind man's place in the animal hierarchy, given in simplified form in table 5.

Table 5
Man's Place in the Animal Kingdom

- Kingdom Animalia (animals)
  - Phylum Chordata (chordates)
    - Subphylum Vertebrata (vertebrates)
      - Class Mammalia (mammals)
        - Order Primates (primates)
          - Suborder Prosimii (prosimians or pre-monkeys)
            - lemurs, lorises, tarsiers and three extinct superfamilies
          - Suborder Anthropoidea (anthropoids)
            - monkeys, apes, and men
            - Superfamily Hominoidea (hominoids)
              - apes and men
              - Family Hylobatidae (hylobatids)
                - gibbons
              - Family Pongidae (pongids)
                - orangutans, gorillas, and chimpanzees
              - Family Hominidae (hominids)
                - Genus *Ramapithecus* (and other sivapithecines)
                - Genus *Australopithecus* (australopithecines)
                - Genus *Homo* (man)
                  - Species *Homo erectus* (primitive man)
                  - Species *Homo sapiens*
                    - Subspecies *Homo sapiens neanderthalensis* (archaic man)
                    - Subspecies *Homo sapiens sapiens* (modern man)

Fossil primates are scarce. Yet they are so avidly sought after by collectors and so eagerly studied by scholars that years of exploration and study have assembled an impressive body of evidence about them, especially during the last two decades of intensive research. This evidence makes it plain that primates are not Johnnies-come-lately on the mammalian scene. They or their close relatives have been about almost from the beginning of the big, latest-Mesozoic-to-Cenozoic mammalian radiation that began about 70 million years ago while a few dinosaurs were still about. The same evidence also establishes rather clearly that the primate ancestor was a simple, small, shrew-like but highly generalized representative of the primitive insect-eating mammalian order called insectivores. In fact, insectivores are the root stock from which arose most or all of the placental mammals—those having a placenta from which the young are born live at a relatively advanced stage of development, in contrast to the marsupials, whose young are much less fully developed. Like the early reptiles, we'd be nowhere without bugs.

The oldest generally accepted primate fossils are represented by rare teeth and jaws from earliest Cenozoic (Paleocene) rocks, and

Figure 34. Lower primates. Living female *Lemur* from Madagascar Zoo (above) compared with skeletal reconstruction of 60-million-year-old pre-monkey *Plesiadapis tricuspidens* from Cernay, France. (Illustrations © I. M. Tattersall, American Museum of Natural History; reproduced with his permission.)

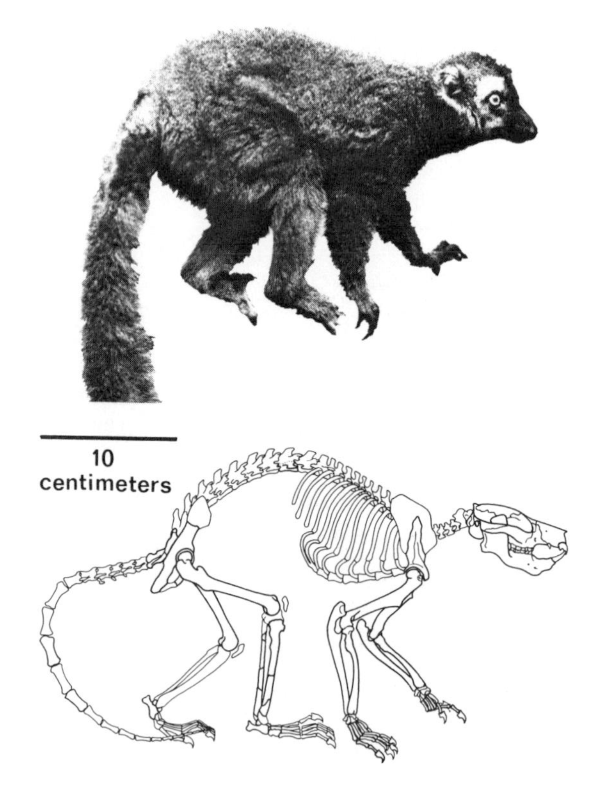

they are indeed very insectivore-like (bottom of figure 34). Insectivores, however, are ground-dwellers. When they went into the trees, perhaps to escape contemporaneous mammalian enemies of other orders, they became primates. The adoption of an arboreal habitat, coupled with the supplementation of their former diet of bugs and grubs with fruit and nuts, led to changes in skeletal structure and teeth that distinguish these little tree-dwellers from their ground-living insectivorous ancestors. Early in the development of distinctive primate features came the opposable thumb, characteristic of the grasping hand, the key to life in the trees for an organism without claws, and a hallmark of the progressive primate. Equally distinctive of primates, along with the grasping hand and the generalized teeth, is the ability to see stereoscopically (in the round) and presumably in color, as well as a relatively large brain. Such adaptations can also be seen as responses to an active arboreal life. In swinging or leaping from limb to limb a miscalculation in depth-perception could be fatal. Stereoscopic vision and enlarged brains also require distinctive skeletal alterations. Depth perception, which man takes for granted until he loses an eye, is a marvel of adaptive reorganization. It requires overlapping images. Thus the eyes must come round to the front of the head, while the muzzle is greatly reduced or disappears so that the face flattens. A large brain requires a large skull to protect it, preferably one that can continue to enlarge after birth. And, with keen stereoscopic vision and good coordination making up for muzzle-reduction, a sense of smell becomes less important, so that development of the olfactory lobes is not selected for.

By Eocene times, perhaps 50 million years ago, primate remains show that vision and the opposable thumb were already well developed, although these early pre-monkeys still retained many insectivore-like skeletal features. The fossil record suggests that the pre-monkeys were widespread and abundant early in Cenozoic history, and they have persisted until now. They are represented today by the big-eyed but small-bodied tarsiers of Sumatra, Borneo, and the Philippines; the equally stunted lorises of Africa, India, and Southeast Asia; and the long-tailed, cat-like or squirrel-like lemurs of Madagascar (top of figure 34). The lemurs, most nearly like the early

Cenozoic pre-monkeys, are, in effect, living fossils that survive on Madagascar because of a vagary of plate tectonics—the Early Cenozoic separation of Madagascar from Africa before efficient mammalian predators developed. Thus continental drift, and the changing patterns of land and sea, are looked on as isolating mechanisms that can and do play a role in speciation, evolution, and the establishment of evolutionary havens—as for the lemurs of Madagascar and the marsupial mammals of Australia.

True monkeys, both with and without prehensile tails, had already arisen from pre-monkey ancestors by late Oligocene time, perhaps 27 million years ago, after which they expanded in numbers and diversity at the expense of the pre-monkeys, whose niches they competed for, as well as moving into new ones of their own. Monkeys need no description. As noted in table 5, they belong to the primate Suborder Anthropoidea, which includes monkeys, apes, and men. Students of the anthropoids also recognize three main groups in this suborder, but they are different ones and less flattering to man. New World monkeys with prehensile tails and Old World monkeys without them are each assigned a separate superfamily. Men and apes go together in the third grouping, the Superfamily Hominoidea.

Where did the hominoids come from? Evolution in the popular mind is often thought to deal exclusively with biological events and to turn on the question: "Did man come from a monkey?" If it is any comfort, we can now say unequivocally that man did not come *directly* from a monkey. Rather he seems to have ascended from an ape-like line of ancestry that had already separated from that of monkeys more than 25 million years ago. Instead of coming from true monkeys, these earliest apes are now believed to have emerged from pre-monkeys or intermediate forms at about the same time and in the same region as the true monkeys arose and went their separate evolutionary way.

For Further Reading

Bakker, R. T. 1975. Dinosaur renaissance. *Scientific American*, vol. 232, no. 4, pp. 58–77.

Colbert, E. H. 1968. *Men and dinosaurs*. E. P. Dutton & Co. 283 pp.

Olson, E. C. 1971. *Vertebrate paleozoology*. Wiley-Interscience. 839 pp.

Ostrom, J. H. 1976. *Archaeopteryx* and the origin of birds. *Biological Journal of the Linnean Society* 8. 91–182.

Stahl, B. J. 1974. *Vertebrate history: Problems in evolution*. McGraw-Hill Book Co. 594 pp.

IV
MAN

16

THE ASCENT OF MAN

Mankind everywhere is interested in its roots. The question takes many forms. Where did I come from? Who were my ancestors? Or, more deeply, under what circumstances, in what sequence, and when and where did our earliest forebears emerge from their primate ancestry, diversify, disperse themselves over the earth, and eventually evolve to become the species we know today as *Homo sapiens?*

Never has the search for answers to those questions been more active than is the case today. It is hard to keep up with the information explosion, to assess the differences in interpretation of the data among qualified investigators, and to cope with the emotional biases that affect perceptions of human origins. The questions involved, however, are too interesting and too important to be passed over without discussion. I will attempt, therefore, to heed the example of a legendary college president who attributed his success not so much to his knowledge as to the skill with which he was able to negotiate his areas of ignorance. At the same time I will try to state what seems to me to be the main probabilities at this time without losing the reader in a quagmire of qualifications.

A great part of the difficulty in keeping up with events in the study of hominid evolution arises from the very development that makes them so exciting. That is the current surge of interest and of well-financed and highly competitive interdisciplinary exploration and research that has followed the widely acclaimed discoveries in east Africa by the Leakeys (Louis, Mary, and Richard) and which has captured the attention and interest of the world. Despite the relative scarcity of primate fossils, this intensive, multipronged effort is uncovering new data so fast that a recent scholarly review of a published collection of papers presented at an anthropological congress three years earlier comments that "some of what was said then now sounds ridiculous." So fast do hypotheses shift when new data keep turning up. Add to this the recent abrupt shifts in the supposed ages of the basal Pliocene and Pleistocene and uncertainty grows. I try in the paragraphs below to sort out the solid from the shifting ground, but the reader should realize that important aspects of what I say could be changed by new findings—indeed may have been changed between my writing and his or her reading.

So to the ascent of man. But first a paragraph about that insistent bugaboo, terminology—for the sake of readers who might otherwise be confused in comparing what I say here with what they may have read elsewhere. The genus *Homo* (table 5), in sparsest terms, designates an anthropoid mammal that stands and walks upright on its hind limbs and has or had a cranial capacity greater than about 800 cubic centimeters. I will use the word *man* to refer to *Homo* of any species or subspecies and of either sex. The term *hominid* refers to any member of the family Hominidae, distinguished physically from apes by their more erect stance, shorter arms, reduced canine teeth and brow ridges, and other details of teeth, skull-shape, and postcranial skeleton. The brain volume of early hominids was comparable to that of the gorilla (500 cc) or the chimpanzee (300–400 cc). Hominid contrasts with *hominoid*, the inclusive term for monkeys, apes, and men. The most advanced representatives of the genus *Homo*, *H. sapiens*, have a general range of brain capacity of 1,200 to 1,400 cubic centimeters (going as high as 1,600 cc in large males). Keep in mind, however, that *H. sapiens* in the broadest sense also includes Neanderthal man and his kin. Thus some anthropologists distinguish between *archaic man*, the subspecies *H. sapiens neanderthalensis*, with a low skull arch and relatively prominent brow ridges, and *modern man*, the subspecies *H. sapiens sapiens*, with a highly arched skull and faint brow ridges. Here I will use the expression *early man* in a broad sense to include Neanderthal man, the still earlier *H. erectus*, and early representatives of *H. sapiens sapiens* such as Cro-Magnon man.

In seeking to understand anthropoid and human evolution, early investigators were handicapped both by the scantiness of the relevant geologic record and by the fact that they themselves, like current students of the problem, are part of its end stages. It is more than ordinarily difficult to be objective about one's own ancestry. In addition, the general scarcity of anthropoid fossils is probably related in part to the fact that monkeys and the smaller ancestral apes lived in the trees and were generally less likely than other mammals to get caught in floods, quicksands, bogs, or other sites of accidental burial. The record of human fossils before deliberate burial is also generally unimpressive. Where early man apparently practiced cannibalism, as did Peking man (a local population of *Homo erectus)* in the caves of Choukoutien, the disarticulated skeletal remains of his victims became anthropological bonanzas; and where he lived in the same caves for generations, as did Cro-Magnon man, the prospects of preservation improve greatly. Otherwise burial, mummification, and other modes of whole-skeleton preservation had to await the evolution of a high level of conceptualization, accompanied by reflections on origin and fate.

Because of the rarity of accidental burial, the finding of anthropoid skeletal remains has always been a chancy business. It doesn't ordinarily suffice to march off to some far corner of the globe with a Ph.D. in anthropology and a crew of bearers or even a steam shovel. One needs to dig where the fossils are. That means that one needs luck, extraordinary persistence, and a good paleoecological and paleogeographical perception of where early man was likely to be. The luck element may account for so many of the significant early discoveries having been made by amateurs—a fact that only exacerbated the all too human tendency to seek immortalization by giving new generic and specific names to every fragment found. Over the last couple of decades, however, the situation with early man has changed for the better. Enough lucky finds have been made, as well as many due to insight and persistence, to suggest promising regions for further search and study. And, equally important, paleontologists, anthropologists, physiographers, stratigraphers, and geochemists (in particular geochronologists) have begun to pool their efforts and cooperate with perceptive amateurs. Thus more and better finds have been and are being made of human and ancestral fossil remains. In turn the increased number of well-dated specimens with gradational characteristics and new information about the environments in which they lived is forcing more critical evaluation, more consolidation of nomenclature, and more fruitful interpretations. Thus has the subject of the origin, diversification, and dispersal of earliest man and his subsequent evolution emerged from the shadows of confusion to become one of the most interdisciplinary, most active, and most fruitful areas of modern research. Among the generalizations that have been reinforced by this research are some conclusions long of popular interest. First, man *is* a product of ordinary and universal evolutionary forces involving natural selection. Second, he *is* the descendant of creatures that, if not called apes, would have to be called ape-like. And finally, although closely related to them, he differs in significant ways from the apes.

The Distinctive Features of Homo Sapiens

What is this different kind of animal whose origins we seek to unravel? What does he have besides a big head that sets him apart from other animals, including his anthropoid first cousins, the apes? Meaningful inquiry into the nature of man begins with the recognition of his biological origins as the only fixed point of departure that a reasonable person can verify for himself. The huge brain, which contemporary man usually regards as the most distinctive human characteristic, attained its present size only after the appearance of other skeletal features that unquestionably represent the genus *Homo*. Looking only at living man, and despite his racial variety, it is clear from the fact that all types interbreed freely and produce fertile offspring, as well as from other similarities, that all men resemble one another much more than they differ. They all share the basic qualities—anatomical, physiological, and psychological—that make us *Homo sapiens* and no other species. Here are some of the more striking human anatomical attributes, essentially as summarized by the vertebrate paleontologist G. G. Simpson in a 1968 paper (included in the book by Washburn and Jay cited at the end of this chapter):

Posture is normally upright.
Legs are longer than arms.
Toes are short, the inner toes frequently longest and not divergent.
The backbone makes a gentle S-shaped curve.
The hands are prehensile, with a large opposable thumb.
Body hair is short, sparse, and inconspicuous compared with that of most other land mammals.
The neck joint is at the center of the skull base.
The brain is uniquely large in proportion to the body, the cerebellum being especially large and complex.
The face is short and flat, descending almost vertically below the front of the brain.
The jaws are short, with a rounded dental arch instead of a U-shaped one with parallel sides as in the apes and lower anthropoids (figure 35).
The canine teeth are usually no longer than the premolars, and gaps are not normally present in front of or behind the canines.
The first and second lower premolars are similar and the structure of the teeth is distinctive of the omnivorous feeding habit.

Given the above characteristics, any paleontologist or anthropologist worth his salt could quickly and surely identify any specimen of *Homo sapiens* whose bones were sent in for preservation or who happened to walk into his office. We, however, who are pondering the question "What is man?" must sense that these anatomical distinctions, diagnostic as they are, still do not respond satisfactorily to our question. Even as regards the lower animals, we are no longer satisfied to form concepts of the true nature of the organism without some knowledge of how it responds to its habitat, including the other organisms within that habitat. In a like manner human anatomy or skeletal structure reflects essential humanness only to the extent that we can relate it to the rest of the animal as it interacts with its environment, including other humans.

Darwin first enumerated the important nonanatomical distinc-

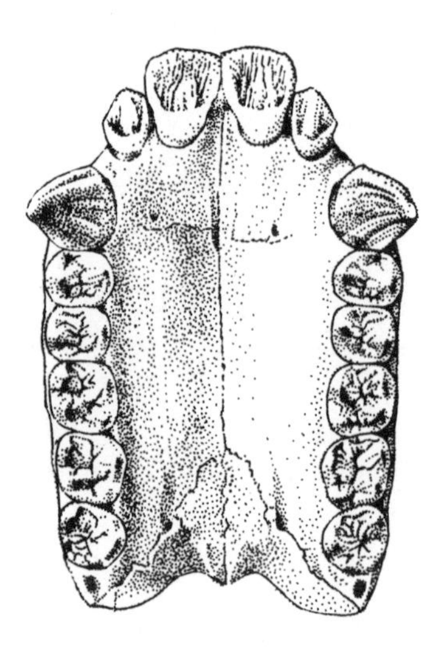

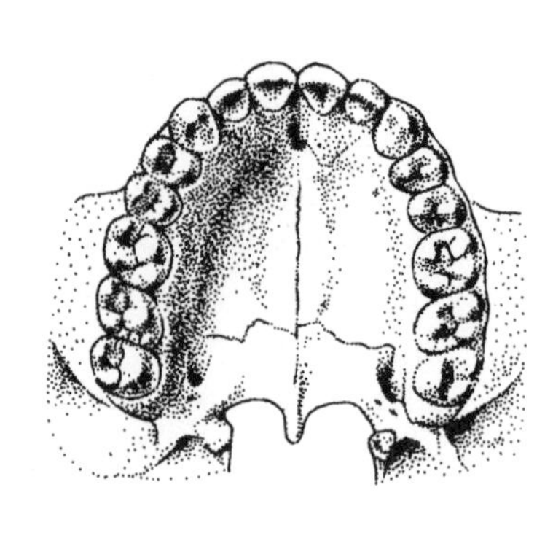

Figure 35. Jaw shapes of ape and man. (Adapted from E. Selenka, 1898.)

tions that are so uniquely characteristic of *Homo sapiens*, and Simpson has refined and condensed them. Similar ideas in different form are expanded far beyond what is feasible here in the book by Watson and Watson cited at the end of the chapter. To paraphrase (and rearrange) Simpson's and Darwin's thoughts on the main features of humanness:

1. Man consciously and systematically both makes and uses tools in great variety, often combining more than one material and element for a single purpose.
2. Man's behavior is more flexible and responsive to changing externalities, less reflexive or instinctive than that of apes or lower animals.
3. Man shares with other advanced animals the complex attributes of curiosity, attention, memory, and imitation, but he has developed them to higher levels and applies them in more intricate ways.
4. Man reasons and adapts his behavior in rational and often far-sighted ways—at least more so than other animals (the chimpanzee, with a brain capacity only one-third to one-fourth that

of *H. sapiens*, seems to come the closest to him in this respect).

5. Man is above all a cultural and social animal that consciously and consistently bands together for mutual benefit and, in consequence, has developed cultures and societies unique in complexity.
6. Man thinks abstractly and develops related vocalizations and symbolisms, chief among which are language and writing.
7. Man is self-conscious and imaginative, reflecting on his origins, previous life, future, and state after death; conceptualizing beliefs about them and shaping his actions according to these beliefs.
8. Men tend to be religious in the sense that their conceptualizations commonly involve awe, superstitution, and belief in the animistic, supernatural, or spiritual. The mind of man probably needs a certain amount of fantasy, which commonly takes the form of local mythologies or fairy-tales.
9. Men are capable of foresight, of foreseeing the consequences of their actions, and of taking actions to achieve an indirect end or avoid an unpleasant consequence; this leads to planning ahead, to caring for the very young and the old, and, in its highest form, to moral sensibility, ethics, a sense of responsibility, and altruistic behavior.

These and other nonmorphological attributes (man is perhaps the only animal that is conscious of the act of killing) are seen as the essence of humanness. Some do and others may have great selective value but they do not fossilize very well. We must still turn to skeletal remains to trace the course of human evolution backward in time. Stereoscopic vision, the grasping hand, erect posture and bipedal gait, omnivorous feeding habits, and large brain all have their skeletal consequences. As noted in the preceding chapter, stereoscopic vision requires that the muzzle be eliminated and the eyes moved to the front of a nearly flat face to produce simultaneously overlapping images. Even the hawk has not such vision. It and other raptorial birds must estimate distance by cocking the head from side to side and integrating two successive images. Fossil anthropoid and prosimian (pre-monkey) faces that have long since lost the power of sight thus proclaim their former ability to observe the world stereoscopically by their flat faces and the locations of their eye sockets.

A grasping hand, of course, means an opposable thumb, one that could articulate in opposition to the other four fingers; and this is easy enough to recognize in skeletal form. Omnivorous feeding habits lead to a distinctive tooth structure of cutting incisors, reduced canines, and grinding molars. Erect posture and a bipedal gait mean that the eyes must look out at right angles from the skeletal axis, so that if the animal tried to go on all fours it would have to bend its head upward at an angle to the backbone in order to see ahead. Such posture and gait also require distinctive rearrangements of the pelvic bones, so that the bony birth canal is narrower in man than in pre-hominids.

A small birth canal means that the fetus must be delivered while the head is still small enough to pass through it. Instead of leading to pin-headed subhumans, however, this seems to have been compensated for in man by relatively immature birth (despite the long gestation interval), of which one prominent manifestation is an open meshwork of cranial bones that allow substantial postnatal brain development before growing shut at a still relatively tender age to form a solid protective skull. Such skeletal and gestational characteristics would seem to be well adapted to facilitate evolutionary enlargement of the brain, a long early learning period, and the evolution of such uniquely human qualities as foresight, rationality, imagination, and altruism—all of which would have been called upon in the onerous child-care needed to assure the survival of relatively helpless young over the long periods before they achieved usefulness to family, tribe, or species. The offspring of parents with such attributes were simply more likely to survive and get progeny of their own, thus channeling human evolution in humanistic directions.

Our Subhuman Ancestors

Let us see how well we can trace the emergence of modern man from the earliest prosimian primates (figure 36). The aforementioned stereoscopic vision, grasping hand, and growing brain, all of which arose among the early primates as responses to arboreal life, now

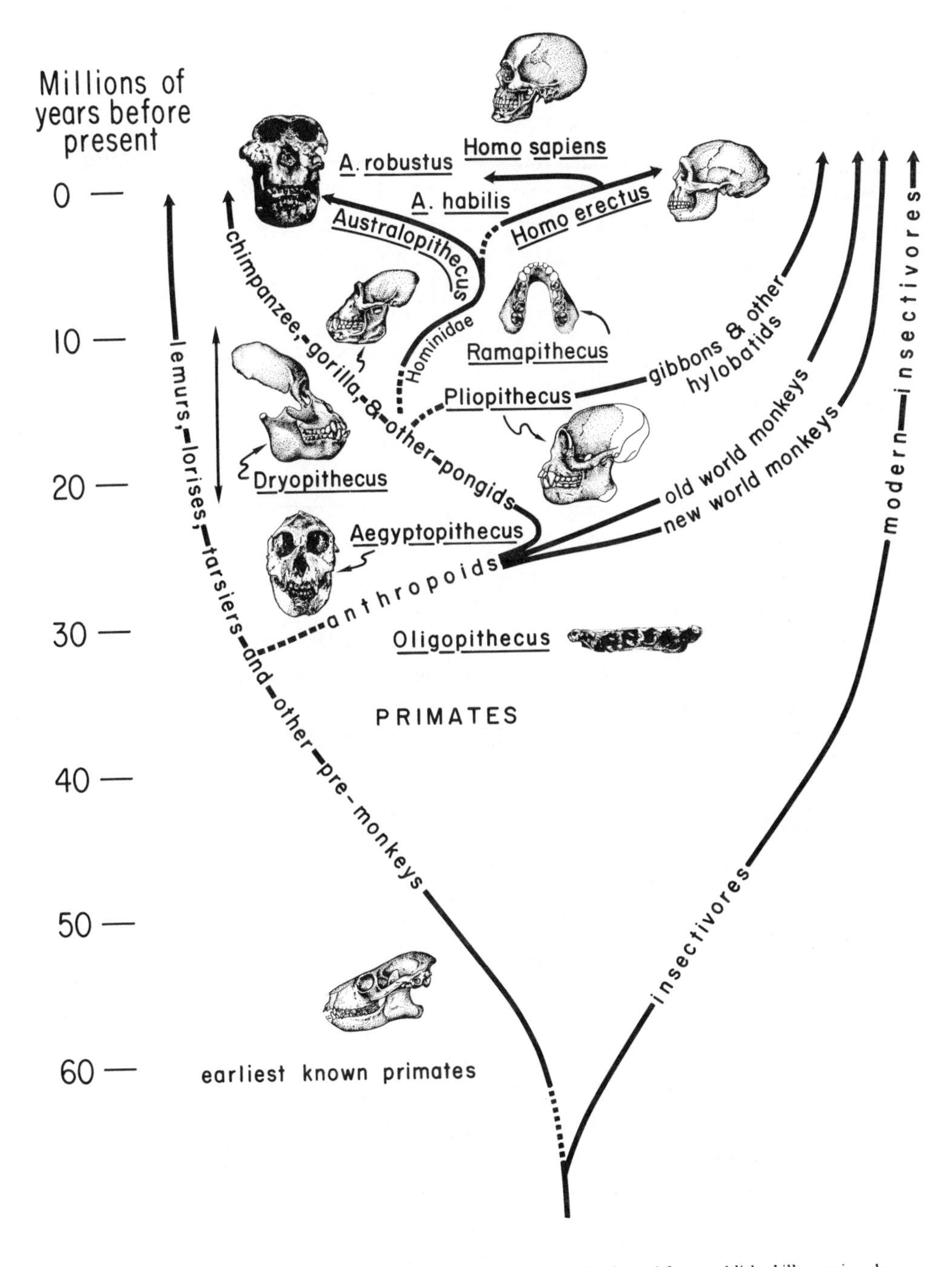

Figure 36. Simplified primate family tree. (Sketches of skulls and teeth adapted from published illustrations by L. S. B. Leakey, D. R. Pilbeam, E. L. Simons, and I. M. Tattersall.)

appear also as preadaptations for man's much later evolutionary success. These were the attributes that eventually enabled him to clutch and use weapons and tools. This was the foundation for his success, first in hunting and the manufacture of tools and weapons, and later in the cultivation of crops and the establishment of settlements.

Primate evolution, however, as in the case of the diversification of organisms previously discussed, needed selective pressures to direct it—pressures generated by such forces as climatic and geographical change, shifting relationships within food webs, population and community structure and dynamics—ecological factors all. There is ample reason to believe that such changes were more than usually active during the time when primate evolution was taking place. This was a time of accelerated continental drift (figure 13). Europe and North America finally separated. North and South America came together, finally closing off the westward drift between them of equatorial waters at the end of the Pliocene and deflecting these waters northward to warm the Atlantic coastal lands and to provide moisture to feed high-latitude glaciers. India moved north from the southern hemisphere to collide with Asia and elevate a chain of mountain barriers. Africa pushed against Europe, raising the Alps and imprisoning the Mediterranean basin, first as a hypersaline lake and then as a deep salt pan, until it was refilled through the Strait of Gibraltar in the late Pliocene. New spreading centers generated lake-filled rift valleys throughout east Africa, opened up the Red Sea and the Gulf of Aden, cut Arabia loose, and set Madagascar adrift. A broad secular cooling, incident to plate-tectonic motions, drainage of the Cretaceous epicontinental seas, and increasing continentality—and interacting with variations in intensity of solar radiation or in Earth's orbital path around the sun—culminated in the great bipolar episodes of Late Pliocene and Pleistocene glaciation beginning perhaps four or five million years ago. The processes involved resulted in waxing and waning of ice sheets at high latitudes and alternating pluvial and dry episodes at low latitudes. Small wonder that Cenozoic history witnessed such a diversity of mammalian radiations or that the relatively large-brained and agile primates climbed so far up the ladder of evolution at this time. Small wonder either that Africa should have been the site of so much of this evolution. Could Darwin have suspected the drift-related African events in 1871 when he proposed that Africa would prove to be the cradle of mankind?

Although the big step from pre-monkey to the anthropoid level of primate organization was relatively slow in coming, upward progress from there onward toward modern man was relatively rapid (figure 36). Data obtained by Duke University's E. L. Simons and his Yale associates in the now desolate but once humid tropical Fayum area of northern Egypt (figure 37) imply that it was not until about 30 million years after the first unequivocal prosimians of Paleocene time that the anthropoid stock branched off from its early primate ancestors. The oldest known genera with anthropoid affinities—the monkey-like apes *Oligopithecus* and *Propliopithecus*—both arose around 30 million years ago. From a 1966 discovery of a complete skull and upper jaw, together with a number of other jaws and teeth, it further appears that only a few million years later, also in the Fayum area, a connecting link to the middle Cenozoic apes of eastern Africa had appeared in the form of a probably arboreal ape called *Aegyptopithecus*. At Fayum, in addition, are Old-World monkeys and other mammals, suggesting a well-organized rain-forest and riverbank community and a roughly contemporaneous origin of apes and true monkeys, presumably from pre-monkey ancestors during Oligocene time.

By early Miocene time (or earlier), a generalized early ape called *Dryopithecus* had appeared in large numbers in what is now Kenya and Uganda. During the rest of the Miocene, from 23 or more to 6 or 7 million years ago, this early ape diversified into several species, while leaving hundreds of teeth and jaw parts scattered over northeastern Africa, southern Europe, and Asia. It radiated into a number of different habitats, including dense tropical rain forest on volcanic slopes, much after the manner of the modern mountain gorilla.

Dryopithecus displays a number of features such as one might expect to find in a common ancestor for apes and men, although it was clearly ape-like and not a hominid. Comparative anatomy of fragmentary remains indicates an animal that ranged in size from a small gibbon to a gorilla, had a small, monkey-like brain, but was advanced in jaw and face structure in the direction of the higher apes and man. The estimated large size of certain individuals may mean

Figure 37. Centers of exploration for early man in eastern and southern Africa.

that some forms of *Dryopithecus* had begun to descend from the trees and spend part or much of their time suberect on the ground.

Now the pace both of hominoid evolution and of fossil discovery begins to quicken and become more complex. Contemporaneous with the later dryopithecine apes lived a more advanced, probably vegetarian hominoid called *Ramapithecus*, together with species of the related genus *Sivapithecus* and the huge *Gigantopithecus*. These forms seem to have evolved out of the dryopithecines about 17 million years ago, as the mostly forested dryopithecine habitat gave way to mixed environments of scattered woods and well-watered grasslands with more open spaces. They may have lived in the fringe of the forest along river-edge grasslands. Originally discovered in the Siwalik Hills along the Pakistan-India border by the paleontologist G. E. Lewis during the Yale North India Expedition of 1932 and described by him in 1934 from fragmentary remains as "the earliest probable hominid," the significance of *Ramapithecus* was long unnoticed or discounted by others. Now, thanks in particular to the work of Simons and David Pilbeam of Yale, the probable importance for human evolution of this fossil and its associated sivapithecids is being recognized, although none can be said with confidence to be the root hominid.

Fragments of *Ramapithecus* or other sivapithecids are known over a wide area in Pakistan, India, Kenya, Greece, Hungary, Turkey, and perhaps China. The best material so far known is from Pakistan and that, unfortunately, is none too good. No cranium is yet known, and until 1976 remains available included only parts of the upper and lower jaws, part of the lower face, and most of the teeth. In the winter of 1975–76, however, another Yale expedition to Pakistan, under the direction of Pilbeam, found a complete jawbone with a full complement of teeth as well as a number of fragments of previously unknown postcranial bones.

Although these remains of *Ramapithecus* show broadly ape-like characteristics, the proportions of the teeth, the vertically oriented incisors, the absence of gaps between the teeth, and the short, deep face are hominid. Until more conclusive evidence is found, it is a matter of choice whether one thinks of *Ramapithecus* as an ape-like hominid or a man-like ape. Yet on balance this fossil seems to be advanced beyond the point of transition from ape to hominid, and so it is provisionally shown in figure 36. The oldest well-dated *Ramapithecus*, a Kenyan form, is 14 million years old, and the species was still around in latest Miocene to early Pliocene time, some 6 or 7 million years ago.

The teeth and bony structure of the sivapithecines are reminiscent of a younger and perhaps descendant group of fossils that is represented by several late Pliocene and older Pleistocene forms in southern and eastern Africa—species of the genus *Australopithecus*. The similarity of these different australopithecines, their overlap of range in east Africa, and their succession in time are consistent with their being interpreted as a branching evolutionary lineage, one branch of which may be intermediate between the sivapithecines and *Homo*, and ancestral to the latter.

In fact the study of hominid evolution in Africa really began with the discovery, in 1924, of the first *Australopithecus* from a lime works at Taung, about 650 kilometers southwest of Johannesburg. The rock containing this specimen, a child's skull, came into the possession of Raymond Dart, a young Australian anatomist then working in Johannesburg. It took him six weeks to remove the skull from the rock, thereby initiating a set of controversies about hominid evolution that still persists. This skull, which Dart called *Australopithecus africanus* (the southern ape from Africa), displays a curious mixture of ape-like and man-like characteristics. Its brow ridges were subdued. Its posture, suggested by the indicated orientation of the skull on the spine, was probably erect. And its teeth and jaws were manlike. Yet a cast of its cranium showed that it had the brain size (440 cc) of an ape. The message of Dart's discovery was that, important though the large brain of man is, its enlargement was not the sole or even the essential feature in the separation of the hominid stock from its hominoid and anthropoid ancestry. Instead stance, gait, teeth, jaws, and general configuration of the skull and brain were equally important and evidently preceded the expansion of cranial capacity in the evolutionary emergence of man from ape.

Dart, however, made little progress in convincing the anthropological world of this until his colleague Robert Broom, in 1936, found an adult *Australopithecus* skull at Sterkfontein, near Johannesburg, showing characteristics similar to the child's. Now many *Australopithecus* bones are known from Sterkfontein, as well as the bones of numerous associated mammals, and the site is generally considered to be of early Pleistocene or late Pliocene age. The average brain size of the Sterkfontein hominids is only 444 cubic centimeters, but they were erect and bipedal, and perhaps tool makers. They probably lived in rock shelters in open grasslands. Since Sterkfontein, additional discoveries have been made by Broom, Dart, and C. K. Brain in the same general area (figure 37), some with brain capacities up to 530 or more cubic centimeters.

The ages of the South African australopithecine material are uncertain because the deposits in which they are found do not lend themselves to radiometric dating, while age determinations based on associated fauna are not precise. Volcanic rocks associated with hominid-bearing deposits in east Africa are conducive to radiometric dating, however. And here, although Richard Leakey doubts ages greater than 3 million years for *Australopithecus*, Simons reports at least 5 million, and the French anthropologist Yves Coppens has claimed an age of 6 million years for a fragment of a lower jaw from the Kerio River basin southwest of Lake Turkana (formerly Lake Rudolph). Such ages, if true, would certainly be considered unequivocally Pliocene. The most numerous and best specimens of *Australopithecus* now known come from a number of localities scattered along 2,000 kilometers of Africa's scenic rift valleys, from the Leakeys' classic site at Olduvai Gorge on the Serengeti Plains in Tanzania almost to the Red Sea. Four species appear to be represented, two emerging from a revised classification of the South African early hominids made by John Robinson in 1954 and two added by the Leakeys from east Africa. The oldest one, *Australopithecus africanus*, is estimated to have been about the size of a chimpanzee. Succeeding and perhaps overlapping it in time was *A. robustus*, almost as big as a small gorilla—about 1.5 meters tall and weighing 40 to 60 kilograms. The third, *A. boisei* (originally called *Zinjanthropus boisei* by the Leakeys) may be an offshoot from or a variant of *A. robustus*. Finally there is *A. habilis*, with a brain capacity of around 600 cubic centimeters or more, considered by the Leakeys to belong to the genus *Homo* and to be a tool-maker and probably a meat-eater.

Australopithecus seems to have been a ground dweller in scattered bushlands with much open space and tall grasses, favoring the banks of permanent streams and lakes such as are so prevalent in the rift valleys above the incipient spreading centers of east Africa. Its limb bones are very like those of true *Homo* and comparative osteology tells us that its posture was slightly stooped but upright. The Olduvai australopithecine population and community lived near the shore of an ancient lake, where their remains are preserved in fine-grained lake sediments associated with volcanic rocks as much as 1.8 million years old. Although these early ape-like men had weak chins and relatively prominent brow ridges, their teeth and jaws were unmistakably hominid. Two australopithecines seem to have lived side by side at this place at the time—the larger *A. boisei*, with a brain capacity as great as 530 cubic centimeters, and the smaller *A. habilis*, with a brain capacity from 600 to 650 or more cubic centimeters (according to P. B. Tobias). The tooth and jaw structure of *A. boisei* are interpreted as indicating that it was an efficient herbivore—an eater mainly of roots, seeds, and fruits. Associated tools and bones implying a more omnivorous diet, including meat, are thought to have been left by its smaller but larger brained contemporary, *A. habilis*. The tooth and jaw structure of *A. habilis* imply that it had evolved away from a strictly vegetarian and insectivorous diet. In addition to seeds, roots, and bugs it was probably eating meat: perhaps it ate carrion at first, but it must have been thinking about catching its own.

Could omnivorous *A. habilis* have been on the line of evolution toward true man? The australopithecines generally give the impression of being an ape-like branch away from the main line of hominid evolution toward man. Yet here is a creature with a brain almost or fully half again as large as the preceding *A. africanus* (with little increase in body size) and substantially larger than the largest measured brain capacity of its big contemporary, *A. boisei*. One possible resolution of the problem, suggested by Pilbeam in 1972, is to recognize two separate lineages within *Australopithecus*. One is seen as branching off to a dead end in *A. robustus* and *A. boisei*, while the other evolves from *A. africanus* through *A. habilis* to *homo erectus*.

The Ascent of Man

In general, then, available hominid fossils display a succession of increasing cranial capacities and related brain sizes. Omitting *Ramapithecus* this begins at around 440 cubic centimeters with *A. africanus*, perhaps 3 million years or more ago, reaches 600 to 650 or more cubic centimeters in *A. habilis* around 1.8 million years ago, and attains a mean of about 940 cubic centimeters in *Homo erectus* at around 1 to 1.5 million years ago. Whether or not this is interpreted as the direct line of ancestry to man, the trend is of the right sort to achieve that end. Perhaps the Leakeys had a point in referring *A. habilis* to the genus *Homo*, for the cranial capacity of this little ape-man is just about halfway between that of other australopithecines and the "magic number" of 800 cubic centimeters taken as the lower threshold for true *Homo*.

What might have caused such a trend toward enlarged brain size? The suggestion has often been made that tool-making and hunting had something to do with it. The oldest known stone tools are about 2.6 million years old. But stones could have been thrown and clubs used before stones were worked. Sooner or later man started making and using weapons to assure his meat-supply and protect himself. In order to hunt successfully he needed to organize, to communicate during and before it, to plan, to conceptualize, maybe even to keep records of some kind. This would confer selective advantage on growth in brain capacity, evolution of language and symbolism, formation of social structures, sharing, foresight, and other human characteristics. The immediate predecessors of man were on their way up.

When and where did true man first appear? *Homo* in Africa has been reported from ever older deposits, as the pace of exploration and competition to establish the oldest record increases. Excitement reigned when Louis and Mary Leakey originally announced the discovery of true *Homo*, now unequivocally assigned to *Homo erectus*, in Bed Two of the Olduvai Gorge sequence, then considered to be 500,000 to 1 million years old. The oldest *reported* man or near-man at the beginning of 1978 was Mary Leakey's 1975 find of fossil jaws and teeth of eight adults and three children believed by her to represent some member of the genus *Homo*, at a site called Laetolil, about 40 kilometers south of Olduvai. The Laetolil find was dated as 3.35 to 3.75 million years before the present by G. H. Curtis of Berkeley, applying the potassium-argon method to crystals from the volcanic ash in which the fossils were embedded. This would seem to support a 1974 report by anthropologist D. C. Johanson of Case Western Reserve University, claiming an estimated age of 3 to 4 million years for remains he considered to be *Homo* from digs in the Afar area of northern Ethiopia. These reports of very ancient *Homo*, however, are both in doubt. Other anthropologists consider them to be australopithecines.

Homo erectus remains the oldest confidently identified member of the genus and a world traveler, wherever he started from. His remains (including those once called *Pithecanthropus* and *Sinanthropus*) are found in sediments and cave deposits not only in east Africa but also in Europe, Java, China, and other parts of Asia. The oldest reliably dated occurrences, however, are from Richard Leakey's digs on eastern Lake Turkana—at two levels, considered to be respectively 1 and 1.8 million years old. Recent studies by Howell even suggest that the older one may be a distinct new and older species of *Homo*. Thus it appears that Darwin and plate tectonics may have done it again. Africa has a better claim than any other area so far known to be considered as the cradle of mankind.

Extensive skeletal remains reveal that *H. erectus* stood upright like modern man and looked like him from the neck down. This primitive man, however, still had a receding chin, relatively prominent brow ridges, and a brain capacity averaging only 940 cubic centimeters (although measurements range from 900 to 1,100 cc and are reported as high as 1,300 cc, overlapping with *H. sapiens*). He was also smaller than *H. sapiens*. Associated artifacts tell us that he made and used a variety of stone tools, including large flint hand axes which showed increasing sophistication of workmanship with time. Disarticulated and broken bones around his campsites reveal that he preyed on animals as small as mice and as large as elephants, including contemporary baboons. He may have eaten meat regularly and sucked the marrow from the bones of his prey in a setting very like that of the present rift valleys and the Serengeti Plains of Tanzania and Kenya.

Half a million years ago, more or less, our beetle-browed but erect ancestors stood on the verge of becoming *Homo sapiens* and at the brink of the last great interglacial interval of Pleistocene history up to the present. Remains of Neanderthal man dating from this interglacial are now known from enough places to make it clear that our species has been around for at least 100,000 and perhaps as much as 350,000 years. *Homo erectus* and *H. sapiens neanderthalensis* probably overlap one another in time, the last known *H. erectus* being from deposits about 200,000 years old. The genus *Homo* may have been literally pushed up the evolutionary ladder by the climatic stresses incident to the onset of a new glaciation, especially in the peripheral parts of his range in southern Europe, where the earliest records of Neanderthal man are found.

The subspecies to which we belong, *Homo sapiens sapiens*, differs from its immediate antecedents mainly in skull morphology. The skull is shorter and the cranium more arched than in either *H. erectus* or Neanderthal man, and there are minor differences in facial proportions and slope. These differences had arisen by about 50,000 years ago. They seem to have appeared more or less simultaneously in different parts of the world, perhaps as a joint function of similar selective pressures, local evolution, and migration. Adult males of one of these local populations, Cro-Magnon man, a relatively tall (average height 1.7 to 1.8 meters) and robust form, surpassed contemporary man in size of brain—1550 to 1650 cc as compared to an average of about 1350 cc for living members of the species.

Before concluding this discussion, I must touch on two subjects that will certainly cross the minds of many readers—early man in the Americas and the Piltdown hoax. Although early fossil primates are known from the Americas, later steps in primate evolution appear to be missing. Our continent and South America were developing as separate landmasses during most of primate history, while Africa, Europe, and Asia were connected. Man seems to have arrived here by way of a land or ice bridge between Siberia and Alaska. He could have walked across on an Arctic steppe or shorefast ice at any time during the last phase of the great Pleistocene ice ages before abrupt climatic warming and melting of continental glaciers created the shallow Bering and Chukchi seas about 12,000 years ago. Although most students of the problem judge his arrival to have been relatively recent—no more than about 12,000 to 20,000 years ago—a few stubborn dissenters (as good scientists often are) have continued to press the search for older records. In early 1975 it was announced by Rainer Berger of UCLA, a specialist in carbon-14 dating, that charcoal samples from an ancient fire pit on Santa Rosa Island, offshore from Santa Barbara, California, had no detectable carbon-14 left in them. As it takes C-14 approximately 40,000 years to decay to levels of radiation that are undetectable by conventional methods, and as the stone tools and charred pygmy-mammoth bones around the fire pit are good evidence for the work of man, it is hard to deny the conclusion that he may have been here 40,000 or more years ago. If you believe in amino acid racemization dating you can add 8,000 years to that. Indeed there are those who have long claimed a duration as great as 50,000 to 100,000 years or more for man in North American (notably Philip Orr and G. E. Carter), although on obscure grounds. It appears that the last word is yet to be said on this subject.

As to Piltdown "man." The finding of his remains in a gravel pit near Sussex, England, was announced in 1912 by the English amateur paleontologist Charles Dawson. Seemingly exhibiting the jaw of an ape and the brain capacity of a man, the "find" attracted wide attention as the then avidly sought "missing link" between man and ape. It was described as such by Sir Arthur Smith-Woodward of the British Museum, who gratefully named the creature *Eoanthropus dawsoni.* Although scientific skeptics were quick to voice their doubts, pointing out the inconsistency between skull and jaw, the will to believe prevailed, as it often does when a few bold souls question hypotheses that have large, uncritical followings. But whoever had fitted together the jaw of a modern orangutan with the Neanderthal brain case, presumably Dawson himself, had done his breaking, filing, staining, and matching so skillfully that skeptics were unable to prove that it was a hoax until 1949. New chemical and physical techniques only then becoming available made it possible for the English geologist Kenneth Oakley to show, as others had suspected, that the bones had been tampered with and stained to match. Dawson died in 1916, without uttering a word to the wise. This amusing escapade has recently been described as one of the ten greatest forgeries of all time by no less an authority on forgery than Clifford Irving himself, who generously ranks it in fourth place, ahead of his own "autobiography" of Howard Hughes.

To return from the bizarre to the germane, what can be said briefly about man's cultural progress upward from his stone-age beginnings? By any criterion man has spent the greater part of his existence as a gatherer and hunter. But somewhat after the end of the last glaciation, terminating in now temperate climes some 10,000 years ago, some discovered that they could live more securely by cultivating plants or herding stock than they could as nomadic hunters or collectors of seeds, wild fruit, nuts, berries, and grubs. New stone tools were invented, marking the beginning of the Neolithic period. This is also called the *Agricultural Revolution*, because of the dramatic growth of populations and permanent settlements resulting from the introduction of agriculture. Such developments in turn led to structured government, trade, standardization of exchange, technology, organized warfare, and eventually to science and industrial society. The machines and other tools of industrial society could only have been invented and produced by a reasoning, tool-using, manually dexterous animal. Man is the only such animal we know.

Some further reflection is in order. For all his unique qualities, achievements, and capabilities, man is as subject to the laws of nature as other animals and no more capable of changing them. Humanity has thus come to a crossroads. The limitations of Earth are pressing upon us. European (including American) influence in the world is waning and a new nationalism grows ever more insistent among developing countries. Post-industrial society waits to be invented. Its shape is shrouded in fogs of rhetoric and conflicting judgment. What will man next make of himself?

For Further Reading

Frazer, Sir James, and Gaster, T. H. 1959. *The new golden bough.* New American Library (A Mentor Book). 832 pp. (Frazer's classic in human conceptualization, abridged, edited, and annotated by T. H. Gaster.)

Pilbeam, David. 1972. *The ascent of man: an introduction to human evolution.* Macmillan Co. 207 pp.

Simons, E. L. 1972. *Primate evolution*. Macmillan Co. 322 pp.

Tuttle, R. H. (ed.). 1975. *Paleoanthropology: morphology and paleoecology*. Mouton Co. (The Hague). 453 pp.

Washburn, S. L., and Jay, P. C. 1968. *Perspectives on human evolution I*. Holt, Rinehart & Winston. 287 pp. (with papers by Simpson, Simons, Leakey, and others).

Watson, R. A., and Watson, P. J. 1969. *Man and nature*. Harcourt, Brace Jovanovich. 172 pp.

17

SIGNALS FROM SPACE

Self-consciousness and curiosity about one's relation to the before, after, and yonder are peculiarly human traits. Wherever there is surcease from the immediate business of staying alive, whether around the sheltering campfire after the day's hunt or taking ones ease after work in a modern city, our minds are wont to turn to other things. What lives in the next valley, over the mountain range, across the sea, or in some imaginary faraway place? Are there people there? Do they look like us? What language do they speak? Would they be friendly? If we were able to go there might we be able to bring back something valuable? And finally, are we alone in space?

As man came to realize that some of those specks of light out there in the night sky are Earth's sister planets and that there might be other planets beyond his ken, his soaring imagination peopled them with little green men, or troglodytes living in shady cracks, or canal-building civilizations, or what have you. Fantasy is not to be denied merely because there is no rational basis for it.

If we wish seriously to tackle the question of whether there is anybody or any living thing in space, however, we must begin by asking what would be taken as evidence for the presence or absence of life beyond Planet Earth. That is the subject which some call *exobiology*—a subject with, as yet, no known subject matter. I prefer to speak of *cosmobiology*, which includes at least Earth's biosphere as a point of departure. That sample provides the basic material from which we can attempt, if not to formulate an answer, at least to define the question: "What is life?" If we don't know what we're looking for, how can we agree on whether we have found it or not, especially when our search must be conducted by remote detection or from automated orbiters or landers? What signals from space might tell us of life out there and what kinds of sensing devices might be designed to receive and test such signals?

A start could be made on other planets of the solar system by looking for geometrical patterns that might be distinctive of the activities of organisms. For example, patterns of regular intersecting linear streaks were systematically plotted on a map of Mars by G. V. Schiaparelli in the late nineteenth century; and he called them *canali*. Later observations of Mars by Percival Lowell at his observatory in Flagstaff, Arizona, led him to believe that he was seeing the same canals and that they were too straight and too systematic to be the product of anything except the planned activity of intelligent and civilized living beings. Thus Lowell peopled the junctures of his Martian "canals" with Martian cities, inhabited by intelligent beings who used these routes to bring water from the polar icecaps to the parched equatorial plains. Even Schiaparelli, who did not regard his *canali* as evidence of life or civilization, thought that they contained water. Other competent astronomers were unable to see the alleged canals and attributed them to optical deficiencies in the telescopes used. Also, close observations made from the Viking spacecraft in 1976 and more distant images from earlier spacecraft reveal no such features. They do, however, show giant volcanoes, the impact craters of large meteorites, looping, dry channels that closely resemble those of river systems, large areas of sand (or dust) dunes, and desert landscapes of volcanic rock and windblown particles that look for all the world like parts of some terrestrial deserts except for the absence of visible life forms. Such a landscape is shown in figure 38.

Inasmuch, then, as man is no longer limited to Earth-based telescopic observations but can now actually send space-probes to other planets, how should he set out to seek useful evidence for or against the presence of life on them? By May 1981 more than a dozen space-probes will have visited no fewer than five extraterrestrial planets and several planetary satellites (beginning in December 1973). These unmanned spacecraft have visited or will visit not only Mars and

Figure 38. Two views of Martian surface, summer 1976. *Above:* view east from Viking-1 lander in early morning light of 3 August shows dunes on rocky landscape, indicating wind from upper left; large boulder at left about 1 by 3 meters. *Below:* view from Viking-2 lander shows surface strewn with vesicular, probably basaltic volcanic rocks, the one at lower right being about 25 centimeters across; apparent slope due to tilt of lander. A thin dusting of condensed water vapor was later observed in the area of the lower image during the Martian winter. (Imagery by courtesy of NASA, Jet Propulsion Laboratory.)

Venus but also Mercury, Jupiter, Saturn, and satellites of Jupiter and Saturn. What kind of information might they send back that would be useful in evaluating the prospects for life on other planets or their satellites?

Suppose they were to send back radar images and chemical data that indicated a water body with a complex set of radiating lines converging on it and fading away from it. That would suggest a watering place with living traffic to and from. Similarly the familiar geometric patterns of city streets or cultivated fields would suggest organized community life—as Lowell deduced for his imaginary canals. And, of course, pictures of little green men, or one-eyed giants, or cows, or even trees, might satisfy us of the presence of life without demanding more sophisticated evidence. No such things are likely to be seen, however, because the known physical conditions on other planets of the solar system are not conducive to the evolution and support of an oxidative type of metabolism such as appears to be uniquely capable of supporting the high level of complexity and activity implied by the observations suggested.

A different kind of question must be asked—at least until humans can leave the vicinity of the planet and look around for themselves: what are the common and universal characteristics of living systems and how might one detect their presence or absence by means of instruments? Lacking something obvious, what indirect evidence might be sought to test for the presence of relatively low and perhaps minute forms of life? Is there any way to cope with the problem that as one approaches the lower boundaries of life the distinction between living and nonliving blurs? How shall we separate the signal from the noise?

On Earth the lowest unquestionably living forms are bacteria and blue-green algae (leaving out some peculiar and possibly degenerate forms without cell walls). Such microorganisms are visible as individuals only at high magnifications, and then they may be motionless, nondescript objects—especially if they are fossils. Some even more minute objects, although they contain the genetic molecule DNA or the related nucleic acid RNA, are scarcely more than form-specific organic molecules or motionless crystals that must penetrate the cell wall of a living host before they can reproduce. These are the *viruses*, which cannot replicate in the absence of cellular organisms and whose position on the scale of life or chemical evolution leading to or reverting from it is not agreed upon. Apparently midway between the viruses and the bacteria is a similarly strange group of bacterium-sized parasites called *rickettsias*.

In fact, biologists and biochemists who have seriously studied the problem of the origin of life have found themselves unable to agree on where to draw a boundary between the living and the nonliving. Instead, the difficulties of definition have led to the recognition that the properties of a system that all would accept as living blend into an evolutionary continuum that spans the boundary between the prebiological and the biological. What then are the properties of the simplest aggregate that all would accept as living if found here on our own planet?

In order to respond to that question, I must touch on a few points made in earlier discussions of chemical evolution and the nature of life. To begin with, living things on Earth display a distinctive and universal chemistry. They consist overwhelmingly of hydrogen, oxygen, nitrogen, and carbon, which, except for oxygen, are not among the most common elements on Earth today, but which are the smallest that have obligatory or preferred chemical bonding charges of 1, 2, 3, and 4. Oxygen, nitrogen, and carbon, moreover, are the only elements that regularly make double and triple bonds among themselves and with other elements. They are thus able, in a variety of combinations, to make long, flexible, stable chains without those unfilled marginal charges that lead to continued peripheral bonding and the growth of large solid structures, such as a quartz crystal (silica dioxide). These elements all occur close together in the first and second periods of the periodic table of the elements (see back endpapers). And when we combine them with the 15 other elements that are regularly involved in living systems in small concentrations or trace quantities, it can be seen that all of the elements essential to life are bunched in the first four periods of the periodic table. This strongly supports the idea that a kind of chemical selection played a decisive role in prebiotic evolution leading toward the origin of life.

The elements that regularly make up living things were probably selected exactly because they possessed the properties necessary to produce the kinds of molecules from which living systems could arise and not because they were abundant or handy. Or, put differently, life probably arose because those elements were present and interacted naturally to produce a system having the properties we call "living." We may safely conclude, since the same array of elements is found throughout the universe, that the same principal elements are likely to be involved in living structures wherever such structures are found. Life, wherever it may turn up, will most probably be carbon-based, not because there isn't plenty of silica around but because silica does not meet the chemical bonding requirements of life.

But the basic elements of living matter, although essential to life, do not by themselves signify it. They must be organized by chemical evolution, first into intermediate products such as hydrogen cyanide, aldehydes, and formate (see table 1); then into amino acids, sugars, and nucleotide bases; and finally into the distinctive giant molecules from which living systems can be made. Making these critical molecules involves the bonding together of other large molecules. Amino acids must be assembled into proteins, including the special small catalytic ones called *enzymes*. Special kinds of five-carbon sugars and phosphate must be joined with nucleotide bases made from hydrogen cyanide to form the nucleic acids DNA and RNA. And fatty materials (lipids) capable of making strong, flexible, and permeable cell walls must be incorporated. Many but by no means all of the necessary steps have been taken in the laboratory. What remains to be done to complete the laboratory replication of life is, first, to make a few more parts from scratch and, second and most crucial, to get them all working together, preferably inside a cell wall. Experimental evidence to date, nevertheless, supports the concept of an evolutionary origin of life from nonliving beginnings—given the absence of free oxygen and the presence of a solvent fluid-medium in which the essential chemical transactions could be carried out.

There is good reason to believe that such a solvent medium could be none other than water. Water is the only solvent that is liquid over a suitable range of temperatures, the only one in which all of the seemingly essential elements dissolve readily and which expands on freezing so that its solid phase remains at the surface of liquid pools instead of sinking progressively to form a solid "rocky" (icy) mass in which the essential metabolic transactions cannot take place.

We can feel reasonably confident, therefore, that, *wherever living things exist, they are likely to be carbon-based, dependent on a supply of liquid water, and contingent on either the absence of more than minute quantities of free oxygen or the presence of suitable oxygen-mediating enzyme systems.*

A mere aggregation of organic molecules, however, even within a suitable membrane, is not of itself a living cell or organism. It must be able to replicate itself, to reproduce. In addition it must be able to vary (to mutate) and to reproduce those variations—for variation among individuals of a population is one of the distinctions between a living species and a mineral species. In order to grow, sustain life, and reproduce, living things must also be able to capture and store energy and to transmit genetic information via some kind of hereditary system. *A system that everyone would presumably accept as living can, therefore, be defined as a metabolizing system that reproduces itself, mutates, and reproduces mutations.* Since entropy triumphs in the end, it is also distinctive of life that it dies.

The macromolecules that accomplish all this display some characteristics that can be measured but which are unknown or very unusual outside of living systems, and which, if observed on another planet, would strongly suggest life.

This is well illustrated by the chemical structure of the amino acids and sugars of living systems and the effect of this structure on light when it is transmitted through them. The arrangement of the elements in amino acids and sugars displays a regular asymmetry such that individual molecules can be said to be right-handed or left-handed. It is purely a matter of convention which is called right- and which left-handed, because a different orientation would give a different handedness; but once a reference orientation is agreed upon, the handedness that follows is consistent, and that choice is based on the direction of rotation of transmitted light by the molecule. Chemically speaking, the chances of a molecule of sugar or amino acid being left- or right-handed are equal. Thus synthetic sugars and amino acids show equal numbers of left- and right-handed molecules. That is not the case with sugars and amino acids of biologic origin. Among organisms, living or dead, protein-building amino acids are invariably left-handed and sugars right-handed in the agreed upon reference orientation (some organisms also contain right-handed amino acids but they are not protein-building ones).

This consistent asymmetry has two interesting and important aspects. One is that it is strong evidence for prebiotic chemical selection. The other is that it can readily be determined by passing polarized light through the substances of interest. If the latter are proteins or sugars of organisms, the plane of polarization will be rotated—to the left for left-handed structures, to the right for right-handed ones. Thus a strong *rotation of transmitted polarized light in either direction implies the presence of living or formerly living matter.* As this is a property that could be studied remotely, it might be used in the search for life beyond the earth. In fact, four different combinations of right- and left-handed sugars and amino acids are possible in four different patterns of evolutionary origin, but light rotation of itself is the point of interest here.

Another universal property of organisms that provides a basis for life-detection instrumentation is the dependence of all life processes on enzymes to catalyze biochemical reactions. As catalytic reactions are universal among (but not limited to) living systems, *conversions implying catalysis may be regarded as suggestive of life.*

If, then, on a neighboring or distant planet or satellite, it were possible to detect water or evidence of its past or periodic presence, more than traces of carbonaceous matter, catalytic reactions associated with the latter, and similarly associated strong light rotation, one would be justified in hypothesizing that life exists or has existed in that place. Water and carbonaceous matter alone would be inconclusive, but would spur the search for better evidence. Add catalytic reactions and hope grows. Strong light-rotation would greatly heighten probabilities. But one would want to see the materials that display such properties under a microscope (in a space or Earth lab) and to analyze their chemical structure, physical properties, and catalytic reactions before removing the last question mark. For the really critical aspect of the search for life in space involves the same principle as any extension of knowledge beyond the well established. The central question in this instance is: Could observations made possibly be explained by nonvital processes, and, if not, can contamination unequivocally be ruled out?

A characteristic of life that deserves emphasis, if for no other reason than that it is so obvious it may be overlooked, is morphology. It is characteristic of living things that they have distinctive shapes and ranges of size. Even though the shapes may be simple and may repeat in morphologically simple organisms that differ from one another only biochemically (like bacteria), particular shapes and size-distribution patterns may still strongly indicate life or former life. At the most general level, evidence of distinctive cellular differentiation would be conclusive, and certain shapes and patterns not known to be produced by nonvital processes would be highly suggestive, especially if they were found in large numbers of limited size-range and had the right chemical composition.

The life-detection systems that have been installed on or are being planned for space-probes to Mars and other planets all involve some means of testing for one or more of the above "universal" characteristics of life. One thing that seems highly probable is that, if there is or ever has been any life at all on other planets, microorganisms will be among the most numerous, the most ubiquitous, and the most size-limited forms. Most designs therefore involve some system for picking up or entrapping particles of dust or soil in sterile containers where they can be chemically analyzed, tested for metabolic activity or light rotation or both, observed and measured under a microscope, or some combination of these things.

It remains to see what the results may be. Most cosmobiological enthusiasts agree that, if life beyond Earth is to be found in our solar system, it is as likely to be on Mars as anywhere. Although the diameter of Mars is only twice that of Earth's moon, its mass is ten times as great. Thus it could retain an atmosphere. Parts of Mars, like our moon, are also extensively cratered by meteorite impacts. But recent space probes have revealed a more complicated history. We now know, from observed dust storms and from the observations of the first two Viking landers (landed on 20 July and 3 September, 1976) that Mars does have a thin atmosphere containing small amounts of nitrogen. We also now know that the polar ice caps of Mars are mainly water-ice thinly covered with frozen carbon dioxide and dust. Combined with their observed waxing and waning, this seems to imply that small amounts of liquid water might occur seasonally as the ice caps undergo their periodic waning cycles.

The trouble is that Martian atmospheric pressure is so low that winds there must travel over 160 kilometers an hour just to pick up dust from the surface to produce the observed dust storms. Such a thin atmosphere also means that water would boil away about as fast as formed. Thus it is improbable that any long-lived bodies of liquid water exist on Mars today. Yet the numerous unequivocal river channels seen in the detailed surface pictures (not photographs, but computer-enhanced imagery) of Mars are clear evidence that large quantities of liquid water existed at the Martian surface when they were cut. How could that be? One suggestion is that there may be deep permanent ice (permafrost) on Mars that is occasionally melted and released as liquid surface water by eruption of the observed volcanoes or by giant meteorite impacts. But that hardly suffices to explain great fossil river systems. It seems more likely that atmospheric pressures were higher at some time in early Martian history, before that planet's weak gravitational attraction lost its grip on whatever gases made up the atmosphere beneath which the Martian rivers flowed, causing them to vaporize and disappear.

Such conditions, if they included long-lasting water bodies and endured long enough, *could* have fostered prebiotic chemical evolution of the sort suggested in chapter 11, leading to an origin of life on Mars. If so, and it's a long gamble, we might, with the right techniques, be able to find evidence of it—either in the form of fossils or as living organisms that have managed to hang on somehow.

Needless to say that possibility was not lost on the scientists responsible for planning the first two Viking missions. The Viking landers carried three especially designed pieces of equipment for life detection. One measures change in gaseous components after the introduction of a sample of Martian "soil" into an airtight container (the "gas-exchange experiment"). The second is a similar chamber containing nutrient matter labeled with carbon-14 to which a sample is introduced in the hope that Martian organisms, if present, will convert the nutrients into other products that can be identified by the C-14 in them (the "labeled-release experiment"). The third is a device that introduces a moistened Martian sample into a C-14–labeled atmosphere, allows time for reaction, incinerates it to drive off gases, and tests for evidence that any of the C-14–tagged compounds has been assimilated (the "pyrolytic-release experiment"). Tests made with the first two devices produced negative results. Results with the third are equivocal. Initial experiments of uncertain implication were repeated by Norman Horowitz and his associates at Caltech under varying conditions of moisture and temperature without resolving the question. Although positive reactions have been observed, the possibility that they are the result of unexpected nonbiological reactions cannot be excluded. Thus the possibility of microbial life of some sort on Mars can be neither affirmed nor excluded by the results of the first two Viking missions. We may have to wait until a sample is returned to Earth or to an Earth-orbiting laboratory before the question can be resolved.

Negative results from Mars would be a heavy but not necessarily fatal blow to hopes for life on other planets in our solar system. Mercury is too hot and results from the Soviet Venera probes and landers indicate that Venus probably is too. As planetologist Carl Sagan of Cornell University has noted, however, some kind of aerial life in the clouds of Venus does not seem to be completely ruled out by data now available. Yet, after Mars, the best prospects for life might be on one of the larger satellites of the outer planets—perhaps a satellite of Jupiter. For Jupiter (and its satellites) seems to be a kind of self-heating mini–solar system within our larger one, a kind of little sister to the sun and her planets.

But suppose, as is likely, that Earth proves to be the only life-bearing planet in the solar system. Does that mean that our planet is the only seat of life in the universe? The proper response, of course, is no, although that is a question to which there may never be a conclusive answer.

Support for the idea of life outside the solar system is, so far, only statistical and theoretical, but it is not negligible. It turns on something you have probably wondered about—the number of stars in the observable universe and the probability that some of them have planetary systems. It has been estimated that about 100 billion

(10^{11}) galaxies float within view of our larger telescopes and that the population of stars in an average galaxy such as our own is roughly 100 billion. That implies about 10^{22}, or ten sextillion, stars in the universe. In addition, modern studies of planetary origins suggest that, far from being a unique feature of our own solar system, planets are a normal product of stellar condensation and should be associated with a large number of main-sequence stars (see figure 6)—more than half of all the stars in the visible universe. Although we have no direct evidence of any planets outside the solar system, a few stars do seem to be associated with smaller dark companions that are apparently stellar satellites and thus planets.

So what does the statistical argument say? If only one in a million of the 10^{22} stars in the universe had associated planets there would be 10^{16} stars with planets. If only one in a million of those planets were of the right size and distance from a central sun of the right luminosity to have undergone an evolution similar to that of Earth, there would be 10 billion (10^{10}) planets similar to the earth somewhere in the universe. If life evolved on only one in a million of these Earth-like planets, there would be at least 10,000 inhabited planets. Then if, as suggested in chapter 11, the appearance of living things is a normal and highly probable event in the evolution of an Earth-like planet of sufficient age, the hypothesis of 10,000 inhabited planets must be considered a conservative estimate. Indeed, more selective evidence suggests that something like half of all stars (spectral types F, G, H, and perhaps M) are old enough and of about the right range of size and luminosity that any planets associated with them could be close enough to maintain life-supporting temperatures without being so close that their rotation was halted by planetary tides. Thus the statistical probability of there being many planets broadly similar to Earth may be considered substantial.

From this it would appear that life of some sort very probably does exist elsewhere in the universe, although the prospects for finding it on other planets within our own solar system are very small. It also seems likely that such life, wherever it may be, will be carbon- and water-based and will include light-rotating molecules. That it will be capable of oxidative metabolism is much less probable. And still less likely is the prospect that it will be intelligent.

From experiment and observation we have strong grounds for believing that the processes that could lead to prebiotic and early biological evolution are chemically probable and statistically likely. But the possibility of a repetition or parallelism of the evolutionary program and ecological staging that led from the first life to intelligent beings on Earth seems extremely small. For that reason it is not scientifically justifiable to forecast similarly or more intelligent life elsewhere until some concrete evidence may be found. "Thought experiments," to be sure, are always permissible, but their value lies precisely in their ability to distinguish between potentially fruitful avenues of inquiry and probably unproductive speculation.

The result of my own thought experiments with respect to intelligent life in space is aptly expressed by Loren Eiseley's variation in *The Immense Journey* on the words of the Nez Perce's Chief Joseph: "Of men elsewhere and beyond, there will be no more forever." Yet, if the universe is truly infinite, even that seemingly safe assumption could be invalidated. Thus the improbable search for intelligent life in space, manlike or otherwise, has a touch of inspired madness about it. It expresses in an intellectual way qualities of conceptual boldness and operational daring reminiscent of Columbus's voyage to the Americas or the Apollo missions to the moon. Granted the needs of man for a degree of fantasy, it would be hard to think of a more entertaining or even potentially more rewarding sort of fantasy. For even if no life is found, no contacts made with intelligent or superior societies, interesting and valuable things may be discovered in the course of carrying out the search. In the tradition of the great voyages of exploration throughout history, it is above all the supreme expression of human curiosity and hope, in which the thing that counts is not so much the practicality of the mission as it is the quality of the searcher.

How might one set about conducting such a search? Inasmuch as the nearest stars beyond the sun are several light years away, there is little prospect of visiting them or even reaching them with spaceprobes within a human lifetime, if ever. A space crew for such an exploration would have to be a completely self-contained heterosexual crew, for it would be en route for several or many generations.

Now that man has discovered radio, built radio telescopes, and detected radio transmissions from other stars, however, he has a new and more practical kind of space-probe. He can listen for radio signals from space which cannot be explained by normal physical radiation. And, if he detects such signals, he can analyze their structure and try to decode them for messages—assuming that civilizations comparable to or more advanced than ours have discovered radio communication and have built and are broadcasting with radio telescopes.

We now have radio telescopes that can hear signals from the limits of the visible universe and some that can pick up and send out the types of radio and radar transmissions likely to be useful in contacting conscious intelligence from as much as 100 light years away. Within that distance of Earth are perhaps 10,000 stars, many of which may have planets. It is possible that some form of life exists on a planet within this range, and it is not completely impossible that it could have evolved to a level of intelligence comparable to or greater than ours. In the latter event it might even be sending out and listening for radio signals, as we, in fact, are doing.

Although often opposed on Earth, those two uneasy giants, the United States and the Soviet Union, have been drawn together by the stars. In 1971 in Yerevan, not far from the biblical Mt. Ararat and the good vineyards of Armenia, delegates from the American and Soviet academies of science met to discuss the subject from which has arisen the acronym SETI—search for extraterrestrial intelligence (or, by some, CETI, for communication with ETI). They did have something to discuss. Eleven or twelve years earlier, in his short-lived Project Ozma I, Cornell University astronomer F. D. Drake had listened for signals from space with the 26-meter radio telescope at Green Bank, West Virginia. But that was the Model-T of SETI. More recent is Ozma II, under the direction of Patrick Palmer and Benjamin Zuckerman of the universities of Chicago and Maryland. Working at the National Radio Astronomy Observatory, they scanned 640 sun-like stars within 80 light years of us, again without encouraging evidence. Now, with the mighty 450,000-watt transmitter and the 300-meter-diameter radio dish at Arecibo, Puerto Rico, it is possible not only to listen over a wider area, but also literally to "shout" specially coded messages into space. And, even though it may take hundreds of years for such a message to reach its unknown destination, and an equal length of time for a response to be received, such listening and "shouting" is now being tried. Perhaps such an exercise in long-range planning will help to put man in a frame of mind to undertake similar planning in matters more germane to the continuation of intelligent life on Earth.

Not to be outclassed, Soviet astronomers and cosmophysicists have designed, and are now putting into effect, two overlapping ten-year listening and sending plans. The first, SETI-1, to last from 1975 to 1985, includes an array of eight ground-based stations, two space stations, and a system of ground-based antennae intended to make a preliminary survey of the whole sky as well as of near galaxies. In SETI-2, to run from 1980 to 1990, an expanded satellite system will continue to monitor the entire sky. In addition, two new space stations with semirotatable antennae, having one-square-kilometer sensing areas, will search for specific objects and analyze selected radio sources. Meanwhile dreamers in the United States talk of something they call Project Cyclops, which they perceive as involving an antenna array several kilometers in diameter and capable of listening for and sending to civilizations 100 to 1,000 light years away—fully aware of the fact that it may take 3,000 years to search the heavens comprehensively to a distance of 1,000 light years. Alternatively there is talk of carrying out a similar search from one or more radio telescopes in orbit around Earth or on the moon.

Improbable as such a search may seem to sober Earth-bound men and women, the imaginative and very bright visionaries involved in it are quick to point out that the three stars of Alpha Centauri (they appear as one to the naked eye) are only 4.3 light years away, which means that a message could be received by someone there and responded to within 8.6 years. Sirius is barely 8.7 light years away (as compared to eight light minutes for the sun). And two very sun-like stars, Tau Ceti and Epsilon Eridani, are but 11 light years away. Radio telescopy, meanwhile, is bringing us all kinds of other information about the stars and interstellar space (see, for example, table 1), and the big radio telescope at Arecibo can also be used to "see" and map the invisible surface of Venus with a limit of error of 100 meters, to survey the four larger of Jupiter's thirteen moons, and to study the subsurface structure of Mars. Finally, important astrophysical data can be obtained while listening for signals, and, as the Soviets argue, the exercise in designing and deciphering the computer languages used in these efforts toward interstellar communication can benefit the study of economics, linguistics, and other fields of technology whose practitioners are always seeking to improve their analytical capabilities. Meanwhile there have been summer workshops on SETI at NASA's Ames Research Center since 1971 and another United States–USSR conference on the subject is planned for the summer of 1978, this one to take place in the United States.

Perhaps the real meaning of SETI lies in the fact that it is an expression of man's need to reach toward "the infinite meadows of heaven," be there forget-me-nots there or no. It's a long chance, for sure, but it may be our least unlikely prospect of discovering whether there really is anybody out there beyond our own solar system. If there is, and if we hear and decode a message, the next question will be: "Should we answer?"

For Further Reading

Avener, M. M., and MacElroy, R. D. (eds.). 1976. *On the habitability of Mars.* National Aeronautics and Space Administration. 105 pp.

Pittendrigh, C. S.; Vishniac, Wolf; and Pearman, J. P. T. (eds.). 1966. *Biology and the exploration of Mars.* Natl. Acad. Sciences—Natl. Research Council, Publ. 1296. 516 pp.

Ponnamperuma, Cyril (ed.). 1972. *Exobiology.* American Elsevier Publishing Co. 485 pp.

Ponnamperuma, Cyril, and Cameron, A. G. W. (eds.). 1974. *Interstellar communication.* Houghton Mifflin Co. 226 pp.

Shklovskii, I. S., and Sagan, Carl. 1966. *Intelligent life in the universe.* Dell Publishing Co. (a Delta Book). 509 pp.

Shneour, E. A., and Ottesen, E. A. 1966. *Extraterrestrial life–an anthology and bibliography.* Natl. Acad. Sciences—Natl. Research Council, Publ. 1296A. 478 pp.

18

SPACESHIP EARTH TODAY

Our reach toward the stars has given us a new vision of Earth. We have looked through the camera's eye from space and seen the whole planet in one field of view—floating against the empty blackness beyond like a multicolored oriental lantern (figure 39). Some lantern! Some view! Mother Earth will never seem the same again. No more can thinking people take this little planet, bounteous and beautiful though it be, as an infinite theater of action and provider of resources for man, yielding new largesse to every demand without limit. Born from the wreckage of stars, compressed to a solid state by the force of its own gravity, mobilized by the heat of gravity and radioactivity, clothed in its filmy garments of air and water by the hot breath of volcanoes, shaped and mineralized by four and a half billion years of crustal evolution, warmed and peopled by the sun, this resilient but finite globe is all our species has to sustain it forever.

One of the goals of this book is to give substance to the concept, now beginning to pervade the general consciousness, that all parts of the human habitat, reaching right around the planet, are interconnected in complex and delicate ways. Those interconnections extend throughout the existing global ecosystem and all the way back to the beginning of time on Earth. We are a part of that ecosystem and a product of that history. And the little sphere on which we ride through space is equally a part of the cosmos and a product of cosmic history. Figure 40 encapsulates this view in time perspective.

Man can never escape the fact that he is a piece of the biosphere,

Figure 39. View of Earth from the moon, 390,000 kilometers away; eastern edge of lunar surface in foreground. This view of the rising earth greeted the Apollo-11 astronauts as they first came from behind the moon after entering lunar orbit. (Photograph supplied by and published with the permission of NASA.)

along with the lower primates whose ancestry he shares and with other more distantly related animals and plants. Although he stands high in it, he is not above it. The *biosphere* is a single, interwoven web of life, comprising not only the thin veneer of living things that is found everywhere at the surface of our planet, but reaching into the greatest depths of the sea and outward into the atmosphere beneath the ozone screen. It consists of about one and a half million described species, and perhaps once again or twice that number not yet described, each living in its own distinctive and limited place and way—its *ecologic niche*: the dog for the flea and the global sea for the whale. These niches, these organism-environment contracts, overlap with one another, and all are linked together into larger *ecosystems* and finally into the *global ecosystem*, comprising the entire living space of the earth and all of the life within it. Our global ecosystem is the product of 3.8 billion years or more of interacting biologic, atmospheric, hydrospheric, and lithospheric evolution. It will probably continue to evolve for another 4 or 5 billions of years into the future before our parent star leaves the main sequence on its way to becoming a red giant and who knows what thereafter.

It is characteristic of this global ecosystem that events taking place in one part of it interact with those in other parts and that these interactions are far more extensive than is apparent at a glance or over a short interval of time. There is a connection between sunspot activity and the luxuriance of the meadow, between the ragweed in the meadow and the prevalence of urban tears. There may even be a connection between the iron in the vehicles with which the urbanite flees the ragweed pollen and the sunspot activity of two billion years ago. Indeed these interconnections are so extensive that it is fair to characterize the system as indivisible. One cannot be sure beforehand how any action in any part of it, whether accidental or deliberate, whether well or ill intended, will affect some other seemingly isolated part of it—if not immediately, then at some future date. Although the past is prelude, although it contains the seeds of the future, it is difficult to know exactly what will grow from those seeds, especially where man takes a hand. All we can be sure of is that the future will be different. Nothing is so permanent as change.

As for the effects of man's actions, some have been good, others bad, many a mixture of good and bad. Control of malarial mosquitoes, disease-causing viruses, and pests of various sorts has enabled man and his domestic animals to live more comfortably, reduced infant mortality, increased crop production, and extended life expectancy for most people. The inventions of clever technologists, interacting with and contributing to a sustained flow of scientific knowledge, have eased and enlightened our lives, given us highly efficient transportation and communication systems, facilitated education, and provided an unprecedented variety of goods and services for an equally unprecedented number of people. Despite all the glaring, indeed intolerable, imperfections of today's

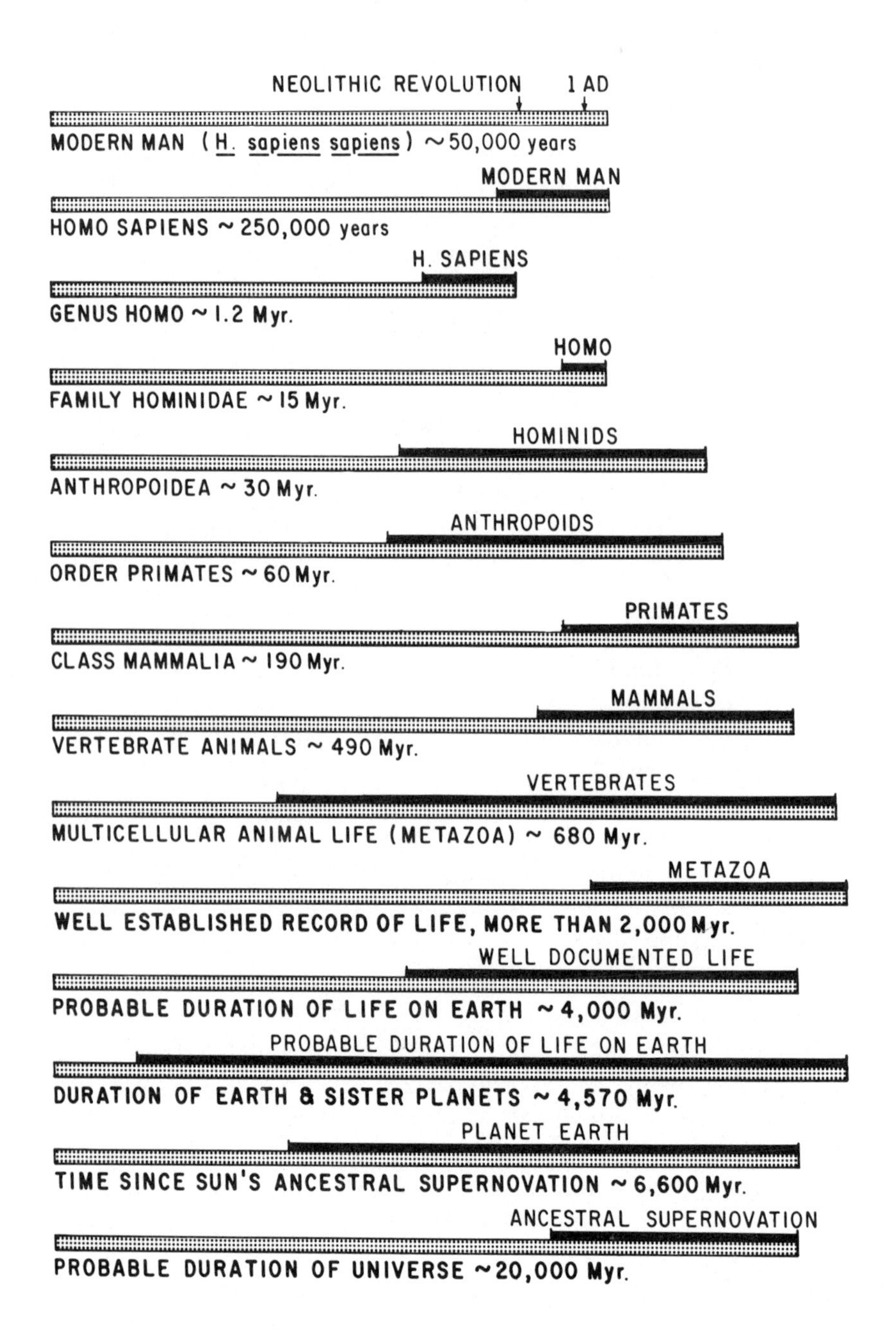

Figure 40. Human history in cosmic perspective. The short wavy lines to the left of the numbers signify approximations.

world, more people live better than has ever before been the case in the quarter-million or so years since the origin of *Homo sapiens*, and they do so because rather than in spite of science and technology.

Yet along with all this has come a growing sense of frustration and longing for simpler times, as unforeseen side effects have arisen to temper our joy in the technical products of our growing knowledge. The same DDT that saved countless lives in the jungles of World War II, when it turned up in Arctic penguins and brown pelicans, caused thinning of the shells of their eggs and well-justified concern about their survival. Insecticides that boosted crop production in the fields of Utah found their way up the food chain to brine shrimp in Great Salt Lake, thereby impairing the progress of fish culture in the coastal lochs of Scotland. Irrigation projects in Africa simultaneously led to an increase in land under cultivation and the spread of infectious and degenerative bilharzia, or schistosomiasis. Mining for the raw materials basic to industrial society makes a mess, and their smelting and fabrication add to it. Seemingly well-designed city centers turn into slums. Suburbs sprawl. Our nostrils and eyes sting from the smog produced by the effort to escape the city centers. Clair Patterson of Caltech has shown that all mankind has become polluted to some degree by lead; yet we have no measure of how debilitating the effects of that pollution may be because there is no unpolluted race of people anywhere in the world to provide a reference point. Even the remaining bits of wilderness suffer from too much loving by too well equipped urbanites.

The work of nature also brings out surprising side effects. The origin of the blue-green algae was apparently largely responsible for cleaning up the initial smog beginning some two billion years ago. The breakup of the consolidated late Paleozoic continent of Pangaea, followed by the drifting apart of the pieces to form the present continents, set in motion a complex of processes that led to a general moderation of Mesozoic continental climates. Late Mesozoic changes and climatic deterioration that contributed to the extinction of the dinosaurs prepared the way for mammals to inherit the earth. Plate tectonics, the opening of well-watered rift valleys over incipient spreading centers in east Africa, and the spread of ice sheets at high latitudes may be related to the emergence of *Homo sapiens* and his transformation into modern man. The 1883 eruption of the Indonesian volcanic island of Krakatoa affected weather, crops, and animal populations on a global scale. A warm sea current that appears episodically off the coast of Peru causes fluctuations in the Peruvian fisheries and thus in the price of chicken and pet foods in the United States.

We cannot prevent changes from taking place on our evolving planet. But we can become more conscious of their possibilities and of the extent to which they may be brought about or affected by man. And we can put more effort into trying to understand, anticipate, and ameliorate them where the prospects are that they will be unfavorable. We need to pay attention in particular to such changes as may or do affect the continuously circulating systems of atmosphere and ocean. For the global sea is so well and continuously mixed that the percentages of the principal dissolved ions anywhere in it can be as closely calculated from a single determination of its chloride content as from individual analyses of all main components. The air mixes so pervasively and constantly that virtually every breath you take contains some atoms exhaled by every other person who has ever lived—by Abraham, by Confucius, by Cleopatra, by Ghengis Kahn.

Another characteristic of ecosystems is their tendency to be self-regulating. This is often described as the *balance of nature*—sensed if not understood since antiquity, when Herodotus and Aristotle wrote about it. The idea was enunciated more explicitly by the French chemist Le Châtelier in 1888, who observed that if a system in equilibrium is disturbed, it tends to react in such a way as to restore a new equilibrium. Le Châtelier's Rule has great generality.

If you are on one end of a seesaw with another person of the same weight at the other end and one of you leaps off, that disturbs the balance. The weighted end goes crashing to the ground and a new equilibrium is established. If you pour acid on a concrete sidewalk it reacts until its hydrogen ions are neutralized. A clear brook babbling quietly through the countryside when a heavy downpour takes place in its headwaters becomes a raging muddy torrent, scouring its bed and banks and flooding low areas until the flood waters run off and a new stream profile is established. A breakwater built across a longshore current causes an accumulation of sand on its upcurrent side until the sand runs over or around the end of the breakwater, perhaps clogging the channel it was built to protect. Large fluctuations of herbivore populations occur in the wild state in different parts of the world as herds breed up to the full carrying capacity of the land when browsing or grazing is good, then decline in bad times. Nor are people immune. Over a million Irish died and another three million migrated when the potato crop failed in 1845 and again in 1846. Starvation in parts of Africa and the Indian subcontinent regularly accompanies failure of the monsoon rains and thus of crops.

In economics, under all kinds of social systems, if a commodity or other trade item becomes scarce the price usually goes up. That reduces user pressure and nudges supply and demand toward a new balance. If population and per capita use create demand pressures that exceed conveniently deliverable high-grade supplies, the usual result is that prices increase while quality decreases.

Natural communities and trade markets thus tend to establish balances among themselves. Affect one aspect of that balance and all others tend to respond. Let a virus reduce predator populations and herbivores increase. Let a drought reduce the browsing or grazing capacity of the land and herbivores and predators both decrease. It also appears to be true that a degree of diversity improves the ability of a community to respond to change, just as the genetic diversity of a population makes it more adaptable. It is not really an exception to this generalization that a forest or a prairie with a climax community of only a few main kinds of plants may remain in apparently stable balance until struck by fire or plague, or that apparently stable animal communities may display widely varying degrees of diversity. Rather it is the case that an area of forest consisting of a pure stand of chestnut will be harder hit by chestnut blight than a mixed deciduous forest of which chestnut trees are but one component. In fact few natural communities are monospecific, even if one counts only plants; and diverse natural communities are simply less likely to be wiped out by blight or plague than a pure stand of potatoes, wheat, or corn. A large farm with a variety of crops and stock is less likely to suffer from the vagaries of weather, disease, or market conditions than a single-crop farm. A diversified industry or economy, provided it does not become unmanageably so, is more resistant to fluctuations than one that is not diversified. Overspecialization and excessive complexity are both augurs of trouble.

The above examples of the ways in which different systems respond to different pressures illustrate the generality of Le Châtelier's Rule. But this rule is not the only important limiting factor in human affairs. Two others are *entropy* and *exponential growth*.

Recall that, according to the second law of thermodynamics, entropy always increases throughout the universe. This law is sometimes called "time's arrow," because, notwithstanding the local investment of available energy to maintain or restore temporary order, the general course of increasing entropy is irreversible. An aspect of increasing entropy is that things tend to become messier as time goes on. Disorder increases at the cost of order. *Available*, or as it is called *free energy*, is converted to *bound energy*, thereby becoming permanently unavailable. If you try to work against entropy by restoring order, you pay a price in free energy. For example, if you turn the kids loose in a freshly neatened playroom and return for inspection an hour later, you may observe "time's arrow" in operation. Entropy has increased. Disorder prevails. A lot of energy has been expended with nothing to show for it except disorder. More energy must be spent to restore order.

Similarly, if you seek to raise a crop you must expend energy clearing, planting, and keeping the field free of weeds and plant-eating insects. In the end, if all goes well, you capture free energy from the sun in the form of an edible crop. In getting that edible crop, however, you are probably also spending a lot of energy that is not renewable in any sense, energy capital that once spent is gone forever (of course, in the deepest sense, that is true of solar energy as well). If you use a tractor, you pay a high price in nonrenewable free energy, from the burning of diesel oil or gasoline, for the renewable energy you get from the crop. The free energy in the oil is converted to permanently unavailable bound energy in the form of released gases and particulate matter.

If, instead of raising a crop, you decide to become a supplier of aluminum, you must expend free energy to explore for, develop, mine, and convert ores of aluminum to metal—often very large quantities of energy in the last step. In the process you decrease disorder and increase order. Dispersed metal in some natural material is brought together to form the stuff of beer cans. A high-entropy material has become a low-entropy product. It might look as if time's arrow had been reversed. But that is not the case. The apparent decrease in entropy of the material has been purchased at a high and irreversible cost in free energy. The entropy of the whole system has increased. As soon as the beer cans are made and the beer distributed, the dearly bought order begins to decrease again. The cans are dispersed. Some are never recovered. A little aluminum is lost by oxidation. And whatever recycling may be done is at additional costs in free energy for bringing the cans together again and melting them into solid aluminum ingots for further processing.

It might seem that if only we could produce enormous quantities of inexpensive energy all would be well. But it costs both energy *and* materials to produce energy in available states and bring it to bear on useful work. The work that is done by the machines of industry is performed at a cost in free energy that includes not only the energy directly used to turn the machines, but all of the energy that went into finding, mining, transporting, and smelting the ores and fabricating and transporting the metal from which the machines are made. Further energy costs include those of building and operating the plant that produces energy in the form used; finding, producing and transporting the energy raw materials, whatever they may be; and the energy losses involved in transforming energy-rich raw materials into useful states and conveying this energy to the sites of intended application. To this must be added the environmental costs of mining, production, and coping with solid and thermal wastes.

It seems that we cannot win against entropy, and in truth we can't. What we can do is to use those remarkable conceptualizing devices and computers we all carry around on our shoulders to minimize both man-induced increases of entropy and environmental degradation—to achieve a kind of tolerable balance with nature wherein the speed of flight of time's arrow is not unnecessarily accelerated.

The biggest obstacles we confront in moving toward such a desirable state are those created by the prevalence of *exponential growth*. Exponential growth is like compound interest on a loan. It gets bigger much faster than you might expect. The easiest way to think of it is in terms of *doubling times*. If annual rates of growth are known, it is easy to estimate doubling times by a rough rule of thumb—just divide 70 by the percentage of annual increase and that will yield the approximate doubling time. If, for instance, you borrow a sum of money at 7 percent interest and fail to pay on it for 10 years, you owe twice as much at the end of that time. At the end of 20 years you would owe four times as much, and so on. If present rates of increase of world energy consumption were to continue for about 240 more years, energy would then be being used at a rate equivalent to the world's total receipt of solar energy. If demands continue to double every generation, as they have been doing, no amount of minerals on or in the Earth can meet such drains for long.

Exponential growth has a peculiar property: although absolute increases are slow at first they mount rapidly in the final stages. Thus, if something being consumed at a steady exponential rate is finally one-fourth gone, only two doubling times are left before it will all be gone. If quantities being consumed are very large and doubling times short, as they now are, the lead time available for action between a general perception of impending supply problems and their actual onset may be too short for effective countermeasures. This is what brought on recent and continuing energy shortages, concerning which recent polls show 48 percent of Americans unaware that we must import oil to meet demands. *Similar relationships will continue to generate episodic shortages in energy, in fresh water supplies, in a number of mineral products, and in food supply in regions of continuing population growth; and related environmental deterioration will continue until such time as world population and rates of material consumption come into balance with other aspects of the man-altered global ecosystem.*

As in steering a large vessel through a narrow passage, the need for a change in course must be foreseen and acted upon well in advance of the response sought if trouble is to be averted. There is so much inertia in the human ecosystem that substantial lag times must be allowed if corrective action is to have the desired effect. Although water from the Aswan Dam, for instance, has increased the arable land of Egypt by about 50 percent, parallel population increases have kept pace with the increase in food supply, allowing no more per capita. Traditional economic indicators tell us about these problems eventually but not soon enough any more to allow us to cope with them properly. Increases both in numbers of consumers and in rates of consumption have reduced the lead time available to cope with shortages, coincidentally with increases in the lead time required to find, develop, and establish delivery systems for new resources. This means that better resource indicators and monitoring systems must be developed and their warnings heeded and incorporated into long-range advance planning. Also needed is a greater emphasis on more conserving use, reduction of demand, the search for new sources of supply, and the protection and restoration of the environment.

All of the above and related issues are dealt with effectively and in detail in works suggested for further reading at the end of this chapter. My aim here is to identify and highlight a few of the more critical issues. I see those issues as centering on the levels and rates of growth of population and per capita material consumption as they bear on the capability of the global ecosystem to supply food, mineral raw materials, and energy; to withstand and recover from environmental stress; and to sustain man morally against a failure of nerve and collapse of the human spirit.

First because of simple sanitation, and then because of medical advances and chemical pesticides, mankind during the latter part of the industrial revolution has achieved a substantial measure of death control. Because advances in death control were not accompanied by comparable advances in the application of birth control, however, the world's population increased for some decades at rates that have led to the alarming growth summarized in figure 41. Birth rates have now decreased slightly, giving a current doubling time of 42 years for the world population as a whole, but with local doubling times as short as 20 to 23 years in a number of poor countries. How shall we feed these new mouths when a 1976 study by economists of a conser-

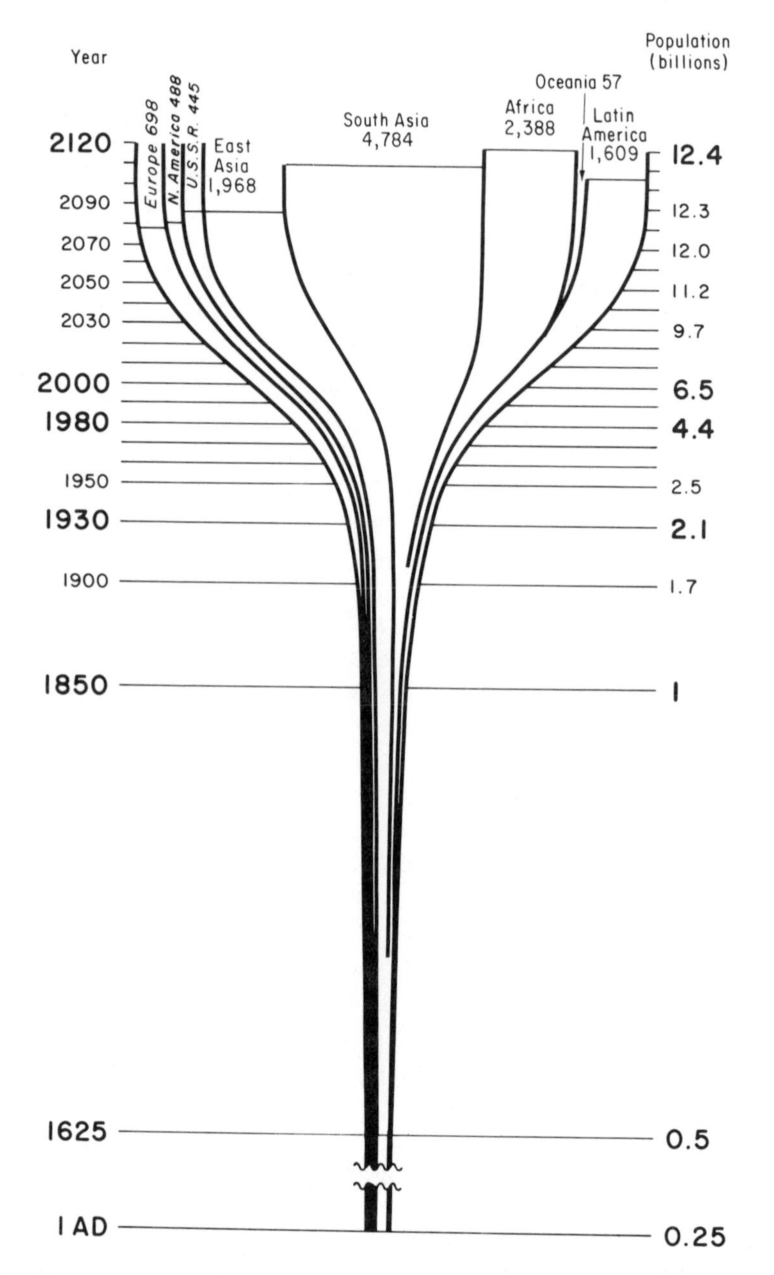

Figure 41. Past and estimated future populations of the world by regions. Numbers at right indicate total past world populations and estimated future ones in billions for years indicated at left. Horizontal lines and numbers at top indicate year at which it is estimated that, with luck, population growth might end and population in millions expected at that time. (Adapted from a drawing by Margaret R. Biswas, U.N. consultant, 1976.)

vative organization like the World Bank finds one-quarter of the present world population already undernourished, and where the fraction who fail to receive a really adequate diet may now be as great as one-half or even two-thirds?

Of the world's potentially arable land, about half is already being farmed and much of what is left is marginal. The north temperate regions are the most intensively cultivated. In Europe about 90 percent of the potentially arable land is in production, in Asia (including subtropical lands) about 85 percent, in the USSR about 67 percent, in North America about 51 percent. The equatorial and Southern Hemisphere lands are the least cultivated for several reasons: first, because of severe problems arising from the rapid formation of a brick-like surface crust (laterization) on cleared lands; second, because much of the land is too wet, too dry, or too salty for conventional crops; and, third, because of other problems of tropical and desert agriculture. For good reason the figures for the percentage of possibly arable land under cultivation in Africa are only about 25 percent, South America about 12 percent, and Australasia scarcely more than 2 percent. To extend cultivation in such areas would almost surely be more difficult than to increase the productivity of present lands. Yet, outside of India, such extensions of land under the plow would help most to feed the countries where the needs are greatest. Not only is this desirable for encouraging self-reliance, but in addition it could provide a more diversified diet than can be supplied by shipping wheat from temperate regions. The problems and some possible solutions are concisely and amusingly discussed by the British biologist N. W. Pirie in his book *Food Resources, Conventional and Novel,* referenced at the end of chapter 19.

Exponential growth, stimulated by the industrial revolution, led to enormous .and unprecedented expansions—in populations, in consumption of raw materials; in production of waste; and in pollution of air, water, and the solid earth. After perhaps a quarter of a million years, the global population of *Homo sapiens* finally reached the half-billion mark around 1625. Spurred by the pressures of industrialization, it doubled to a billion by 1850. With the above mentioned advances in sanitation and disease control it doubled again in 80 years to slightly over two billion in 1930. And it attained its next doubling, to four billion, during the American bicentennial year. *If present birth and death ratios continue, the next doubling, to eight billion, will occur by the year 2018.*

In the United States whose bicentennial population was close to 220 million counting illegal aliens, we have used our heritage of mineral raw materials at such rates that the country long ago ceased being a net exporter. Of 36 major mineral commodities monitored by the United States Bureau of Mines, 1975 supplies of all were to some extent imported; 12 were 80 percent or more imported; and 20 were more than 50 percent imported (figure 42). Man's appetite for energy and materials worldwide has seemingly become insatiable at the same time as Earth's ability to absorb and recover from pollution shows signs of approaching limits. Meanwhile, world population is all too slow in moderating its dizzying upward spiral, let alone leveling off, and industrial man spreads his litter to all corners of the earth. Greenland has a telephone directory. Tourists travel to the Canadian Arctic islands. There is regular traffic to and from a dozen permanent bases on Antarctica. And I have a depressing collection of photographs of abandoned industrial trash taken in various corners of Earth once thought to be remote. The sequence of actions taken over the past 200 years of industrialization, mostly well intended, has had not only good effects but also cumulative, unsought and increasingly undesirable bad effects.

Despite some glib assurances to the contrary, the maintenance of a favorable balance between man and the rest of nature on Earth today is threatened in unprecedented ways. Never before have there been so many consumers, consuming so much, at doubling rates that are so short both for numbers of consumers and quantities consumed. That leads to the above-mentioned shortening of warning times between general perception of crises and their onset—times which may in the future, if present trends continue, become too brief to allow for effective counteraction.

What will tomorrow bring? Will deprivation continue side by side with plenty? Will the gap between rich and poor continue to grow? Will we be able to feed the 70 million additional mouths (over 100 million births minus deaths) that still crowd to Earth's austere dinner table yearly? What foreseeable natural or man-induced variations in

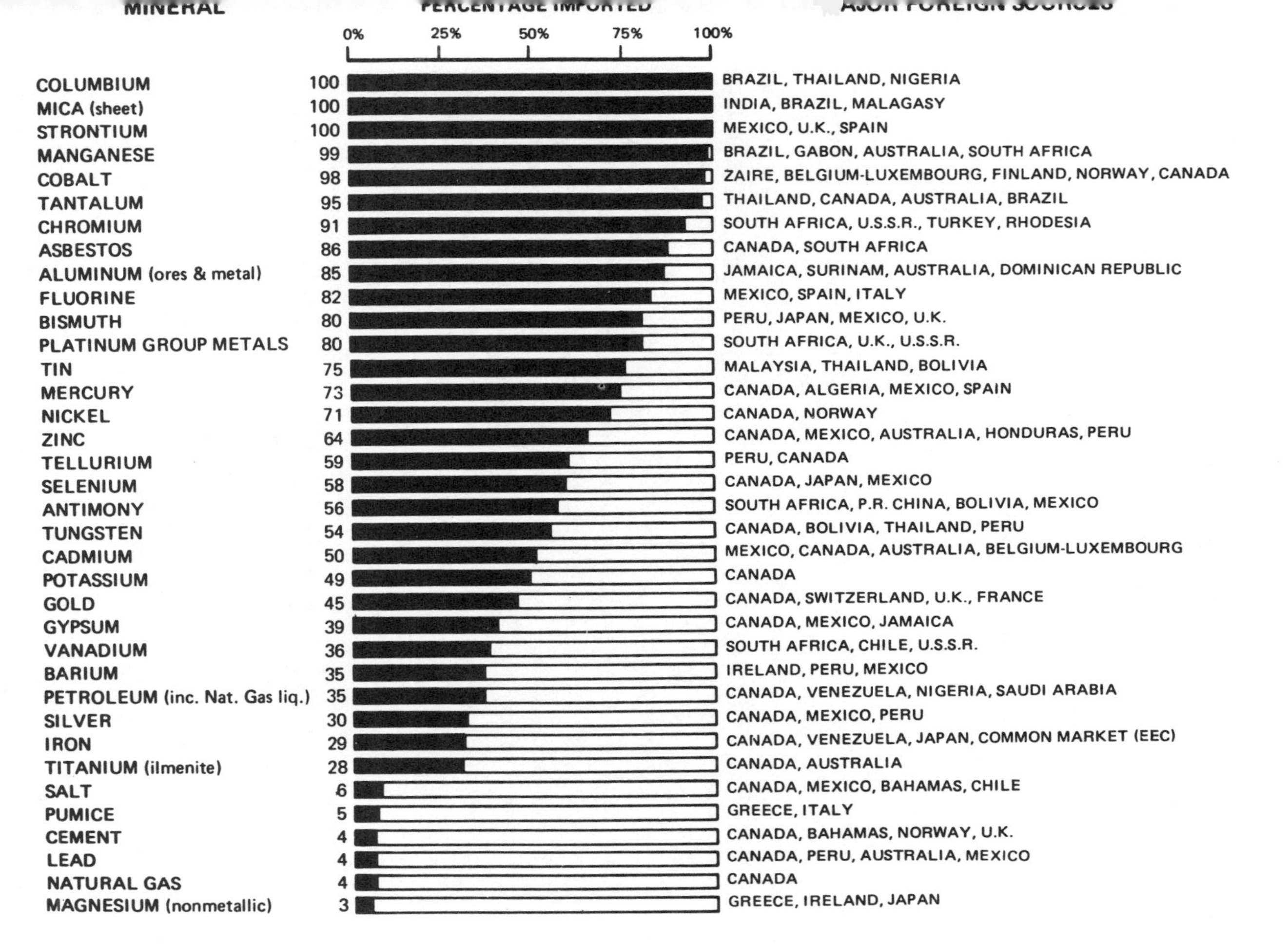

Figure 42. Percentages of U.S. mineral requirements imported during 1975. (Source: U.S. Bureau of Mines, Bull. 667, 1975, p. 12.)

climate, rainfall, and the like may affect our ability to do so? Will world population continue to grow in spite of limitations on agriculture, mineral production, space, and clean air and water? Will we meet the challenge of man's future?

Inevitably our evolving Earth will continue to evolve. Crustal plates and the continents they carry will continue to drift, sea level will rise and fall in rhythm with the plate motions, solar activity and Earth's orbit will vary, global atmospheric circulation will shift, climate will change and agricultural patterns with it, rainfall patterns will differ, deserts may bloom, breadbaskets may become deserts. Man also will continue to evolve and to affect Earth's evolution willy nilly, if only by breathing and undertaking the minimal activities needed to endure. Thus, to do nothing at all is equally to decide something important about the future of the planet and ourselves. If the future of civilized man is to be more than a flash in the pan, if *Homo sapiens* is to outlast the dinosaurs and to do so with grace, living generations must grasp the fact that better management of the global ecosystem and its resources is called for and called for most urgently, and they must act more effectively upon this realization. This is demanded, not only in the best interests of Earth's present tenants, but also, and especially, by that most basic of human rights—the right of those brought into the world without their consent to inherit the planet in good condition and with a maximal variety and flexibility of significant options for the future. *Laissez-faire* may have had advantages in a thinly populated world where supply greatly exceeded demand and clumsy, wasteful, error-filled, even willfully polluting practices are tolerated by a forgiving environment. But it is no longer tolerable on our tight little globe. To subordinate the rights of posterity to selfish and undisciplined perceptions of freedom and democracy is a betrayal of freedom and a perversion of democracy. For they, in their most basic sense, must grant and protect the access of the now underprivileged and the yet unborn to a fair share of Earth's bounty.

For Further Reading

Bates, Marston. 1960. *The forest and the sea*. New American Library (a Mentor Book). 216 pp.

Ehrlich, P. R.; Ehrlich, A. H.; and Holdren, J. P. 1977. *Ecoscience: population, resources, environment*. W. H. Freeman & Co. 1051 pp.

Lauda, D. P., and Ryan, R. D. (eds.). 1971. *Advancing technology–its impact on society.* William C. Brown Co. 536 pp.
Murdoch, W. (ed.). 1975. *Environment–resources, pollution and society* (2d ed.). Sinauer Associates. 488 pp.
National Academy of Sciences. 1967. *Applied science and technological progress.* U. S. Government Printing Office. 434 pp.
Turnbull, C. M. 1972. *The mountain people.* Simon & Schuster. 309 pp.

19

POSTERITY'S WORLD

Come what may our colorful Chinese lantern of a planet and its motley crew of habitants will continue to evolve. The lithospheric spreading centers will continue to spread and the positions and outlines of the continents to change (figure 43). And those who inherit the earth will view it and their responsibilities to it and to one another in different lights, depending on who they are, who preceded them, and when and where they live. Meanwhile the livelihood of industrialized man has improved so markedly, and the hours that must be spent in gaining it have decreased so substantially, the he has more time than ever before to reflect on the future. Despite this leisure for reflection, however, little heed is given to matters that might affect that part of posterity more distant than our children or grandchildren.

Who then will speak for posterity in today's world? The industrialized regions of the world react primarily to market pressures, while the theory of competitive markets, in addition to excluding community assets such as air, water, and scenery, assumes all participants to be fully informed and free to choose. But posterity cannot participate in its own behalf. It has no information and no choice—not even the choice to go unborn. There is no term for it in the economic equation. The third world, for its part, must struggle too hard to shelter and feed itself to think about children as much more than a source of gratification, complication, cheap help, or social security. We are prone to dodge the issue of posterity's rights with the complacent judgment that, after all, no one can foresee the future,

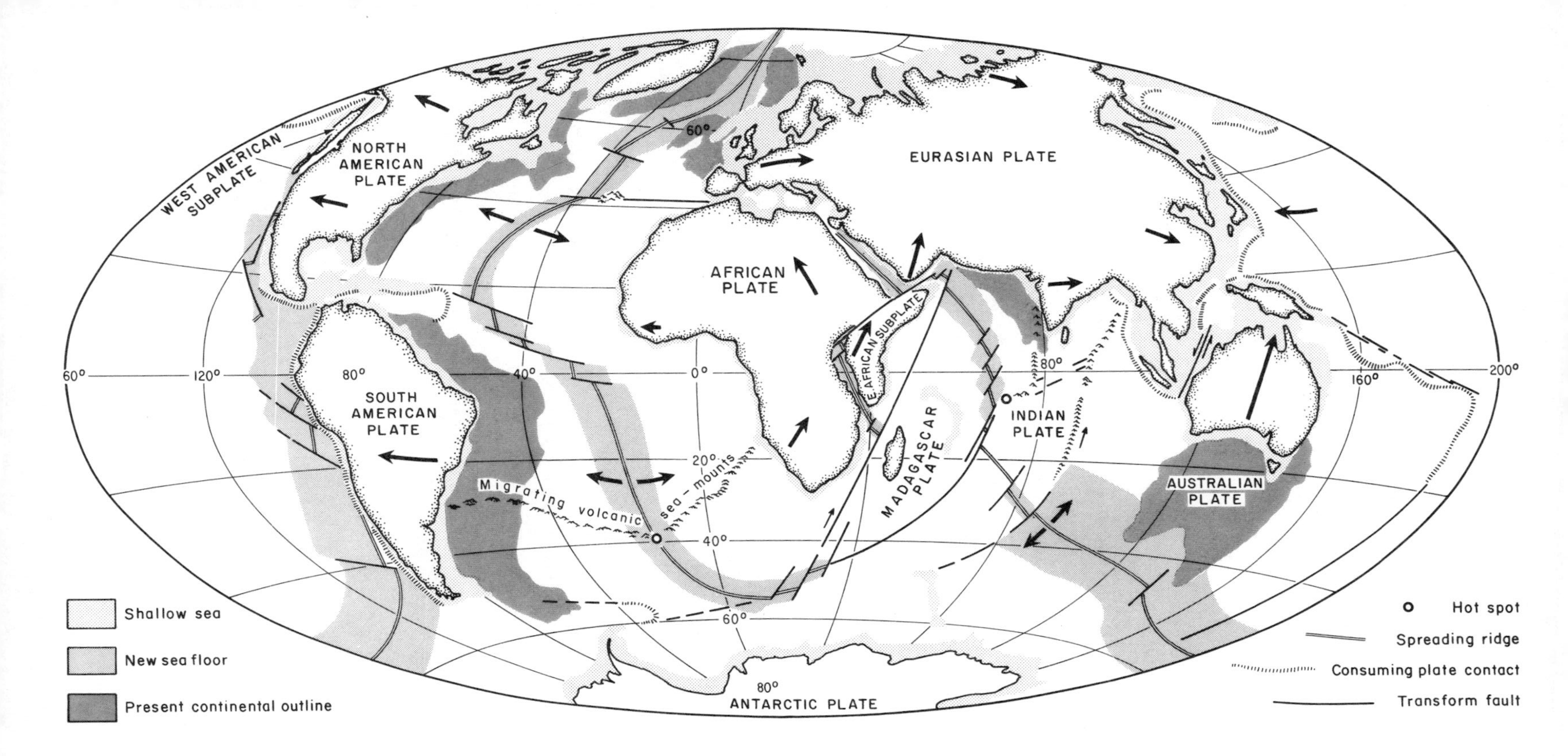

Figure 43. World geography 50 million years from now assuming lithospheric plates continue present rates and directions of motion. See also figures 11 and 13. (Slightly modified from R. S. Dietz and J. C. Holden. Copyright © 1970 by Scientific American, Inc. All rights reserved. Reproduced with permission of the authors and the publisher.)

and that each generation must therefore look after its own needs as they come along, with whatever means avail.

There is some truth to that view, but more escapism. It is true that all we can be sure of about the future is that it will be different, that only change is certain. It is equally true that neither complacency nor panic nor any intermediate state of mind will, of itself, expunge the relevant variables from which the future evolves. Our attitudes and judgments, nevertheless, will determine how we respond to them. The variables are not unalterable. Only the constants cannot be changed—only the laws of nature remain fixed. We are not mere pawns of fate. We can to some degree foresee the consequences of our actions and take heed not only for our own welfare in years to come but also in the interests of future generations. All our yesterdays need not have passed in vain. Our knowledge of past events, current trends, biological and societal processes, and natural laws can all be brought to bear in attempts to anticipate the future and to exercise some control over it. From such considerations arises a responsibility to posterity that cannot be set aside merely because we cannot see exactly how to fulfill it in all particulars.

How will it be, then, in 2003, or 2033—for the next generation and the one following it? And on to 2073, when those born in 2003 will be dying, and beyond? What do the trends suggest? How might actions we take or fail to take today affect the outcome? What are the costs and benefits of different courses of action or inaction, and to whom? What should we who are now living try to do differently? After all, as all smokers will affirm, it is hard to forego what may appear as a present good to avoid a future evil, especially where it is not certain that a specific future evil will in fact materialize, or whether it may beset some other place or later generation and not ourselves. What, after all, the cynical may ask, has posterity ever done for us?

Population, Food, Forest Products, and Climate

The trends clearly imply that by far the gravest problems for mankind in general in the decades ahead will be continued population growth, food supply, and related issues such as peace, pollution, and climatic change. Hungry people dream only of food. Starvation suppresses our finer instincts. People become desperate when they have only their chains to lose. Such problems, to be sure, may appear remote to the average citizen of an affluent society. He has other problems—also arising from population growth, but in a heavily consuming society. Fuel shortages, inflation, smog, launching the children on a career, or finding a quiet place for a vacation may so distract him from the realities of the larger world that he fails even to notice them. But we must notice these realities, even if only to build defenses against them. In the end the affluent minority and the underprivileged majority must either work together to achieve both population control and a more equitable distribution of the world's goods or else sink separately. The distinction between steerage and first-class vanishes when the ship goes down. It is well also for the affluent to bear in mind that even rich and prosperous societies can fall on hard times when population growth continues beyond the supporting capacity of the land. India, once rich beyond belief by the standards of its day, now tops the sick list of the nations where hunger stalks the land and the people perish.

Let us, then, turn our eyes, our hearts, and our minds to the larger world. What is to be done if world populations continue to increase, as they give every sign of doing? Suppose we simply disregard the connection between scarcely bridled population growth and hardship and set out to feed the daily addition of hungry new mouths? What is the most optimistic scenario one could write for the next scene in the industrial act?

Suppose that the amount of land under cultivation could be doubled by increases mainly in the equatorial regions and the Southern Hemisphere, where the need as well as the areas of uncultivated land are greatest. And suppose that all these new lands could be brought up to the same average level of production as presently cultivated areas. On the highly improbable further assumption that all this could be achieved in 40 years, that would just about feed the somewhat more than 8 billion people to be expected in the year 2018 if existing rates of population growth continue. They would be fed, however, at the levels of malnutrition presently prevailing in much of the world unless a far more equitable system of distribution than now exists could be achieved in the meanwhile. Of course such growth of agricultural land would also threaten or exterminate the politically voiceless remaining hunter-gatherers so beautifully described by Colin Turnbull in his book *The Forest People*—a cost that a world claiming compassion for its minorities should balance against the "benefits" for barely sustained, ever denser concentrations of ill-fed humanity living lives at best short and brutish.

But what about the second doubling of populations? To feed such numbers would call for a different set of actions. Suppose now that massive and continuing flows of fresh water are somehow made available for irrigation, that desalination of lands and water is carried out on a large scale, that more richly productive new crops are introduced and most domestic animals eliminated in favor of expanded growth of cereal grains, and that there is continuing genetic improvement in bearing-quality and disease resistance, expanded production through ever greater use of mineral fertilizers, and ever-improving control of insects, vermin, weeds, and viruses. Suppose that, through such measures, it were possible again to double the food production of an earth on which the only lands not cultivated were simply too rocky, too steep, too salty, too frigid, or too arid to bear crops. Such herculean achievements—again assuming they came to pass in the 42 years beyond 2018, and leaving out of consideration unsought environmental side effects—would barely feed all the 16 billion or so mouths that continuation of present rates of growth would place on Earth by the year 2060—at a food production four times that of 1976. For present purposes I assume that an adequate supply of those amino acids that are found naturally only in animal proteins would be provided synthetically by food chemists.

For the third doubling you must really let your imagination soar. Consider the full play of technology—utilization of leaf protein and calories from woody plants, algal food production, artificial photosynthesis using water from the sea and carbon dioxide from limestone, bacterial conversion of vast quantities of nonrenewable coal to food, feeding of what stock remains with fodder from wood pulp and urea derived from atmospheric nitrogen, and other tricks. Could world food production be doubled once more in another 42 years to eight times present production? Maybe. But now we are talking about feeding something more than 32 billion people around the year 2100 at 1976 levels. That would mean, assuming it could all be achieved within 120 years, either continued marginal diets for the great bulk of the world (and continued malnutrition and episodic starvation for a minimum of one in four), or an enormously improved distribution system with much leaner diets for the now well fed. Either solution involves agonizing questions of numbers versus quality. And, of course, all problems and responses would be changed were trends now visible to bring on massive starvation or global military genocide.

I have not, as some readers may be thinking, forgotten either the sea or other planets. But new surges of knowledge about the potential aquatic harvest, the failure of recent efforts to increase it significantly, and the limited regions of sufficiently shallow depth and degree of enclosure for aquaculture do not support a cornucopian outlook for food from the sea. Although aquatic sources can, if properly treated, continue to be important providers of much needed high-quality protein at near or somewhat above present production levels, the harvest can probably not be greatly increased. Nor can the sea contribute significantly to the world's caloric budget. Students of world fisheries such as William Ricker of the Fisheries Research Board of Canada, Hiroshi Kasahara of the United Nations Development Program, and John Ryther of the Woods Hole Oceanographic Institution doubt that the sustainable annual crop is likely ever to exceed double the present one, and that adds but little to the very rough and sanguine computation of *maximum* possible world food-productivity from the lands which was given in the preceding paragraphs. As for other planets, they may serve interesting functions as sites of small, expensive space stations, but extraterrestrial colonization or agriculture on a large scale belongs in the realm of science fiction rather than that of serious discourse.

Supposing is a game that can be played in other ways. If, instead of the world's attaining that postulated eight-fold increase in food production and an equitable distribution system, famines such as have been spreading in Africa and India were to persist and extend to other parts of the world, what then? For one thing, an increase in death rates for any reason has the same effect as a decrease in birth rates—a lessening of the rates of population growth. It is one way of solving the problem, although not a way we like to think or talk about. It would be better if world literacy (now barely or less than 30 percent in large segments of the Southern Hemisphere) and awareness were to lead to further reductions in population growth and eventual stabilization. That, were it to happen soon enough, would make the problem manageable. The odds against its happening soon enough are great, however. And it will certainly not happen without continuing and increased effort, including moral, technical, and financial support from the now literate and comprehending.

One big difficulty is that the present age-structure of global populations is such that numbers would continue to increase even after births reached the level of bare replacement (2.2 children per couple in the United States in the late 1970s). That is because such a large number of the women now living who will bear children and continue to live thereafter have either not yet reached childbearing age or have not yet borne children. Given the attainment and maintenance of bare replacement rates of birth worldwide at present age structures of world populations, demographers estimate that the eventual increase would be about 35 percent for the United States and 50 percent for the world as a whole. That means that if bare replacement rates had taken effect worldwide on American Independence Day 1976 and were maintained thereafter, the population of the United States (leaving out immigration) would finally level off at around 300 million and world populations at around 6.4 billion just before the middle of the twenty-first century—both perhaps manageable numbers but far exceeding the ideal.

If attainment of bare replacement rates is delayed and war avoided, eventual balance is achieved at correspondingly higher levels and later dates. As in the case of a nervous tightrope walker, *the balancing act becomes harder to manage the longer its initiation is delayed*. The forecast indicated in figure 41 of a final leveling off of global population at about 12.4 billion around the year 2110 is probably as realistic as any. Earlier leveling off at less alarming numbers, however, could take place either because of unified global effort toward that end, or because of massive starvation, collapse of the world economy, global war, or simply failure of the human spirit under the growing psychosocial burdens of a massively overpopulated and undernourished world, leading to the collapse of society and global anarchy. Horrible as the thought is, that might well appear as a retrospective blessing to our twenty-second–century descendants.

Assuming for the sake of discussion that by some combination of miraculous agricultural achievements and continuing generally beneficent climate, the growing world population could be fed and supplied with fresh water, how would their demands for material goods be met? And at what cost to the environment, to societal structures, to world harmony, and to individual freedoms?

Later I will focus on mineral raw materials and on energy from all sources. But first a brief word on forest products and other natural fibers. Forests are renewable resources that convert the sun's energy into forms useful to man for building materials, fuels, and even food. Timber now supplies about one-fifth of the nonfuel industrial materials. If forests are properly managed and utilized they may continue to meet important fractions of many industrial demands. And modern wood technology has created more new and better forest products than have been known before. Not only are forest products good for building houses, barns, and bowls, but the organic wastes from forests and farms can be converted into plastics and liquid fuels, and stock feed can be made from leaves, fruits, or even wood.

The limitations are several, however. Trees take a long time to grow and must be managed and harvested according to long-range plans to get the best production. The price of wood is high—a sure sign that this is a commodity for which demand is crowding supply. Metal, concrete, and ceramics increasingly substitute for wood, not the reverse. If man is successful in his dubious venture of seeking to bring the Amazon Basin and other now forested, submarginally arable lands under cultivation, destroying the forest people in the process, those lands will not be available for forest products. Liquid fuel and plastics from forest and farm wastes, even if efficiently and completely converted, can meet only small fractions of the current demands. Finally, removal of forests, with exposure of thin forest soils on steep slopes or in areas of lateritic soil, can produce new wastelands, especially in semiarid regions, as has already happened in many parts of the circum-Mediterranean to name but one locale.

Finally, the prospect of climatic change is a Damoclean sword hanging over the world's hope for increases in production from fields and forests. Evidence of this threat appeared in the late 1970s in the form of high coffee prices resulting from freezing of the Brazilian coffee crop, failure of the European potato crop because of drought, upheaval in world wheat prices as a result of droughts in the USSR, and so on. Here, to be sure, I speak of weather, whereas climate is more properly considered as the statistical average of the weather. But weather is a manifestation of climate, and long-term weather records support the view that current world agriculture developed under unusually favorable climatic conditions which have prevailed since the end of the so-called "little ice age" about 250 years ago. Even under such favorable conditions, normal weather cycles seem to be causing us increasing problems, because, like the small-brained herbivores whose populations increase in good years and die back in bad ones, we have built our expectations on the good years. How long will they continue to recur? What is to be done about the episodic drought cycles? Suppose there were a global or hemispheric deterioration of climate?

The British meteorologist Derek Winstanley has pointed out that a general weakening of zonal wind-circulation over the Northern Hemisphere seems to have had a number of effects: changing rainfall patterns in Europe, crop failures associated with variations in the timing and areal extent of the monsoon rains in parts of Africa and the Indian subcontinent; and the accelerated southward march of the Sahara Desert. The same weakening of zonal circulation may well be responsible for changing precipitation patterns and droughts in North America as well. Based on wind and rainfall records in England, together with correlations and anticorrelations between the weather there and elsewhere, Winstanley finds several long-period cyclical variations in Old-World, African, Middle Eastern, and Indian rainfall, which he associates with variations in the strength of zonal winds.

Evidence for the cyclical nature of climate comes also from the Harvard-Smithsonian astronomer John Eddy, who observed that variations in sunspot activity correlate with carbon-14 anomalies in tree rings. This provides a tree-ring record of intervals of solar irregularity since the late Bronze Age having durations of 50 to 200 years. From this and related data, Eddy concludes: "We have lived our lives . . . during solar conditions that have applied but 10 percent of the time, or less, in the longer run of history." In related studies J. M. Mitchell, Jr., of the National Oceanic and Atmospheric Administration finds a 22-year cycle of droughts in the western United States, coinciding with every second 11-year minimum of sunspot activity (the so-called double sunspot cycle).

Still larger climatic variations could be in the offing. Concern has been expressed from time to time that increases in atmospheric carbon dioxide resulting from the clearing of forests and from large-scale industrial activity based on the burning of coal and oil might, in turn, lead to retention of solar heat and thus to global temperature increases. It has been estimated by geochemist W. S. Broecker of the Lamont-Doherty Geological Observatory, for instance, that the next century of industrial burning could double atmospheric CO_2 content, leading to a temperature increase of 2° to 3° Celsius. That would cause climatic belts to move poleward, shifting present agricultural patterns and increasing the area of low-latitude desert. It would also induce melting of polar ice, leading to a rise of sea level and perhaps to drowning of the world's major seaports. Opposing such effects the University of Wisconsin geographer Reid Bryson sees a long-term cooling trend in the Northern Hemisphere that may be shifting hemispheric climate in the direction of a new ice age. And still others point to records that show a cyclical succession of warming and cooling episodes since the beginning of the industrial revolution that seems to ignore the steady increment of manmade carbon dioxide. What may one believe? What can or should be done? If climates shift who will be harmed? And who might benefit? What consequences would accompany a shift of the global "breadbasket" from the north central United States to Saskatchewan and Siberia? Should we refrain from burning fossil fuels so as to prevent overheating the atmosphere or burn them as fast as possible to counteract the suggested cooling trend, or just wait to see what happens next? To be considered also are natural buffering effects and the numerous other feedback loops of such—for example, the beneficial aspects of increased atmospheric CO_2 on agricultural productivity. And what can possibly be done about the variation in solar activity other than to take note of it and attempt to foresee and prepare for its consequences?

In truth the causes and probable directions of climatic variation are but dimly perceived. Although a major effort to assess these problems is currently under way, the variables are so many and their implications so poorly understood that it will be some time before we can seriously hope to achieve useful accuracy in climate forecasting.

Meanwhile, in counting how many people the world can feed, or in estimating the future role of forest products in the world economy, mankind would do well to keep the climatic uncertainties in mind and under surveillance.

The Material Basis of Industrial Society and Its Implications

The average resident of the industrialized quarter of the world is more likely to be impressed with the problem of maintaining a "way of life" than with malnutrition or starvation in faraway places or even in neighboring lands. His way of life is a product of industrial society and it can be sustained only if industrial society is sustained. He is all too prone to be complacent about his material affluence, to be blind to inequity and insensitive to the adverse environmental and social consequences of over-industrialization.

Yet, within limits, industrialization has been a good thing. It brought with it surcease from much drudgery, lengthened life spans, increased leisure time, and unprecedented cultural opportunities to larger numbers of people than ever before. It has been the most civilizing influence since silver from the mines of Laurium ignited and sustained for a time the Athenian age of enlightenment. Indeed in many ways it has been the most civilizing influence ever, for it abolished many forms of slavery and compulsory servitude and permitted a larger fraction of the world than ever before to live to advanced years in relative comfort and improved health. The benefits of industrialization, however, need to be made more widely available, concurrently with more effective limitation or elimination of its unpleasant, dehumanizing, and hazardous aspects.

What is required to sustain industrial society at appropriate levels where it exists and to extend its benefits more widely? The productive capacity of this social system is utterly dependent on natural resources, mainly geological. It needs minerals and fibers for making machines, tools, structures, and products and energy to drive the system, including both mineral fuels and the food products that energize the brains and muscles that design and serve production lines. Recall now an earlier discussion of entropy—that quality of disorder and decline in the availability of energy that increases as the universe runs down. All of the essential minerals for industrial society, without exception, are produced, beneficiated, and transformed into well-ordered low-entropy products at a cost in available energy. All of the energy utilized is concentrated and brought to focus on useful work at a cost in materials. All of this energy is irreversibly transformed into unavailable, high-entropy states. All of the low-entropy metals and other mineral products utilized become disordered high-entropy waste with time. They can be restored to states useful to man—recycled that is—only with further inputs of energy.

When we think of minerals we are likely to have in mind things like gold, silver, copper, rubies, diamonds, turquoise, and so forth. Indeed such things were among the earliest mineral products used, and for very practical reasons. Gold, silver, and copper, being among the few metals that are found in elemental form in nature, could be seen, along with gems and gemstones, as separate components of the rocks in which they occurred. None of them required smelting to become visible. Geologists define a *mineral* as a naturally occurring substance, having a characteristic chemical composition and definite physical properties. Many minerals in this sense are capable of yielding metals or other elements of commercial interest that are not visible within the untreated rock or mineral, but which appear on smelting or the application of other modes of mineral separation. The term *mineral* in a colloquial sense refers to almost anything that comes out of the ground except fossils—to metals, gemstones, salts, cement, soil nutrients, water, oil, natural gas, and so on. In the following discussion, however, I take only passing notice of water (a huge and vitally important subject by itself) and I emphasize metals. I will also refer to mineral resources other than water as *nonrenewable resources*, even though, except for the fossil fuels (coal, oil, gas), many can to some extent be recycled. *Renewable resources* such as agricul-

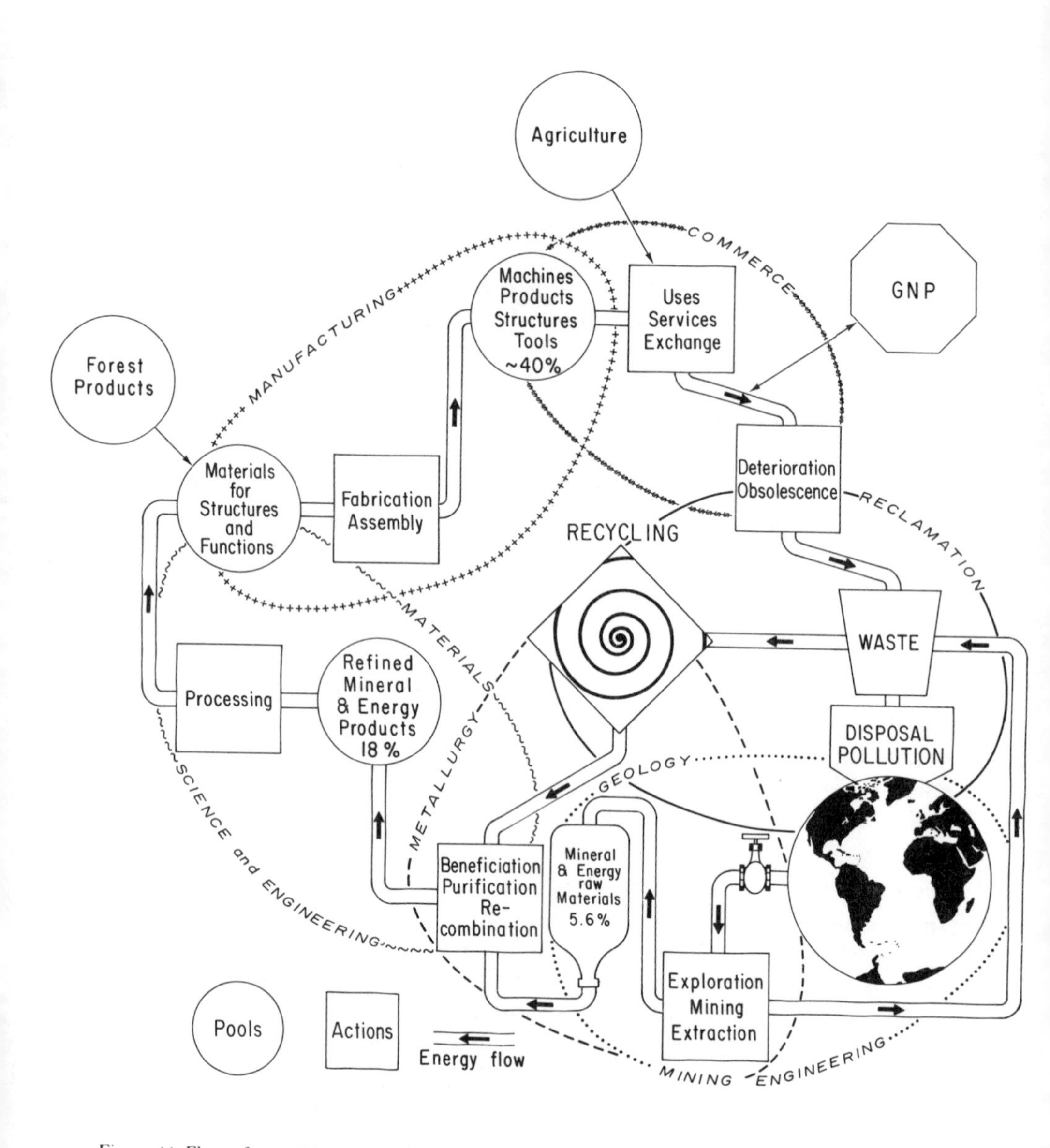

Figure 44. Flow of materials and energy in an industrial economy. Percentages given are of 1975 GNP for the U.S. (After P. Cloud, 1976, U.S.G.P.O., 78-653, p. 56.)

tural products, wood, and fresh water can, under appropriate circumstances, be reharvested at short intervals from the same sites. All are gifts from the sun. However, nonrenewable resources, once taken from the ground, are not replenished naturally at the same places or at rates significant for human purposes.

Mineral raw materials, then, are the useful metals, chemicals, raw fuels, mineral-containing concentrates, and nonmetallic substances that result from mining and extraction (before further processing to produce a refined product), or which are used directly, in the manner of stone, sand, and gravel.

The importance of mineral raw materials for industrial society cannot be overemphasized. They have an enormous multiplier effect on industrial economies. Nonenergy raw materials move through the system, driven by a sustained flow of energy to ever higher forms and uses, until they finally end up either being recycled or becoming a waste-disposal problem. This process is illustrated in figure 44. In the United States during 1975, for instance, the initial dollar value of mineral and energy raw materials used was only 5.6 percent of the *Gross National Product (GNP)*, of which 1.4 percent was imported and 4.2 percent was domestic. But as refined products they represented 18 percent of the GNP, and by the time the stage of finished products was reached the combined value of materials, plus inputs of energy, plus cost of manual and managerial skills, accounted for some 40 percent of the GNP. Moreover the products of mineral fabrication, supplemented by agriculture and forestry, underpin all the rest of it. Everybody "consumes" raw materials in the form of transportation and communication, housing, work space, and daily operations.

Mineral raw materials are produced from *mineral deposits*. It is important in thinking about minerals, therefore, to have some idea as to what mineral deposits are and how they form. Familiar kinds of mineral deposits include, for example, placers, where heavy and durable materials like gold and diamond are concentrated in streambeds or beaches by the winnowing action of water and gravity. Others are residual aggregations produced by weathering at Earth's surface, such as the ore of aluminum called bauxite. Just about all other mineral deposits, however, are formed by local concentration of metals or other elements from fluids of some sort. Many of these fluids are warm, salty, aqueous solutions that move through rock, depositing minerals chemically just beyond points of temperature change or constriction, in response to boiling or fluid mixing, or because of chemical reactions with particular adjacent rock types. Others are actual rock melts that produce chemical or gravitational segregation of particular minerals during crystallization. And still others are vapors, special kinds of open-water bodies, fluids that were buried along with ancient sediments, rainwater moving through permeable rocks near Earth's surface, or some mixture of these things.

Because of reactions within these different solutions, melts, and vapors, or between them and the rocks through which they flow, different elements are locally concentrated over long intervals of geologic time to levels far above their normal abundances in Earth's crust. Such concentrations are rarely in elemental form; instead they combine with other elements to form crystalline or amorphous compounds. The shapes of such mineral deposits are often highly irregular and are difficult to discover and develop. And of course the metal elements, which are of particular interest, display a wide range of normal geochemical abundances in Earth's crust, depending on rock type.

Some metals are so abundant and so widespread that it is fair to expect that technological advances and market pressures will assure a continuing supply sufficient to meet all demands of a sane world for thousands of years to come. They include iron, aluminum, and magnesium. At intermediate levels of abundance, but not so universally available, are manganese, chromium, and titanium, all with crustal abundances greater than a hundredth of a percent by weight. All such elements are found as essential constituents of minerals. Together with various rock-forming minerals, they make up 99.23 percent of Earth's crust and provide a fall-back position for industrial technology and civilization—one that should already be the object of study and analysis.

All other metals, those with crustal abundances less than a hundredth of a percent by weight, are relatively scarce. They tend to occur by atomic substitution in compounds of different composition rather than as essential components of the minerals that contain them. The ore bodies they form are commonly irregular and highly localized. For a number of such metals, scarcities now threaten. Their increasing scarcity or practical exhaustion would force changes in technology or institutional structures. Should they not, therefore, be the subject of strategic research and planning focusing on the principles that govern the natural concentration of the elements, mineral exploration, extractive technology, and materials science and engineering?

How shall future demands for mineral resources be met, especially for the scarce ones? To what extent will industrial man be successful in finding ample new deposits, and to what extent will he, in fact, be forced to turn to ever lower grades? What kinds of problems will arise with regard to energy requirements, waste disposal, and environmental protection as we try to produce more and more metals from ever larger volumes of ever leaner grades of ore? Will there be cutoff levels in grade of ore below which the energy or other costs of mining will escalate to intolerable levels? To what degree can pressures be moderated by recycling, substitution, or other conserving practices?

Geologists and ecologists, impressed with natural balances and limitations, are likely to point out that resources although large are finite; they cannot withstand all conceivable pressures, especially where such pressures grow exponentially. Others, mainly economists and engineers, impressed with the forces of technology and the past record of relative success in stretching the limits, are likely to cite the *first* law of thermodynamics—matter and energy are neither created nor destroyed but only recycled (although available energy *is* converted to bound states). They point out that the quantity of matter and energy available on Earth is very great and susceptible to substitution and reorganization by clever *Homo sapiens*. These views are not irreconcilable, but controversy tends to become a game and enthusiasts at both ends of the spectrum wage it with gusto. That is good, as long as candor prevails, for it gets the facts and the degrees of uncertainty before the public. And even those who seek a balance do not always agree on the exact rates of increase or decrease of population, per capita consumption of material goods, and mineral and energy production that are either hazardous or desirable, or on the time scale at which they become so.

In fact, no reasonable, informed, and sober person will argue that the quantities of mineral raw materials in Earth's crust are not truly enormous. Rational and informed opponents of further material growth will concede that more is possible. At the same time they will point out that substantial further growth is inevitable in any case because of lag effects and the legitimate demands of the underprivileged. The point they are trying so desperately to make is that it is time to start applying the brakes, and hard, if the machine is to be stopped before it crashes. Similarly, rational and informed believers in the power of technology and economics will concede that exponential growth cannot go on forever. But they will point out that versatile and imaginative man has overcome restraints in the past and that there are natural braking effects in market systems that tend to reduce pressures in areas where they become unprofitable or visibly threatening to the public welfare.

The issue, thus, is not whether the exponential growth of new mineral production can continue indefinitely, for all agree it cannot. The issue, rather, is when and how shall mineral production and consumption be brought into a state approaching balance with what is possible on a sustained basis in a tranquil world. Questions of levels and rates of population growth and per capita consumption, as well as questions of environmental, socio-political, economic, and military consequences, are central to such considerations.

The problem, then, for the next few decades is to find, separate, and produce an optimal quantity and selection of those minerals and mineral products that can enhance human welfare without imposing unduly adverse effects on living systems or the physical environment. As they say in the Caterpillar tractor ads, "We must have minerals . . . but we must also have an inhabitable environment." As those minerals come, however, from ever leaner ores by moving ever larger volumes of rock, the cost in both dollars and the energy units needed to mine and process the minerals, move them to market, and safeguard the environment is going to go up. Such costs will be transmitted to the user in the form of higher prices. Higher prices for energy and raw materials, in fact, are not a bad idea. That they have been priced so low, considering their enormous significance for industrial economies, explains why they have been used so profligately. The benefits from higher prices, induced by imposing clean-up costs and depletion taxes instead of allowances, could include not only more efficient, more conserving uses of materials but also support of needed research on mineral deposits and reduction of other forms of taxation. Posterity would benefit from such practices with respect to its minimal basic right to inherit the earth in good condition and without foreclosure of options that should be left open.

Posterity, however, is of all colors, nationalities, and economic states. Add to other considerations the idea of somehow closing the ever-growing gap between rich and poor during the next generation without lowering anyone's material standard of living and you have a problem of the first magnitude. Consider the goal expressed by black statesmen of bringing all of Africa to the 1975 standard of living of Western Europe by the year 2000, for instance. That would require a sustained real growth rate of 13 percent a year. Can this be more than a naive dream or a negotiating point? Will the now industrialized countries be willing to wait for Africa to catch up? Can the third-world countries ever catch up, even if the industrialized world is willing to forego further increases in its own material growth? And suppose the whole world were able to attain and maintain rates of consumption equivalent to 1975 rates in Western Europe—what would be the consequences for the global ecosystem and the human habitat of more than 8 billion people consuming at such rates in the year 2018? Or of 16 billion in 2060?

The problems raised by these questions boggle the imagination. They are not wholly rhetorical and certainly not unimportant. Yet I can only ask them. I have no panaceas to offer. But I do call for a new and greater concern among the affluent for posterity, on the one hand, and for a genuine striving for greater equity among races, nations, and economic sectors on the other.

I do have very grave reservations about the wisdom of attempting to industrialize the entire world, even assuming that all people want this for themselves and that it could be achieved. Industrialization involves the large-scale production of goods, machines, and service systems from raw materials. If the whole world is doing this, who will buy the product and who will supply the raw materials? How can serious conflicts, leading perhaps to war, be averted except by international covenants of diversification? I also have the impression, substantiated by survey data, that many migrants to urban centers would really have preferred to stay on the land and that they have gone to industrial centers only in search of employment, charity, or amenities. What I perceive then as the ideal world of the future would be a world that is both increasingly equitable and healthily diversified; one where the supplier of raw materials, food, fiber, and labor is compensated for his efforts at levels at least approaching those of manufacturer and middleman, allowing for the greater cost of living and recreation for the latter. This would be a world where the basic amenities and the means of livelihood were extended somehow to rural areas throughout the globe, but where those who preferred simpler life-styles were not forced or seduced into accepting unsought amenities. It would be a world of honesty in advertising. It would be a world that allowed for national, societal, and individual variety in goals and styles. And the way to it would start with constructive new action by both rich and poor. Overdeveloped nations would stabilize and eventually decrease both populations and per capita levels of consumption, while simultaneously extending *appropriate* educational and technical assistance wherever it was needed and would be beneficial. Underdeveloped nations would strive to reduce family size to bare replacement levels or less, not only by family planning programs but also, with necessary help from the rich, by the introduction of social security, rural education, and rural electrification.

In the following paragraphs, however, I will confine my attention to the matter of what would be involved in attempting to maintain current worldwide trends of growth in the production and use of mineral and energy resources, as an illustration of the need to pause, take heed, reconsider our goals, and redirect our energies.

It deserves emphasis, to begin with, that the central and limiting cost of mineral resource production is not dollars so much as it is energy. Returning to considerations of entropy, the mining, extraction, and processing of ores to produce metals represents an increase in order (metal) from disordered materials (ores) at a cost in available energy. About 16 percent of all energy used in the United States in 1975 was expended on a domestic mineral production that fell 25 percent short of demand. Energy costs, moreover, increase as the concentration of metals in the ore decreases to some grade below which the rate of energy demand per unit of metal produced increases rapidly.

Figure 45 illustrates this relationship. It shows how, as the grade of ore decreases toward the left beyond certain critical values, energy costs climb dramatically. This in effect establishes *cutoff grades* below which metals cannot be produced from those ores at tolerable energy costs with existing technology. Nor is it easy to foresee how technology could drastically reduce such energy demands. Energy costs, of course, include not only those of actually producing and processing the ores, but also those of transportation to the market and a suitable fraction of the energy costs of all plant equipment and all facilities involved in production, processing, and transportation. The 16 percent of the United States energy budget that presently goes for mineral production will increase substantially as the nation seeks to reduce its dependence on foreign sources. As the price of energy goes up, the hard-currency cost of nonenergy mineral procurement will also increase, with consequent braking effects on the economy.

This follows inexorably from the entropy law and the decreasing grades of new primary ores. Economics enters the picture only to the extent that the growing energy investment is reflected in rising prices, which, in turn, will stimulate the geological search for higher grades of ore and the technological search for substitute materials and technologies. A question to consider is why we must wait for economics to tell us what appropriate science can foresee now, when the time needed for the development and introduction of new materials and technologies is likely to be so long as to introduce unnecessary stress where we wait for economic signals.

I do not mean to deny that economics is a critical factor. But an economic philosophy rooted in the concept of ever-increasing material growth as the basic or even sufficient ingredient in human welfare has dominated economic and political thinking in the Western world to the point of conceptual bankruptcy. It is time to balance the decision-making process by introducing a better mix of economic

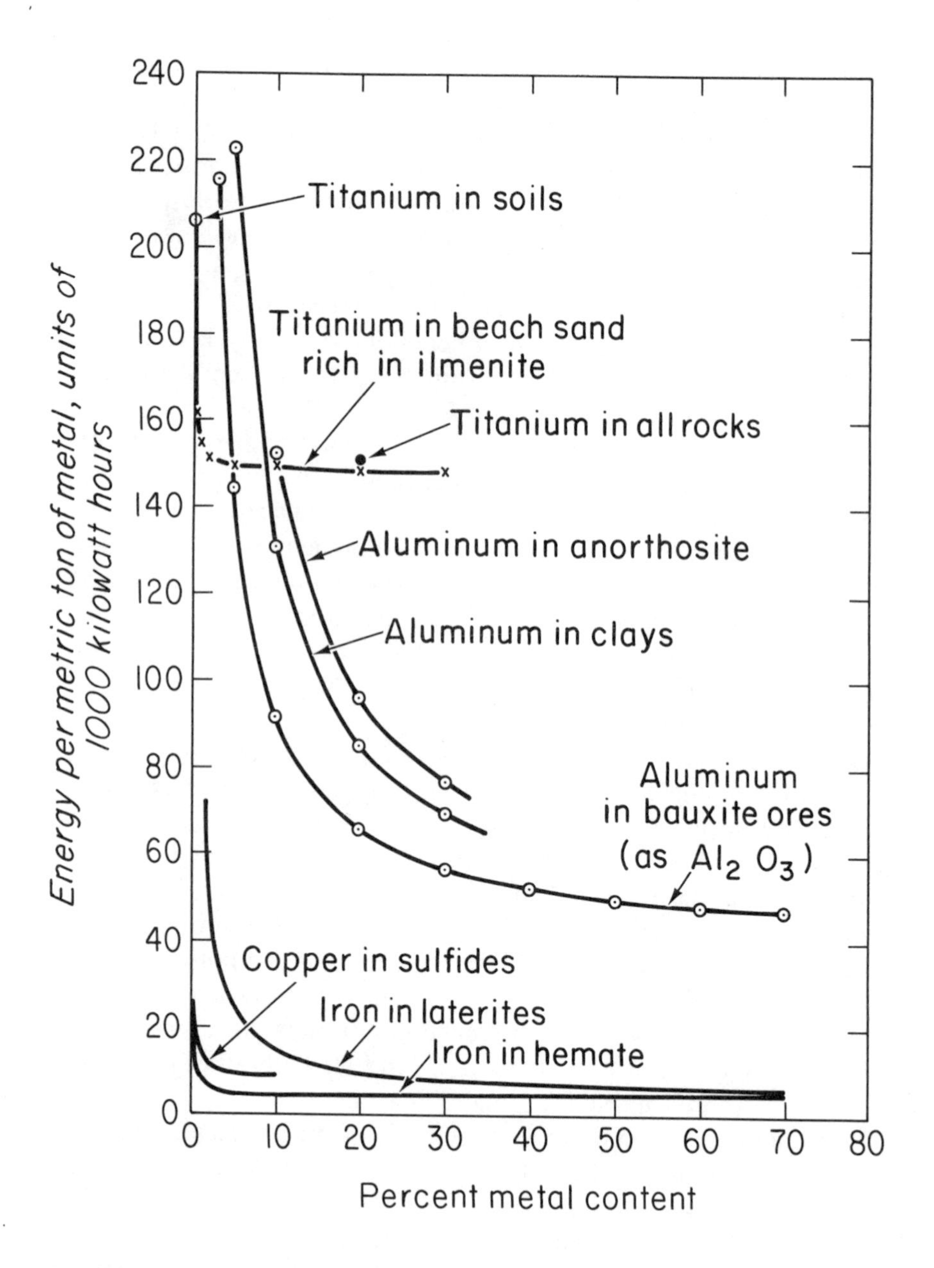

Figure 45. Energy costs of metal production. (Simplified from N. J. Page and S. C. Creasey, 1975, *Journ. Research U.S. Geological Survey,* vol. 3, no. 1.)

considerations and philosophies and by paying comparable attention to other factors, including the long-range benefits that arise from an economical use of most materials.

The narrowly economic approach to problems of mineral supply is epitomized by the arguments of two economists who not long ago advanced the superficially logical but practically nonsensical conclusion that we can never physically run out of mineral supplies because the whole earth is made of minerals. That reminds one of the argument of Vannoccio Biringuccio, who in 1540 argued that: "Miners are more likely to exhaust the supply of ores than foresters the supply of wood needed to smelt them," because "very great forests are found everywhere." If the whole earth is made of minerals, then isn't it all one big mineral deposit? Such an interpretation is in fact implied by the suggestion, often made, that as grade of ore decreases we will turn to mining "common rock." As with many other reassuring oversimplifications there is both an element of truth and food for dangerous complacency in this one. Indeed, to the extent that we obtain lead and zinc from limestone and copper from sandstone we mine "common rock" now. But the idea is misleading to the geologically uninformed in the same way that the picture of an earth made up of minerals is misleading, albeit a fair literal approximation. The zinc-rich limestones and copper-bearing sandstones are, in fact, very unusual rocks. *It is the uncommon features of a rock that make it mineable.* It is the local concentration of elements to commercially exploitable levels, beyond, and usually far beyond, their normal abundances in Earth's crust that makes a mineral deposit. Even if the whole earth were accessible to mining, which it is not, other factors would limit the exploitation of it—above all energy accounting, but also considerations of public health, safety, and environmental quality. Thus, when geologists think of mineral deposits, they think in terms of reserves, resources, and the total stock in Earth's crust, of which reserves and resources are a part.

Reserves are mineral deposits that have been discovered, defined as to volume, and shown to be commercially exploitable under conditions determined by existing technology, transportation networks, and market values. They differ from *potential resources*, which include probable undiscovered deposits, and those that are known to exist but are of unknown, unproven, or submarginal economic feasibility. Potential resources can be thought of as possible future reserves, and both reserves and resources are but small fractions of the total stock.

On the probably secure premise that man is unlikely to recover minerals from below the outer 10 to 40 kilometers that generally comprise Earth's rocky crust, the quantity of an element within that crust can be thought of as the *total stock* of the element. It is easy to estimate total stock. The mass of Earth's crust is known within narrow limits and the average crustal abundances of different metals and other chemical elements are reasonably well established from many analyses of different rock types. If you multiply mass of the crust by percentage abundance you get the total stock.

Similarly, reserves are not difficult to estimate. Basically it is necessary only to get reserve estimates from all mines and mineral districts in the nation or the world and add them up. Within the United States Department of the Interior, the Bureau of Mines publishes reserve figures regularly, along with thoughtful analyses of their implications. There is little variation between different sets of published figures for reserves or total stock. The big uncertainty and the big differences come in estimating resources—that is, how much of the total stock for a given element is likely ever to enter the category of reserves and thus eventually to be mined and transformed into materials for structures and functions.

Reserve figures, together with census data and a variety of economic indicators, provide information useful for short-range planning only. Reserve estimates can be thought of as minimal quantities. Unless made by a promoter, they are almost always conservative, owing to economics, local tax laws, and other factors that make it disadvantageous for a real mining company to develop or reveal the existence of larger reserves than are needed to meet its own short-range projections. What one would like to get at, therefore, are reasonably reliable figures for potential resources that could provide a basis for strategic intermediate and long-range planning. Such estimates are more difficult to compute, but they can be made within broad limits of error. One may start with the recognition that most of the world's ore deposits appear to be limited to the upper few kilometers of the crust. That substantially narrows the target. In addition, the earlier described general model of the earth provided by *plate tectonics* gives clues as to the best places to look within that outer crust. A new theory of ore-finding is emerging which relates ore deposits to relatively metal-rich provinces and epochs that are determined by times and types of plate motions, events within plates, and past atmospheric and weathering conditions.

Meanwhile crustal abundances provide some clues for the estimation of potential resources. Resources that may some day become reserves, however, are many times less than might seem to be implied by crustal abundances alone.

I like to illustrate the distinctions among total stock, reserves, and resources with the story of the economist, the geologist, and the engineer who found themselves wrecked on a desert island with only the clothes on their backs. When a case of beans from their wrecked vessel floated ashore, the geologist and the engineer had an animated discussion over how they might get at the beans with no tools of any sort. The economist, however, proposed to solve the problem by assuming a can opener. Given a can opener, the case of beans would become a reserve. All the cases and cans of beans that had been aboard their ship and might eventually float ashore on their island would constitute potential resources. And all beans from all ships that had ever been wrecked would make up the total stock. It is easy to see both that the potential resource can be only a small fraction of the total stock and that ultimate reserves will probably be substantially less than the potential resource.

A more practical example is that of copper in Earth's crust. It has an average abundance in crustal rocks of 55 parts per million. The mass of the crust is 25 quintillion (or 2.5×10^{19}) metric tons, or slightly less than half of 1 percent of the mass of the earth. The total stock of copper, therefore, is about 4.5 quadrillion (or 4.5×10^{15}) tons, which is a lot of copper. But less than 1 percent of Earth's crust is accessible to exploration by any likely means, reducing the quantity of copper to about 10 trillion tons. It is very doubtful that as much as 1 percent of that 10 trillion tons is present at concentrations above the cutoff grade at which energy costs become intolerable (figure 45), which brings the quantity down to somewhat less than 100 billion tons. A realistic guess about how much of that copper might eventually be discovered is probably somewhere between 1 and 10 percent. Thus the ultimate recoverable resource of copper in Earth's crust is probably somewhere between 1 and 10 billion tons, which is between 2.5 and 25 times the 1975 United States Bureau of Mines estimates of copper reserves; or, averaging and rounding, perhaps 10 times known reserves.

Independent estimates of ultimately recoverable resources based on crustal abundances have been made for a number of industrially important elements by the Yale geologist Brian Skinner and by R. L. Erickson of the United States Geological Survey, using a method pioneered by V. E. McKelvey, also of the Geological Survey. Skinner also shows why sea water and the deep ocean basins are of little interest for *most* metals, so that future metal production must come mainly from continental types of rocks.

These estimates of ultimate potential resources by Erickson and Skinner rather closely parallel one another for most elements, although Erickson's are generally higher. They set probable outside limits on mineral production. And they provide important clues to where future mineral supply problems are most likely to arise, under optimistic assumptions about ultimate supply. On balance, and *assuming a strong and well-financed program of research and exploration, good advance planning, and no major energy problems*, it seems that difficulty with global metal supply in and beyond the first third of the twenty-first century is to be expected mainly with respect to copper, gold, lead, mercury, molybdenum, silver, tin, and zinc. A similar difficulty with the domestic supply of the United States threatens for the same metals plus nickel, platinum, and perhaps others—not to mention oil and natural gas, whose domestic production has peaked and is now declining, or helium, which will essentially disappear with natural gas early in the twenty-first century (except insofar as it may be stockpiled or recovered from the atmosphere). Nor should the other nonmetallic mineral resources or the material costs of energy be overlooked. A surge of drilling in search of new domestic oil, for instance, would generate great demand for the heavy mineral barite, the critical component of oil-well drilling muds needed to keep uncased parts of the holes open, retard blowouts, and carry cuttings to the surface. Thus barite, in limited supply, could become a bottleneck in energy and mineral production unless and until suitable substitutes or new resources are found. It must be stressed also that we have no assurance that undiscovered potential mineral resources will be discovered, or will be discovered in the quantities estimated by Erickson and Skinner, as they have made quite clear. Estimates of potential resources by the United States Bureau of Mines for instance, based on different techniques from those of Skinner and Erickson are very much smaller.

It is clear that mineral statistics can be presented in various ways, depending on the compiler's previous experience and training and on what he or she may be trying to prove. If you want to know the lifetime of a particular commodity, it would seem simple to divide reserves by current rates of use. Such simple arithmetic is often used to suggest either very long lifetimes or imminent disaster. In fact, if one wishes to estimate the real lifetimes of known mineral reserves one must divide the reserve figure by a rate of use that allows for projected growth of demand. When the resulting doubling times are taken into consideration, apparent lifetimes become much shorter. But lifetimes of known reserves are not likely to be the same as lifetimes of producible resources. Here we must allow for a number of variables which are sure to increase reserves but whose effects are hard to foresee and compute. Among these are new discoveries, as well as price increases and technological advances that allow lower grades of ore to be mined at a profit, more ore to be recovered, or distant sources to be brought to market. And lifetimes may be increased by more conserving use, substitution, and recycling, although substitutes may sacrifice performance, and *even 100 percent recycling would yield only half of what is needed to meet each doubling of demand.* In addition, the greatly increased use of aluminum and plastics in place of copper, steel, and glass containers over recent decades has not appreciably reduced the doubling times for consumption of copper, steel, or glass. That is just one more aspect of exponential growth.

Thus, it is uncertain what the real lifetimes of various metal elements or other mineral commodities may turn out to be. It is possible, however, to make some informed guesses about lifetimes based on knowledge of crustal abundances, geological processes and conditions, the localization of certain types of mineral deposits, and the probable environmental and other consequences of production, use, and disposal. For most metals, given appropriate economic incentives, hoped-for technological advances, bearable environmental costs, and sustained effort in research and development, reserves can probably be substantially increased above those known in 1974. Abundant commodities, such as iron, aluminum, magnesium, and the silicate minerals, have potential resources so great that, again given appropriate stimulus, research, technology, and effort, plus sufficient energy, they can sustain heavy demand pressures for a very long time. The question here is, will the required advance planning be done? Nor should we overlook the environmental consequences of sustained heavy production from ever-decreasing grades of ore, requiring ever-greater energy input and generating ever-larger quantities of waste.

Figures 46 and 47 show apparent lifetimes of a number of key mineral commodities for the United States and the world, based on 1974 reserves and estimated doubling times of consumption. These same illustrations compare estimated lifetimes of known 1974 reserves with those of reserves five and ten times as great and with similar lifetimes implied by the United States Bureau of Mines resource estimates. They suggest what the situation could be for industrial society in the year 2000 and beyond if present trends continue. To look on the rosy side, my own guess is that ten times present reserves or more may be obtainable for most commodities shown and much more for the few so indicated. During the next half century, given the early initiation of intensive exploration, research, and environmental-protection programs, and assuming a continued access to foreign supplies while neglecting equity, the industrial world might experience nothing worse than recurrent supply problems, some serious shortages, fluctuating but generally increasing prices, and nagging environmental deterioration. Following that, continuation of present trends can be expected to lead to increasing shortages, still higher prices, practical depletion of some important metals except to the extent they can be recycled, and dramatic environmental consequences.

Long before then, however, measures should have been taken to shift industrial economies toward the more abundant and cheaper minerals, to curb the appetites of industrial societies for material goods, to reduce inequity, and to bring the number of consumers into balance with what there is to be consumed over the long term.

It is only prudent to attempt to shift the burden of demand as much and as soon as practicable to abundant and common minerals. Such a shift would conserve the rarer mineral resources for essential applications, provide cheaper and more assured sources of supply for other uses, probably reduce unfavorable environmental consequences, and tend to reduce sources of international tension. The trouble is that if these more abundant materials had been the best ones for the various tasks to which the rarer minerals are now applied, they would have been so utilized long ago. "Infinite" substitutability without loss of performance is fantasy. No material can substitute well for gold, helium, copper, or mercury in all their applications. But a good deal of substitutability should be possible, given appropriate incentives for the creation of new alloys and composites by materials science and technology.

It is by no means premature to begin asking what kind of an industrial or postindustrial society could be created using only abundant and widely distributed natural substances in new combinations. In emphasizing industrial society, of course, I consider as negligible the prospect of mankind generally and willingly returning to an agrarian or hunter-gatherer existence. This may eventually come to pass, perhaps through negligence, but such modes of life are capable of supporting only a tiny fraction of the people now alive, let alone those expected. To the extent that man is humanitarian, he has no choice at this point in his long and mostly upward evolution except to sustain as best he can those industrial societies now functioning, to try to extend their benefits to others insofar as desired by them, and to work to moderate the consequences. This does not mean, of course, that it is proper to accept continued population growth anywhere, or to condone the environmental deteriorations and other shameful side effects of ever-increasing per capita material consumption by the already affluent.

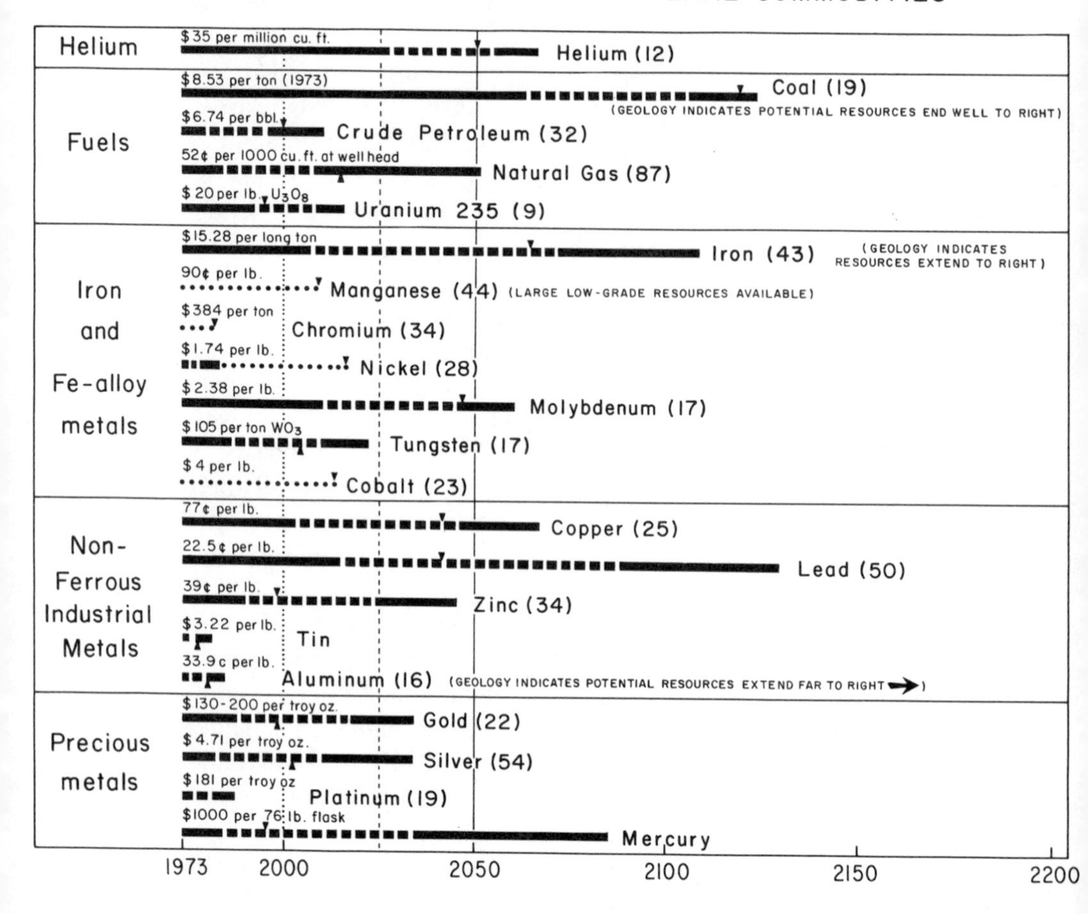

Figure 46. Apparent lifetimes of 21 key U.S. mineral commodities compared with lifetimes of hypothetical reserves 5 and 10 times as great and with estimates of potential resources. Solid bars at left represent known reserves. Short-dash lines in middle extend lifetimes for 5 times known reserves. Solid lines at right extend lifetimes for 10 times known reserves. Dots indicate potential resources only, no or scant reserves. Solid triangles above or below denote lifetimes implied by U.S. Bureau of Mines resource estimates. Bracketed numbers at right are years for projected doubling of demand. Prices in 1975 dollars. (All data from U.S. Bureau of Mines as arranged by P. Cloud, 1976, U.S.G.P.O., 78-653, p. 70.)

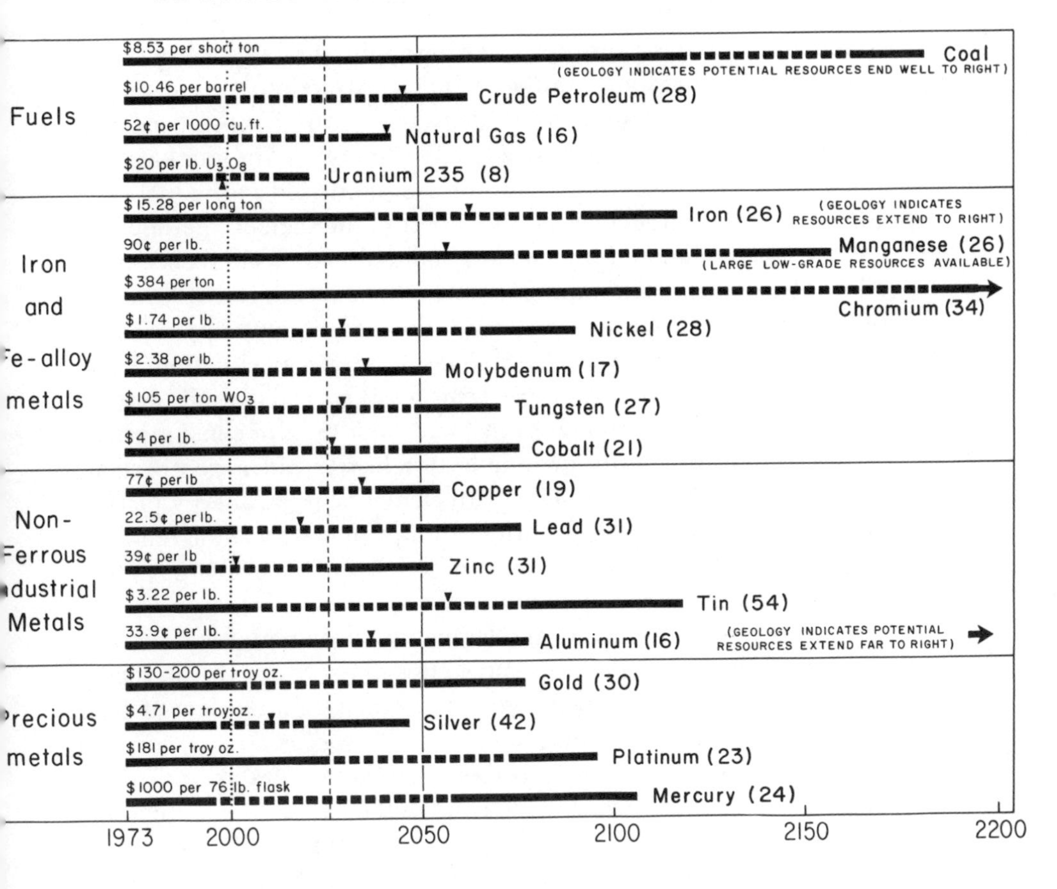

ure 47. Apparent lifetimes of 20 key global mineral commodities compared with lifetimes of hypothetical reserves 5 and 10 es as great and with estimates of potential resources. Solid bars at left represent known reserves. Short-dash lines in middle end lifetimes for 5 times known reserves. Solid lines at right extend lifetimes for 10 times known reserves. Solid triangles above elow denote lifetimes implied by U.S. Bureau of Mines resource estimates. Bracketed numbers at right are years for projected bling of demand. Prices in 1975 dollars. (All data from U.S. Bureau of Mines as arranged by P. Cloud, 1976, U.S.G.P.O., 653, p. 70.)

Alternatives in Energy Supply

What now about energy, earlier highlighted as a limiting factor in mineral production? Mining, mineral extraction, and materials science and technology convert dispersed natural substances into useful materials at a cost in available energy. Automation, the working of ever-declining grades of ore, increased individual consumption rates of materials and energy, inefficient methods of energy-conversion and use, waste, and planned obsolescence have all resulted in large and rapid increases in the rate of transformation of available to bound energy and the consequent approaching depletion of petroleum and natural gas. The energy equation, however, involves a great deal more than fossil fuels (coal, oil, natural gas, and oil in shales and tar sands) and their rates of consumption. Other sources of energy include hydroelectric, tidal, geothermal, solar, and nuclear power. The main elements of the energy picture are outlined in figure 48, where I attempt to clarify some relationships—in particular the several components of solar energy, often considered as if independent of the sun.

The fossil fuels, in fact fossil solar energy, fix our attention because they are overwhelmingly the main source of energy at the present—now being consumed in the United States at a rate equivalent to more than 2.3 million constant megawatts a year (a megawatt is a thousand kilowatts). That is equivalent to about 300 slaves working around the clock for every man, woman, and child in the nation. In addition, per capita consumption of energy has grown with time, and current rates of increase give a doubling time of about 14 years for both the United States and the world at large, which, at present, uses only three times the United States consumption of energy, or about 6.8 million constant megawatts a year. Indeed electrical energy consumption is increasing even faster than general energy consumption, and there is little prospect of a voluntary reduction in either of these uses at any time soon, least of all outside the United States. Use of electricity had been doubling once a decade for several decades until about 1973—increasingly from the burning of fossil fuels. Thus the fossil fuels account for an overwhelming proportion of current energy use, either directly or via their grossly inefficient conversion to electricity.

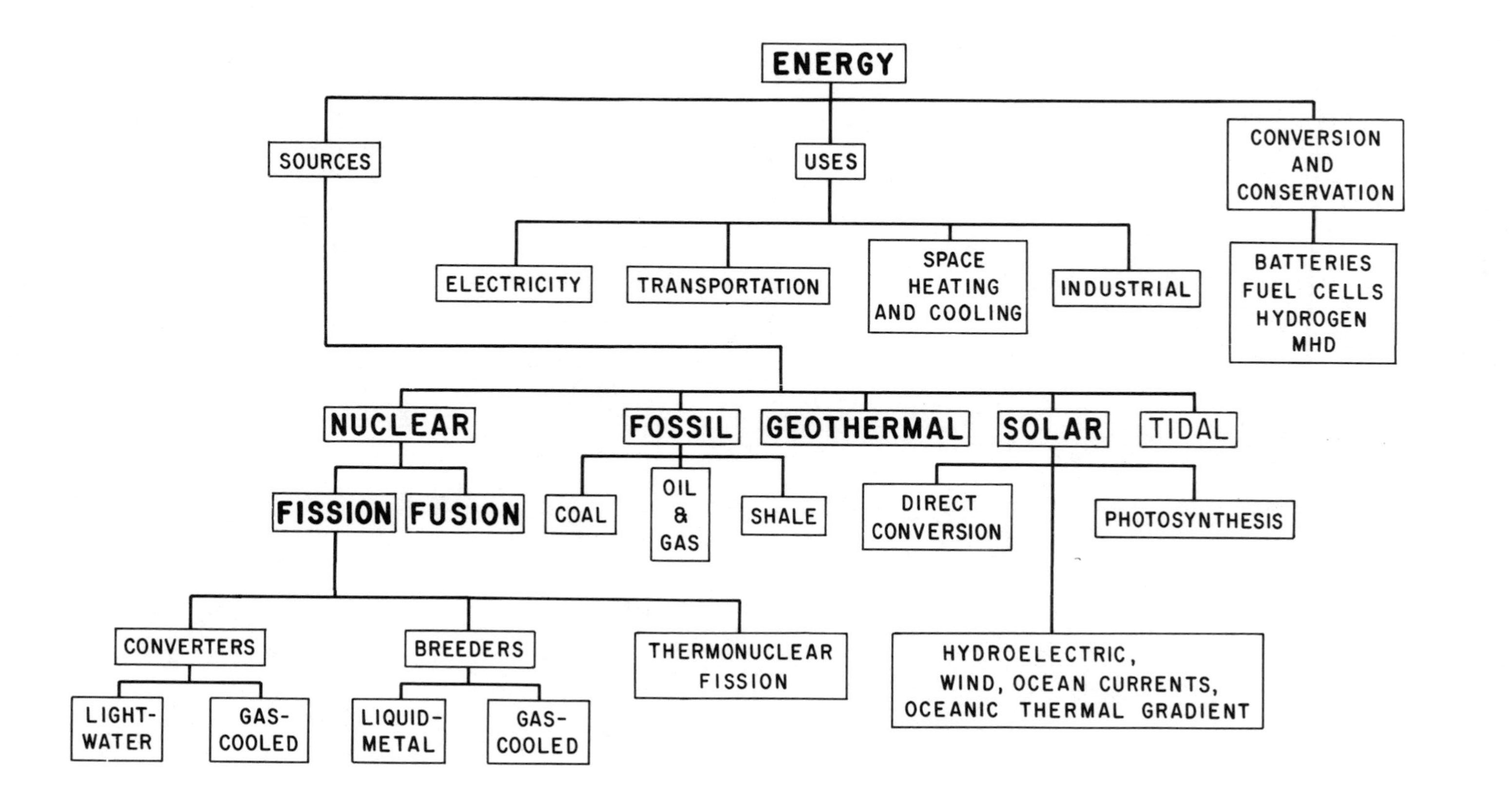

Figure 48. Main elements of the energy picture. (After P. Cloud, 1976, U.S.G.P.O., 78-653, p. 73.)

Because oil and other fossil fuels occur in sedimentary rocks whose volumes are reasonably well known, the probable quantities of petroleum and natural gas that will ever be produced are agreed upon within relatively small limits. They will not suffice to meet global energy demands very far into the next century, and good management calls for their conservation to serve as petrochemical feedstocks for plastics and the like. Coal could last for hundreds of years at *current* rates of consumption, but not if called upon to bear the whole or a large part of the energy burden for expected rates of increase. Oil shales and tar sands contain vast quantities of potentially energy-yielding hydrocarbons, but the necessary technology is not perfected and the price is not yet right. When the price of oil gets high enough tar sands may become an important industry in Canada, and an oil shale industry may be born in the United States. It would take years, however, for either to become an important fraction of the fuel market, and we cannot be confident that they would much more than double the energy available from the total historical supply of crude petroleum—thus extending the lifetime by only one doubling time, if current rates of growth continue. Nor should we lose sight of the above noted climatologic problems that could arise from the continued increase in atmospheric carbon dioxide incident to the continued burning of fossil fuels.

Turning to some indirect sources of solar energy now in use, if hydroelectric power from flowing water were totally developed at all practicable sites it would be nearly equal to present world power consumption. The undeveloped sources are mostly in the Southern Hemisphere, which is in many ways a boon, but they are of little consequence in meeting the projected doublings of demand for energy worldwide. Tidal energy at usable sites, if fully developed, would meet only about 1 percent of present world power demands. Other indirect sources of solar energy, the efficient combustion of solid wastes and the generation of gaseous and liquid fuels from garbage and bodily wastes might meet another 6 to 15 percent, a non-negligible fraction of *current* demand but a fraction, nevertheless.

Conventional (and exhaustible) sources of geothermal energy from hot water and steam would be about equivalent to tidal power if consumed over a fifty-year interval. If, on the other hand, suitable technologies could be developed for tapping such unconventional geothermal power sources as volcanoes and areas of high general flow of heat from Earth's interior, much larger quantities of power could become available. Efficient technologies for such purposes, however, do not now exist and may not be developed. Assuming they can be developed, it is still too early to evaluate realistically the promises and hazards of large-scale geothermal energy production. It would probably not be nonpolluting. Conventional thermal waters, for instance, generally contain large quantities of dissolved salts which can contaminate surface and ground waters and corrode installations and equipment. As conversion efficiencies are low, thermal pollution is a problem.

All in all, however, conventional sources of energy, although clearly finite, are not inconsiderable, and taken together they appear in no *immediate* danger of running out despite current and expected shortages. The problems, rather, involve technology, public health, safety, and environmental protection. The question remains open as to what is the best mix and scheduling of energy options for the future, in particular the part to be played by those now seen as unconventional.

Prominent among such unconventional sources is the direct use of solar energy. Energy flows to Earth from the sun at a nearly constant rate of 173 billion megawatts, or nearly 26,000 times the present rate of global energy demand. The total conventional nonrenewable terrestrial energy inventory is equivalent to only a few days of solar energy. Some of that solar energy is already used and more will be used indirectly as indicated in figure 48. We cannot harness much of it directly; but, since it is essentially nondepletable, we can keep on using it at whatever rate we can tap it for at least a few more billions of years. The problems in utilizing direct solar energy lie mainly in its dispersed state and local prevalence. It must be concentrated during the daytime from very large areas of collection where the sun shines enough hours to make it useful. Or it might be beamed into solar furnaces from orbiting satellites with very large collecting surfaces. There is renewed interest in the technology but little progress as yet. From what is now known, it seems that direct solar energy, given a determined effort to convert to its use, could meet a substantial fraction of space heating and cooling requirements within 10 years. With redesign and modification of existing structures, it could eventually meet nearly all of the roughly one-quarter of the United States energy budget that now goes for space heating and cooling. Large-scale industrial application of solar energy, however, is probably still many decades in the future.

Thus man has turned to nuclear energy as one of his main hopes for the intermediate as well as the long-range future, if not the main hope. As indicated in figure 48, the energy of the nucleus is available from two main sources; nuclear *fission* and nuclear *fusion*. Fission is the energy of the so-called atom bomb; fusion of the hydrogen bomb. The trick is to control the reaction so that it releases energy slowly instead of catastrophically, and to contain its noxious products. The first is done in a *reactor*, of which several kinds are in operation and more are theoretically possible. Fission reactors are of two main types. Those now in commercial use are burner or *converter reactors* that use an actively fissionable rare isotope of uranium (U-235) and also do a non-negligible amount of breeding. In nuclear parlance *breeding* is the process of bombarding either ordinary uranium (U-238) or thorium (Th-232) with high-energy neutrons (the active ingredient of the neutron bomb) to make actively fissionable plutonium-239 and uranium-233. *Breeder reactors* produce their active fuel from ordinary uranium or thorium in this way. They are still in the developmental stage, and their development has focused on the plutonium cycle. Both France and the USSR claim working plutonium breeders, but United States breeder development has experienced many technical difficulties and delays.

The great problems with fission reactors are the long-lived radioactive wastes they produce and the radioactivity of the reactor materials and fuel rods. The most threatening of the radioactive materials is plutonium-239. Microgram quantities are carcinogenic in experimental animals. It has a half-life of 24,500 years. Because 20 half-lives is the time required to degrade to mainly harmless end products, this means that all plutonium produced must be safely excluded from contact with the biosphere for half a million years if risk of increase in cancer incidence is to be avoided. Five kilograms of plutonium suffice to make a plutonium bomb capable of causing great destruction and many deaths wherever people congregate. At 10,000 dollars per kilogram, plutonium-239 is a tempting target for theft or hijacking. Any reasonably intelligent person or group who can read the published instructions, possesses mechanical skill, and is sick enough, hateful enough, or desperate enough can make a plutonium bomb, given enough plutonium-239.

The fuel rods of an ordinary water-cooled, 1,000-megawatt converter reactor are reported to accumulate about 220 kilograms of unburned plutonium-239 per reactor year of operation. These fuel rods are periodically removed from the reactor and sent away for "cleaning." So far none appear to have been hijacked or stolen, but that source of plutonium-239 is not tempting. It would be a tough and hazardous job to concentrate it from the rods. Of course there is also a large supply of plutonium-239 in the hands of the armed forces—well guarded, we may hope.

If and when uranium breeder reactors go widely into production—and the uranium rather than the thorium breeding cycle is the focus of most effort toward an operable breeder—quantities of plutonium-239 in transit will increase greatly and it will be conveniently concentrated. According to published government data, each 1,000-megawatt breeder reactor is expected to require about one metric ton of plutonium to start up and will contain anywhere from one to three metric tons of plutonium at any given time. Plutonium might then be in frequent or constant shipment in concentrated quantities of many kilograms to various parts of the country or the world where it would be stored or used. Even "reactor parks" would not exclude some transshipment or eliminate the hazard of unauthorized acquisition and use.

How to safeguard an always imperfect civilization against the deliberate or accidental misuse of nuclear force would be a central and growing problem in a world less divided and more trusting than ours. Although in 1969 I joined in a recommendation that "the development of high-neutron-economy reactors be accelerated, including an efficient and safe type or types of breeder reactor," I have since come to have grave misgivings about the advisability of such a course of action. Real and permanent safety in connection with the production and movement of plutonium-239 and in the isolation of noxious by-products with shorter half-lives, assuming this is technically feasible, calls for a degree of care, order, and permanence in world society which is unprecedented in human history and unwarranted as an assumption. While I lay no claim to expertise in matters of nuclear energy, I share the concerns of a few who do understand it well that fission enthusiasts may wish to take us down that road faster than we need to go.

A glance at figure 48 shows that there are a number of possible variations to the mix of sources and technologies with which a less hurried world might choose to meet its energy agenda. And there is less pressure of time on resolving legitimate problems of energy supply than there is with regard to many related problems about which lesser public stirrings are heard. I need only mention public health and safety, environmental protection and restoration, continuing growth of populations among the poor and material consumption among the affluent, looming materials shortages, and the need for technological conversion to the art of doing more with less and doing that with cheap and abundant materials instead of rare and ever-costlier ones.

Before pulling all stops on fission energy, leading inevitably to a breeder economy, I would like to see much more attention paid to more widely ramifying and more conserving public transportation systems, to reduction of the energy now being wasted as a result of excessive consumption of metals and other material goods, to the many other possible means of energy conservation, to the potentialities of alternative energy sources not now in conventional use. Among such alternative energy sources I refer especially to accelerated development of direct solar-energy applications, nonconventional sources of geothermal energy, and nuclear fusion.

Fusion energy, the reader should understand, may never be applicable to peaceful ends. The continuing temperature required for a sustained fusion reaction is about 100 million degrees Kelvin—about 20,000 times the sun's surface temperature. No material substance can withstand such a temperature. Thus all designs for a fusion reactor call for containment of the reaction within a "magnetic bottle" of some sort. The maximum sustained temperature achieved so far in such magnetic bottles is only about 6 million degrees. If a successful fusion reactor can be built, it seems that it might be able to run on abundantly available substances with a high degree of safety. If the magnetic bottle fails, the reaction simply stops. Although some long-lived radioactivity would result from bombardment of structural materials with fusion neutrons, noxious products produced are expected to be generally shorter lived and less dangerous than those from either kind of fission reactor. And 1 percent of the deuterium (heavy hydrogen) in the sea, in a reactor designed to use only deuterium, could produce energy equivalent to that in 500,000 times the world's initial supply of coal, iron, and gas. A commercially workable fusion reactor may well be beyond the reach of technology and at best is some decades down the line, but the effort to bring fusion under control as a practicable source for peaceful energy probably deserves more attention than it seems to be getting.

And should we choose to go the breeder route, that should be only after having made a detailed study of the advantages and disadvantages of the thorium versus the uranium breeding cycle and after having exhausted the potentialities of less hazardous mixes of energy options.

At the upper right of figure 48 I also show hydrogen and MHD—both recurrent darlings of the Sunday supplements. The notations refer to the burning of hydrogen as a fuel and the abbreviation for something called *magnetohydrodynamics*, which is a big word for a potentially highly efficient energy-conversion mechanism. Both depend on other forms of energy as a primary source and thus are not primary energy supplies but rather means to use them more efficiently or in different ways. Neither calls for amplification here.

To summarize what I have said about energy for the future: *we need not panic or rush headlong into "solutions" that may be premature or unnecessary such as massive proliferation of converter or breeder reactors.* Thanks to other sources, we have some decades of grace during which to seek more conserving ways to use and transform our energy, to explore and develop solar energy systems, to take a harder look at nonconventional sources of geothermal energy, and to continue research toward a practicable fusion reactor before we may be forced to make decisions with such potentially far-reaching adverse consequences as the generation of vastly larger quantities than already exist of plutonium-239 or long-lived radioactive wastes.

It should be emphasized also that, although alternate energy sources of vast potential exist, there would be little useful energy without metals to build the machines and superstructures needed to capture it, convert it to useful forms, transport it, and apply it to the performance of work—including getting more energy to produce more metals and other mineral products. *As energy restraints limit access to materials, so material restraints limit access to energy*. It would be unfortunate if either were to dominate our thinking to the exclusion of concern for the other.

The Outlook

How should one react to such a tabulation of difficulties and hazards? A common reaction is to point out that although Malthus and others from early times until now have cautioned against various potential limits imposed on man's existence, human ingenuity has so far been able to stretch or overcome such limits in some or many parts of the world—although that was probably small consolation to the million Irish who starved in 1845 and 1846 and may be of little comfort to those starving today in Africa, India, and other places. The problems mankind now faces, according to many self-professed "optimists," are no different from those faced and overcome before. As he has controlled or overcome pollution, disease, famine, and materials shortages in the past, so—with the help of technology and the free market—they see him continuing to overcome such nuisances into the indefinite future. This viewpoint is expressed with unusual candor in the words of a distinguished adherent to it and a cherished friend of mine: "You can't count on it but you can bet on it."

These are vigorous and stirring words. The human spirit responds to confidence, determination, and imagination, and approval flows to those who display such admirable attributes. Cautionary advice is hardly ever popular, let alone heeded. The name of Cassandra, sister of Hector, who combined the gift of accurate prophecy with the curse of disbelief by others, came to be considered pejorative because she warned against admitting the "Trojan horse" with its trick cargo of armed Greek warriors. Her words might well be paraphrased: "Beware of seers prophesying plenty."

I too believe that man *can* overcome his growing complex of problems, but only with early and effective global population control combined with substantial and conscious effort toward the other ends mentioned, and beginning soon. Neither technology nor market forces alone will set everything right, although both can help. Reliance on the technological fix is a copout. My rejoinder to my buoyant friend is: "If you can't count on it, and if the future of man is involved, don't bet on it," which is not to say that I propose to discard either discriminating technology or perceptive economics. My confidence in the judgment of an alert and informed electorate, however, is too great for me to adopt the view that the fate of society is secure only in the hands of an elite guard and that the society itself, through its elected and appropriately supervised representatives, is incapable of the responses needed to shape the problems to manageable proportions. Indeed, although small elites may and probably will play disproportionately important parts, only the collective actions of society as a whole can achieve the goals of human survival and progress.

We can begin by recognizing that the problems man now faces and will continue to face for some time *are* different from those he has faced in the past. They are, above all, different in scale. Never before have the quantities involved in the growth curves been so large. Never before have there been so many consumers, consuming at such high per capita levels, with such high expectations for the future. Never before have so many different elements and mineral commodities already been in use, with consequent diminution of opportunities for substitution at similar or better performance levels. Never before have annual demands for metals and other minerals been so large. The reason these differences are so important is that the quantities involved have become so large and the doubling times so short that *the lead time for action between general perception of a threatening situation and the onset of crisis or even catastrophe has become dangerously small.* Market reactions and technology need all the warning time they can get to be effective. Thus it behooves us to set up scientific and technological monitoring and warning systems with respect to mineral, energy, food, and water resources, and to exercise that same power of foresight which makes us responsible for our actions to anticipate where new pressures may arise.

Deficient though the data base may be, enough information is available to warrant five general conclusions that have an important bearing on the future:

1. *The environment is indivisible.* Any action that affects any part of it may affect other parts in indirect, unforeseen, and commonly undesirable ways. Forethought as to the possible consequences of actions contemplated is, therefore, imperative.
2. *Present trends cannot be sustained.* We do not need a computer to tell us this. Even though food can be increased and resources of many materials are large, the growth rates are too high and doubling times too short to be sustained indefinitely by any finite quantity, however large.
3. *A sequence of actions, each defensible on the basis of evidence available when it is decided on, can have unplanned and undesirable cumulative effects.* Such eventualities can be only partially averted by continuing research and advance planning. Even after all due caution is exercised, continued watchfulness is in order.
4. *Early limitation of population growth by the poor and similar limitations on both population growth and per capita rates of material consumption by the rich are essential to survival, no matter what else is done.* No amount of technological and economic creativeness, important though it is, can solve the long-range problem of human survival under tolerable conditions without these actions. Indeed failure to achieve such limitations could reduce human society to a destructive state of gross inequity, desperation, and lethargy long before shortages of any kind became de facto global crises—as failure to limit populations has done in some areas of epidemic malnutrition and crowding today. Students of animal populations find that environmental factors other than limitation of resources may act in unexpected ways to limit populations before theoretical maxima are reached. It may be that further crowding, the inevitable restrictions that

accompany unrelieved dense settlement, and isolation from nature in massive, uniform societies may prove so depressing to the human spirit or so destructive of coherent social organization that theoretically maximal limits will never be reached. Or, like the Ik of the Kenyan border ranges, so chillingly described by Colin Turnbull in his book *The Mountain People*, we may learn to steal from the helpless, abandon the suffering, and live in dread of our neighbors.

5. *Inevitable population increases and the need to raise the level of living of the now deprived nevertheless make substantial increases in both materials and energy resources necessary over the short and intermediate term. Such necessary additions provide the opportunity to convert gradually from a growth-oriented economy to a steady-state material economy* over the interval of the seven decades during which population can be expected to continue increasing even after bare replacement growth has been achieved.

These things must become generally understood and effectively acted upon well before the end of the century if that turning point is to be one to look forward to. Yet the chances of that happening seem fair at best.

Inaction is easy to justify. Futurecasting is a hazardous business; most who have tried it have been wrong in some respect. Some argue that concern is premature, that the problems really center in the developing countries, which may not follow the same patterns of development, use the same resources and technology, or run into the same environmental problems as the developed countries. Others claim that the real problem is with people in the developed countries who are the heavy consumers and polluters, and that it is only meddlesome to worry about population growth somewhere else. And so on. We need not tarry over such cavils, despite the elements of truth involved. When I caution of impending difficulties it is in the hope that precautionary measures will avert or ameliorate them. When I look at what developing countries (e.g., Brazil, to name one with a visible effort) are trying to do, I see little evidence of departure from the patterns of development, resources, technology, or pollution experience observed in the developed countries. It is a temptation to go on, to decry the insanity of pronatalist views, the application of temperate-land methods to Amazonian agriculture, over-industrialization beneath atmospheric inversion layers, and so on. But that would be discursive.

Experience warns that whatever is done, if it is to be effective, must be an expression of the general will, preferably with a popularly supported legal basis. Education and discussion is needed among the literate, literacy (and rural electrification to promote it) among the now illiterate, some form of social security other than surviving sons among the poor, and easily available free or inexpensive means of family limitation among all so that the need for control of populations can be universally appreciated and effectively acted upon. Imaginative and broadly acceptable incentive legislation must be invented and enforced so as to assure a more equitable distribution of foods and limit individual per capita consumption of materials and energy among those who already use an unreasonable share. The costs of environmental protection and cleanup must be charged to the polluter, thereby increasing the costs of raw and refined materials and stimulating more conserving and nonpolluting practices.

Although substantial additional material growth will be needed to care for inevitable population increases over the next 70 years at least, as well as to uplift the now deprived, that growth should not be allowed to escalate for trivial reasons. Planned obsolescence and waste should be legislated out of existence. Outlets and symbols other than material consumption and progeny should be sought to assuage the very human need for a sense of personal worth and pride.

An economy that achieves a steady state as regards materials transactions need not be "stagnant." Economic growth, to the extent that it has social validity, acquires that validity through its use as a means to accepted social goals and not as an end in itself. Industry can be capital intensive, labor intensive, energy intensive, or materials intensive. Materials- and energy-intensive activities should be regulated to essential levels. The contributions of industry must be judged on some basis other than their inflation of GNP. Many modes of enhancement of the human condition are not materials intensive—education, most scientific research, music, art, sports, handicrafts, and entertainment, for example. Such activities could be the focus of our future advance toward a generally more fulfilling human state. Methods other than employment through overproduction must be found for putting the means of livelihood into the hands of all in dignified ways. Unemployment might then even be seen as an opportunity for intellectual or social growth.

International accords and protocols should be sought toward whatever constructive action is possible on the global scene, especially as regards control of populations and pollution. In so doing, the same regard for human dignity and liberty should be observed as at home. That includes the dignity and liberty of hunter-gatherer tribes who may be displaced from traditional but legally unclaimed lands against their will in the effort to sustain by agriculture increasing numbers of mainly urban dwellers somewhere else.

Each individual can do a little in his or her own way, and all who think as I do can talk to others in the hope that 51 percent of the people will come to see the need for constructive personal and societal action and legislative incentives. That would be enough. If the world could achieve bare replacement growth of its populations by 1985, along with a more equitable distribution of goods, it might be possible for science and technology to manage the innovations needed to sustain the eventual populations in a life worth living for a very long time. If such bare replacement reproduction has not been attained globally by the end of the century, the world population problem can be expected soon thereafter to take on truly catastrophic proportions, if it has not already done so. Even today a large proportion of the world's people live short, brutish, hungry lives. As those unfortunates become more numerous, or their lives more brutish and hungrier, civilization will be imperiled and humanity indicted.

The cheeriest note I can command on which to close this chapter is one of simple hope, the hope that if men can soon come to appreciate the reality and immediacy of the problems before them they will do what is essential and accept those lesser reproductive and material constraints that are necessary for the preservation of larger and more important freedoms and the creation of a life of high quality for all. It is within our power to act constructively toward that end. The things we do or decline to do over the next few decades can brighten or prejudice the future of all mankind.

If, to guide our choices, we look for some common denominator to express the various individual and subjective concepts of life-quality, we can perhaps agree that quality of life is somehow related to the variety and flexibility of significant options available. Our goals for the future should include avoidance of the thoughtless foreclosure of significant options and the creation of valid new ones. As we reach toward those goals it is well to remember that to do nothing is equally to make a decision.

For Further Reading

Committee on Resources and Man NAS-NRC (P. Cloud, chairman and editor). 1969. *Resources and man.* W. H. Freeman & Co. 259 pp.

Georgescu-Roegen, Nicholas. 1971. *The entropy law and the economic process.* Harvard University Press. 455 pp.

Holdren, John, and Herrera, Philip. 1971. *Energy.* A Sierra Club Battlebook. 252 pp.

McKenzie, G. D., and Utgard, R. O. (eds.). 1972. *Man and his physical environment.* Burgess Publishing Co. 338 pp.

Page, N. J., and Creasey, S. C. 1975. Ore grade, metal production, and energy. *Journal of Research of U.S. Geological Survey,* vol. 3, no. 1, pp. 9–13.

Pirie, N. W. 1969. *Food resources, conventional and novel.* Penguin Books. 208 pp.

Skinner, B. J. 1976. A second iron-age ahead? *American Scientist* 64:258–69.

20

PERCHANCE TO DREAM

"Perchance to dream: aye there's the rub." Like Hamlet, musing on "the thousand natural shocks that flesh is heir to," prudent men and women give thought to the prospect that their dream of a better world could turn into a nightmare. Often in the past great dreams have proved illusory. Grand plans have come to nought. And high intention has shattered against harsh reality. Perhaps that is why, in times of stress, pragmatists may choose to defer action in the hope that market forces or other balancing processes might lead to a new equilibrium in time to avert hardship. Such processes, of course, are real, and until now the balance between man and nature has not got too far out of line under such policies of studied neglect, although it has often had tragic consequences for some. Yet, as the preceding two chapters have emphasized, the quantities of people and materials involved have now become so large as to call for more foresight in future planning.

As for natural balancing forces, the question is what they turn out to be and when and how they come into play. They could be catastrophic—economic collapse, anarchy, famine, pestilence, war, or some combination of these things. Better that mankind acknowledge the extent to which it already manages the earth and consciously set out to do a better job of it in the future—leaving large ecological buffer zones in the process as a reference base for monitoring activities and in case a new start must be made.

"The beginning is the most important part of the work." The words are Plato's (in the *Republic*), but the idea has been rediscovered by persons of foresight and achievement throughout history. For those who have not yet begun their response to what one wag has called the "popollution" problem, a good beginning is to recognize that mankind does operate within limits, and that it is better to anticipate and do something about adjusting to or coping with them than it is to discover them only by pushing ahead until they are reached.

Infinite is a word that is all too loosely used in reference to human capabilities and resources. It has great poetic appeal, and Longfellow's "infinite meadows of heaven" may indeed be as he says. But it is a word that means literally "without limits." It is a word that can be used only figuratively, for we have no way of knowing whether anything is infinite and some reason to wonder whether infinity is possible. George Gaylord Simpson has observed that to consider even the universe as truly infinite means not only that everything permitted by the laws of nature is possible but that it happens. That would imply that somewhere else in such a universe right now a person exactly like you on another planet exactly like Earth is reading this very line of a book called *Cosmos, Earth, and Man*. That is hard to imagine. Everything we know or can easily visualize has limits, whether or not we can say precisely what those limits are. That includes energy, materials, the carrying capacity of Earth for man, and above all the precious resource of wilderness, supreme in its capacity to restore man's soul. In fact this great nation has so far found it possible to incorporate only slightly more than half a percent of its land area in its National Wilderness Preservation System, as compared to more than four times that amount that is paved over! And even the free energy of the sun will some day degrade to the bound energy of a white dwarf star or a denser state, just as millions of the lights of heaven have gone out before.

When, therefore, in 1945, following a remarkable flood of scientific and technological achievements during World War II, the Carnegie Institution's then president Vannevar Bush called a work *Science, The Endless Frontier*, he did not mean to suggest that there were no limits to science and technology, although that seems to be the expressed faith of some, including some who should know better. Instead Bush was voicing the well-founded conviction that the mind of man is his greatest and most versatile resource, and that no matter how much we think we know about the machinery of the universe there will always be more to learn for as long as *Homo sapiens* endures. That may not appeal to human pride as much as the idea of infinite capability on our part or that of our planet, but one may take some comfort in the thought that both human capability and planetary resources still have large untapped potentialities.

We need not think the cathedral less magnificent because it remains within our sight, the Rembrandt less beautiful because we can take it in at a glance, or the heavens less impressive because there are only 10^{22} stars in the visible universe.

The burning questions before us are: how may the not inconsiderable potentialities of man best be turned to the improvement of the human condition? How shall we cope with the excess of people and their ever-growing demands on the rest of nature? How shall we bring out the best in ourselves? Consider the state of our assets.

At present the human resource is a blighted asset. Among the 100 million or more new babies now being born each year throughout the world, some 25 million or more of them lack the diets needed during the weaning years to achieve their full inborn genetic potential for physical and mental development. The heterogeneity of the human gene pool, however, tells us that the potentiality for genius cannot occur with significantly less frequency among these children than among the properly nourished ones. Such dormant genius is lost to the world where malnutrition blocks its emergence and poverty its recognition. It is the greatest of many great wastes and it is a shame to man that it continues.

The human resource can never be fully developed as long as malnutrition and poverty are tolerated. Malnutrition and poverty are inevitable consequences of overpopulation, illiteracy, and ignorance. Overpopulation, illiteracy, and ignorance are epidemic in many regions of the world. The more of them we have, the more blighted the intellectual potential of mankind. They must be banished before human talent can be fully consecrated to the realization of man's proud dreams. Illiteracy and poverty show a close correlation with overpopulation. It makes little difference which is cause and which effect, for all must be dealt with eventually and probably simultaneously.

The future of all nations and peoples depends on the outcome. Spaceship Earth has become too small and too interdependent for any industrial society to live in isolation. All nations can start by setting a good example at home and continue by making population policy a part of all other policy. Underdeveloped nations and peoples can stop talking nonsense about genocide and observe how hard-won gains in production of food and amenities are wiped out by continuing population increases (see figure 41). City councils and chambers of commerce in developed nations can see how the costs of police protection and other public services escalate out of proportion to population growth above a given size. And industrial nations generally can contemplate environmental deterioration and unrest at home and abroad as products of unrestricted material consumption and the growing inequity between rich and poor.

Yet despite the large and rueful suppression of talent as a consequence of overpopulation, malnutrition, poverty, and inequitable distribution of the world's goods and opportunities, mankind is blessed with a large corps of men and women of high ability. Given education and opportunity, their minds turn eagerly to the growing frontiers of knowledge, whence flows the stuff of new concepts, understanding of complex relations, and practical inventions on which industrial and postindustrial civilization is and will be nourished. Increasingly also good scientific minds turn to questions of how best to cope with the ever-increasing effluvia and other adverse consequences of modern society.

Whatever society's choice may be about how to cope with the future it will need the best these minds can give to address the problems it brings—regardless of whether the choice is consciously made or merely emerges, as most important societal judgments do, from innumerable individual acts of preference, however wise or misguided. Thus it will be advantageous to continue and increase support for education of all types, at all levels, in all parts of the world. For we have ample evidence that an educated citizenry better understands the decisions to be made, makes better decisions, and provides the structure from which special talents emerge to apply themselves to larger problems. In educating such a citizenry we should stress the limitations as well as the potentialities of science, technology, and learning in general. It is necessary to emphasize the limitations simply because failure to do so in the past has led to a widespread misunderstanding, to disappointment, and, as a consequence, to much thoughtless rejection of science and indeed of objective learning in general.

After World War II, a hopeful public, impressed with wartime achievements, fired by the words of Bush, and desiring the good life, turned to science and technology to create it. Scientists and technologists were in general only too eager to respond. Ardent advocates, commonly neither scientists nor technologists themselves, at times promised more than could be delivered, while the latter often failed to warn of limitations or the possibility of unforeseen adverse side effects. A return to the things of life after six years of concentration on death produced an unprecedented crop of babies. Advances in physics and chemistry led to the production of all kinds of sophisticated instrumentation for measuring things that had never been measured before or in quantities smaller than had ever been measured before, leading at times to premature alarms like the one about mercury in swordfish. The drive to feed more people and combat disease produced a rash of seemingly miraculous pesticides which added fuel to the population explosion and produced a number of inadvertent consequences. The splendid achievements of the communications industry put nearly everyone in touch with nearly everyone else on a short-response basis and showed the downtrodden and impoverished just how downtrodden and impoverished they were.

The mixture was understandably explosive. To the age-long problems of man were added those arising from higher concentrations of humanity in larger aggregations than had ever existed before, a growing but superficial awareness of environment on the part of people who had more leisure for reflection and more detachment from the environment than ever before, an unprecedented solid-waste problem that could feed only on the equally unprecedented affluence of Western societies, environmental pollution on a huge scale, and a worldwide wave of rising expectations. Technology was blamed for the most visible difficulties and science was castigated for creating technology, while failing to produce the millennium. Let it be admitted that science and technology in truth have much to answer for in terms of their frequent indifference to the possible adverse consequences and misuses of their discoveries. At least an equal burden, however, must be borne by the social and political systems which failed to keep the fruits of science and technology in exclusively constructive channels—which too often demanded and got palliatives instead of cures.

The world does not need whipping boys. It needs appropriate social and political institutions to assure that the productions of science and technology are focussed exclusively toward the benefit and enlightenment of mankind. In addition to technology transfer and technology assessment, we must also have a kind of technology information and control system. This is required to keep society informed about what is happening, to look into possible side effects, to evaluate the likely costs and benefits, and to devise legal and other regulatory mechanisms for curbing or preventing pollution, litter, waste, excessive noise or thermal variation, misuse of technology in general, and other insults to health, happiness, aesthetics, and culture. The United States Environmental Protection Agency has moved in this direction, but more comprehensive, more analytical, and more scientifically informed action is needed, and similar or better systems should be extended worldwide.

To achieve all of these things, to limit "popollution" by acceptable means, and simultaneously to maintain a continuing and equitable flow of material and energy resources without unacceptable environmental consequences is a large order. It can be done only with more, not less, good science and technology and appropriate sociopolitical developments—including a strengthening of the behavioral and political sciences and their redirection away from almost total preoccupation with urban systems, power politics, and material growth to more comprehensive views of life and progress.

But disillusionment lurks in over-expectation; and despair, like complacency, is counter productive. Much must be ventured in order to discover what may be achieved. The laws of nature, including human nature, cannot be abolished or suspended. Gravity will endure. Entropy will prevail. The finite will not become infinite, no matter how devoutly wished. Plutonium-239 will not become benign, man omniscient, nor societies permanent. Perpetual motion machines will continue to evade the cleverest inventors. Granite will not be converted to diamonds nor lead to gold. Nothing very useful will happen without effort. Some good ideas will come to nothing. Many worthwhile discoveries will not find immediate applications. Many miraculous seeming developments will spring from old ideas—as gigantic crushing plants and big magnets are used to get iron from the once "worthless" but now vital taconite ores of the banded iron formations. Technology may reach a plateau, as it has done in the past—perhaps even a plateau from which it can rise no further.

The unembittered disillusionment of the mature mind informs us that nothing secular will ever be completely perfect and that no great advances are achieved without trial, risk, and often failure.

Science is not magic. It is a modicum of inspiration and a maximum of application and stubbornness about getting the facts straight. Some springs from the mind of genius. Most, however, is carried on by ordinary men and women with ordinary ambitions and failings, but often with extraordinary discipline and persistence. The scientific endeavor to expand knowledge and understanding of natural systems is a deeply civilized, intensely humanistic, and ultimately useful human activity, comprehensible in its broad terms to all who are willing to pay attention, if not always successful or immediately applicable. When it is successful it always adds to the significance of the larger human endeavor and is sometimes materially improving of the human lot. It is nearly always gratifying. And it is often beautiful.

But don't expect miracles. And do insist that decisions about the *uses* of science and technology be made consciously and deliberately. There must be adequate understanding, as well as public information and discussion of relevant variables and possible consequences. We cannot return to some mythical Garden of Eden but we can make a better world.

Anyone who finds no cause for concern in the present condition of the human species can have in mind only some narrow and privileged enclave and assume that it will be able to maintain its favored position against the rising pressures for equity. He or she cannot be viewing the world as a whole or considering the prospects for posterity. In the larger outlook no less a prize is at stake than the duration and quality of human society worldwide. Any program that realistically seeks to improve the human condition will recognize that without population control and restraints on material consumption all else is in vain. To be constructive, therefore, programs for national, regional, or global management should include the following twelve steps:

1. *Move society toward the voluntary two-child family norm as quickly as possible*—making use of any combination of education, family-planning services, social security, public proclamation, and home improvement that is appropriate to that goal under circumstances obtaining. The recommendations of the Tokyo International Symposium on Population of April 1977 offer a good starting point.
2. *If efforts toward the wholly voluntary two-child norm are not successful, supplement them with legal incentives and social pressures,* evolved through public information and discussion and democratically agreed upon for the common good—in particular the good of the yet unborn. Nations and social sectors slow to respond to the primacy of this goal should be made to feel the pressure of the contrary view—but insofar as practicable in ways that do not work further hardship on unwanted minors.
3. *Once reproduction is at or, better, below bare replacement growth, take available democratic measures to hold it at that level until zero population growth is reached.* It is likely that in many parts of the world the population will then be substantially larger than can be supported without massive importation of food and materials. Thus it will probably be advisable to encourage the continuation of reproduction at sub-replacement levels until population is reduced to the point where it can be sustained indefinitely by local resources at a dignified level of living. That population in much of the world is likely to be substantially less than at present.
4. *Seek concurrently a more equitable distribution of the world's material goods,* by reducing excessive consumption on the part of the 33 percent who now consume 90 percent of the world's

resources and increasing the flow to the 67 percent who now get by on 10 percent of the world's resources.

5. *Strive through scientific research and technological innovation to increase the production and more conserving use of food, forest products, and mineral raw materials as needed to support inevitable population increases at a decent level of living.* Keep this up until such time as population levels permit stabilization or decrease, following the most conserving practices indicated by appropriately inclusive cost-benefit analyses. Research in tropical agriculture and materials science would be important aspects of this policy.
6. *As essential material production increases, regulate it as needed with a view to resource conservation, the earliest possible stabilization, and reduction of hazards to public health, safety, and environment.* Nonrenewable resources should be regarded as a public heritage. Mining and energy production might be regulated by some kind of sliding tax scale that would encourage the recovery of a wide range of grades and qualities while the mine or other source is operable. Depletion permits might be sold to regulate rates of exploitation, as economist H. E. Daly of Louisiana State University has urged. Solar energy should become the main source of energy for space heating and cooling, and other demonstrably real energy needs should be met by an appropriate mix of the most hazard-free available systems.
7. *Emphasize materials-conserving modes of enhancement of life*—education, research, music, art, amateur sports, handicrafts, and so on.
8. *Encourage diversity, for diversity is the wellspring of adaptability, the absorber of economic shock, the spice of life.* Diversity serves these purposes in crops, products, nature reserves, life-styles, employment, and the human habitat. The seemingly most efficient techniques for producing goods and services are not invariably the most desirable or even the most effective. People may work better under other conditions and different people under still different conditions. A sterile and uniform efficiency may lessen satisfaction in tasks

performed and the sense of personal fulfillment in creative enterprise that all people need. Efficiency is not the whole goal of existence. If a fraction of the population produces all the goods and services needed, the rest are likely to be unemployed, irrelevant, unhappy, and troublesome. Assembly lines and mercantile chains should be curbed in favor of wider and more creative individual involvement, including home and village handicraft and artisanship. Police forces, psychiatric and welfare services, and unemployment might be cut if more people had the opportunity to participate in society in ways they perceived as significant and gratifying.

9. *Establish strategic research and planning centers for natural resources*, to be staffed by appropriate mixes of scientists, technologists, and others, including, but not directed or dominated by, economists and lawyers. These centers should have appropriate authority to propose action at top executive, legislative, and international levels. They should be as free as possible from the pressures of special interests. In the United States they should work closely with a new *Department of the Future*, charged with anticipating the needs and defending the rights of posterity, just as the present departments of State, Defense, and Commerce are expected to serve and safeguard the interests of living generations. The present Department of the Interior could well provide the nucleus of such a department.
10. *Create legislation and accords limiting the weight and the horsepower or number of cylinders of automotive vehicles* to what is required for the approved use. Large reductions in the consumption of fuels and metals could thus be achieved without limiting any freedoms other than that to disregard the rights of posterity.
11. *Seek regional and world accords* that will stress human goals and reduction of conflict without eliminating the diversity of harmonious and equitable economic and social styles.
12. *Persist.* It took half a century or more for artificially assisted birth control to evolve from a clandestine, socially ostracized or even illegal practice to a way of life in most of the Western

> world; but, when acceptance came, it came rapidly. Limitation of population and material consumption is inevitable in time. Every day that can be gained in attaining these states increases the prospects for a favorable long-term outcome for mankind.

A nation with the intellectual and material resources of the United States can go a giant step further. It has the time, the talent, and perhaps the flexibility to set for itself and to implement a whole new agenda for the future, one designed to anticipate and meet the realities of the coming twenty-first century. Why do we need a new agenda and what are some of the things it should include?

Material growth has traditionally been considered in the United States and other industrial societies as a basic good, a goal in itself. This probably has some relation to the fact that population growth was early seen not only as a response to religious teachings and a way to tame the continent, but also as a means of increasing the labor force and providing a kind of social security. Then growth of production became a means of keeping the labor force employed and of enhancing national (not to mention personal) wealth and influence. Under such an ethic, wherever production to maintain employment exceeds the essential needs of the people plus foreign trade outlets, demand must be created to absorb the overproduction. This has traditionally been expressed by increased levels of consumption, obsolescence, and waste—including enterprises such as arms races which may in turn lead to destructive ones such as war.

I believe that the time has come in the generally upward course of human and societal evolution when, as a part of our agenda for the future, we should be trying to break away from traditional modes of growth. Those modes now tend to shut out the deeper values of life. They are the central cause of environmental deterioration. And they violate the rights of our unborn and thus voteless descendants to have the planet passed on to them in the best possible condition. Indeed I have the impression that the growing awareness of Earth's limitations is interacting with other concerns to generate a new outlook toward growth for its own sake and is crying for new norms to govern man's relation to nature, including his use and abuse of natural resources. We must reconsider what growth is intended to achieve, for it is a cancerous outlook to see it as a goal in itself. A recent Harris Poll, for instance, showed 90 percent of respondents favoring reduction of consumption and waste. Here is a response so nearly unanimous as to constitute a public mandate for abandonment of the growth tradition and the installation of a new economic and ecologic ethic in which healthy balance with nature takes precedence over material growth.

If we need the word *growth* to catalyze human reactions, consider a different kind. Instead of worrying about whether or not that sterile and misleading measure of prosperity, or, more accurately, rate of turnover called GNP continues to increase, let us promote growth in *enhancement of the human condition*. I will call it EHC. It is my way of emphasizing that traditional ways of growth are neither particularly relevant to the last quarter of the twentieth century nor sustainable in the old pattern. They have served their purpose, outlived their usefulness, and deserve a decent burial. Population increases must be curbed, but, in addition, growth in material overproduction, overconsumption, and waste by the already affluent must give way to an increased flow of essential goods and products to the deprived. Despite much public handwringing to that effect by well-meaning demographers and economists of the old school, this does not presage high levels of unemployment, blocking of upward mobility to the young, stagnation of the means of essential production, or a general return to agrarian forms of society. It does, however, call for strategic research and planning involving all sectors of society.

EHC can be achieved by taking steps to put the means of livelihood into the hands of all people; by emphasizing nonmaterial ways of achieving a sense of personal value and standing in the community; by eliminating planned obsolescence and emphasizing the quality and value of working material stock rather than rate of fiscal turnover as a measure of economic well-being; and by denuding nonessential material consumption and waste of their prestige symbolism. It also calls for bringing populations into balance with the carrying capacity of the nation and the planet for lives of high quality; for protecting and restoring the environment; for continuing education; for the decentralization of industry and populations; for legislating incentives and disincentives that will promote these goals;

and, finally, for getting the top people into the top jobs instead of letting those jobs go to willing mediocrities as political rewards or symbols.

Growth in EHC, of which the keeping of world peace is an essential part, should be the central goal of the new agenda. A common complaint against such "impractical" goals is that they are not susceptible to numerical analysis. But I claim that we can put numbers on EHC that are more deeply meaningful in terms of human welfare than those we compute for GNP. We might, for instance, adopt as a measure of EHC the *physical quality of life index*, PQL, developed by the Overseas Development Council of Washington, D. C. PQL is computed for a nation, state, or region in a very simple way. Equal weight is given to measures of literacy, infant mortality, and life expectancy. The average is the PQL, and it gives a fair measure of the level of education, sanitation, medical care, and a variety of other basic needs. Or one might work out an index for EHC, K_{ehc}, using a more complex system of scoring, such as one that measures the standing value of operating capital stock, the area and quality of ecological reserves and recreational areas, educational levels, health services, and other "goods." From the sum of these one would subtract measurements of "bads," such as numbers below the poverty level, population growth beyond replacement levels, numbers of alcoholics and suicides, frequency of crimes of violence, quantities of unrecycled waste, and person-days of respiratory discomfort (and other measures of pollution). Out of such measures, abbreviated to a fine alphabet soup, we could make up an impressive equation. Then the value K_{ehc} for any given year, and whether it was a positive or a negative number, would sum up how well or how poorly societal affairs were being managed.

An urgent issue now is to put the means of livelihood into the hands of all people without excessive production, waste, planned obsolescence, encouragement of prestige consumption, large military establishments, pork-barrel-type public works projects, and other stimuli that we normally think of as means of creating jobs.

Through various social arrangements that now exist, the nation has already approached something that is very like a guaranteed individual or family income. It seems logical to me to go the rest of the way, coupled with incentives to discourage the multiplication of dependents, while at the same time not penalizing those dependents who do arrive. But where will the jobs be, particularly if more women are to have the option of making meaningful careers for themselves outside the home?

One step that would reduce unemployment without stimulating production, while simultaneously reducing problems of technological obsolescence and otherwise promoting EHC, would be a program of *sabbatical educational and advanced-training leaves for all members of the working force*, including housewives and househusbands. During one full year out of every seven, in an appropriate rotational cycle, *everyone* would be free, on salary, to undertake some self-improvement plan. That might involve a return to school or the undertaking of some special, expense-free training or research program in industry, government, a foundation, public service, or an academic institution for the purpose of acquiring new skills or improving old ones. Or the individual might elect to spend his or her time in travel or some other rewarding activity or inactivity. Funding would be worked out between employers and federal and state agencies organized to administer the program.

Universal sabbatical leaves could account for 14 percent of the working force, wiping out current involuntary unemployment. They would also prepare people whose skills had obsolesced to undertake relevant new tasks, and they might improve the outlook and performance of those who had simply grown stale on the job. If employers were required to budget for such training programs and guarantee appropriate positions without loss of seniority to qualified returning employees, USLs might also reduce job-hopping and provide management with a more stable, more interested, and more regularly upgraded working force.

Urban decay and excessive concentrations of people run against EHC. They are largely products of increasing immigration to city centers, paralleled by flight to suburbia on the part of the more affluent. From 54 percent in 1950, the fraction of the United States population living in cities of 50,000 or more increased to 71 percent in 1970. It is expected to reach 85 percent by the year 2000, when some current projections call for people to be concentrated mostly within a few gigantic urban complexes. Problems already plague the larger cities, and they appear to be worsening. If people are forced or determined to live in cities it might be better all around if those cities were smaller, more pleasant, provided with better public transport systems, closer to natural recreational centers, and more widely dispersed. The steps that European countries such as Norway and Switzerland have taken to provide incentives for rural families to remain on the land warrant study, and perhaps emulation.

As for smaller and better dispersed cities, that goal might be combined with materials conservation, decentralization of commercial activity, and the sabbatical-leave plan suggested. I have in mind the creation of a new system of Urban Grant Universities (UGUs), parallel to but with different functions than the existing Land Grant Universities. These UGUs would not be in existing cities but at the edges or corners of large tracts of scenic, well-watered public lands. A substantial block of public land would be granted to the parent state for the foundation and partial support of each incipient UGU. By developing appropriate programs, combined with easy access to recreational lands and other conveniences, as well as by contractual arrangements with industry, private foundations, and government, each UGU would seek to attract light industry, commerce, cultural institutions and activities—and, of course, people—to the region. Support funds would be generated by contracts and the leasing of UGU lands. Thus Urban Grant Universities could nucleate the growth of pleasant, modest-sized, economically viable, and culturally rich new cities while simultaneously reducing the pressures on existing ones and achieving a healthier dispersal of populations. If the new universities draw students away from the present ones, so much the better. Many existing universities are already far too large to perform the educational function at a level that is rewarding to either students or faculty. In addition, the UGUs would be bound by no traditions and therefore would be free to explore, experiment, and innovate, at least until they developed their own traditions.

Utopian? Perhaps. But isn't enhancement of the human condition what society and its use of resources is all about? Aren't conservation and dispersal of the means of production and livelihood part of it? And don't most major advances in societal affairs begin with dreams and flower through education? The problem is to make them happen. For that we need a legislative basis. And that's where politics becomes statesmanship.

The ways in which individual citizens can influence the agenda are legion—for instance by informing their elected representatives of actions they believe need to be taken, by how they cast their votes, by talking to others, by example, by achieving public office and then working to establish the legislative basis for progress, by using all the leverage that exists to raise humanity to higher levels.

As this book draws toward an end I want to remind the reader that he or she is one with nature willing or no. The planet will in some way be different for each sojourner upon it. We live on an evolving earth, on whose future we have an influence. However we choose to exercise that influence, the larger balancing forces will eventually redress present imbalances and restore a new but evolving steady state. No condition is permanent; no plan of action good for all time, all places, and all circumstances; changing conditions require changing strategy. Mankind can be the master of its fate only to the extent that it preserves a broad variety of favorable options, while recognizing the impermanence of conditions and adapting to demonstrable natural constraints. To press too closely to such constraints is to court disaster. To continue to overpopulate the world and to tolerate waste and excess is to condemn ever larger numbers to greater deprivation at some future date. The alternative we still enjoy of establishing and maintaining a dynamic balance with nature under terms favorable to mankind is a revokable one. Were industrial civilization to collapse it might never be able to recover in a world whose easily exploitable resources of industrial materials and energy are all but gone.

Yet great and grave though the problems that confront us are, mankind has a powerful armamentarium with which to engage them—its collective genius and its seemingly unquenchable spirit. That spirit and that genius, nonetheless, can wither and come to naught without both a feasible plan and the will to carry it out. And they can work at cross-purposes where combined with arrogance in seeking the conquest of nature rather than a harmonious balance with it. Instead intellect and spirit must be joined with historical perspective, a keen awareness of limitations, and a clear perception of the relevance of all strands in the web of life in generating a formula for the bounteous, long-term continuance of our species on an uncrowded and ecologically wholesome earth. Better management of the planet and its resources at all levels is called for. That starts with a flexible, legislatively based agenda for action, arrived at by way of a searching and balanced discussion and assessment of alternatives and their consequences. I have sought in these last chapters to contribute to such a discussion. I also suggest some interim actions. But a public will to equal the task at hand is of the essence. And the tides of time are running low.

For Further Reading

Abrahamson, Julia, and others of the American Friends Service Committee. 1970. *Who shall live?* Hill and Wang. 144 pp.

Bell, Daniel. 1973. *The coming of post-industrial society*. Basic Books. 507 pp.

Brown, Harrison. 1954. *The challenge of man's future*. Viking Press. 290 pp.

Daly, H. I. (ed.). 1973. *Toward a steady-state economy*. W. H. Freeman & Co. 332 pp.

Gordon, Kermit (ed.). 1968. *Agenda for the nation*. Doubleday & Co. 260 pp.

Wynne-Edwards, V. C. 1972 (first pub. 1962). *Animal dispersion in relation to social behavior*. Oliver & Boyd. 653 pp.

INDEX